INCLUDES ACCESS TO THE **KHQ** STUDY APP

PRIMER FOR AN
Evolving eWorld

Second
Edition

Michael L. Fox, J.D.

Assistant Professor & MBA Coordinator – Mount Saint Mary College
Assistant Adjunct Professor of Law – Columbia University School of Law

Foreword by Hon. Lisa Margaret Smith

Kendall Hunt
publishing company

Dedication

To my parents, Hon. Mark D. and Jean Amatucci Fox,
and my family, friends and colleagues, for their continuing support and
encouragement behind this updated labor of love.

Foreword

As I write this foreword to Michael L. Fox's 2nd Edition of <u>Primer for an Evolving eWorld</u>, our world is in the midst of the COVID-19 pandemic, and our nation is in the midst of protests resulting from violent confrontations between police and citizens of color. Despite the tumultuous times, or perhaps because of these tumultuous times, the reality of our lives is constantly and increasingly intertwined with the world of electronic communication, also known as the eWorld. It is therefore appropriate that Michael's updated <u>Primer</u> should be published now. With the addition of significant chapters on COVID-19 and other calamities, and teenage-centered social media use, Michael's tome is useful for this time. I am honored to have been asked to write a foreword for this 2nd Edition.

I have reason to know that Michael L. Fox is extraordinarily well-positioned to offer this book, not just for those in the e-discovery world, but for anyone with a keen interest in the electronic world around us. When I was appointed as a United States Magistrate Judge for the Southern District of New York in 1995, I was blessed to have as my mentor, colleague, and friend, the Honorable Mark D. Fox, United States Magistrate Judge. Adding to that marvelous circumstance was that I had the opportunity to get to know Judge Fox's extraordinary and talented wife, Jean Amatucci-Fox, and their wonderful son, Michael, who was then in high school. I shared the Fox family's excitement as Michael selected Bucknell University as his undergraduate choice, as Michael became a member of the University Judicial Boards at Bucknell, and as Michael eventually made the decision to attend law school at Columbia, where he excelled. I have followed Michael's legal career closely, finding it particularly rewarding when he represented litigants in my courtroom. As the child of a sitting Magistrate Judge Michael was not permitted to serve as a law clerk in the Southern District of New York, much to my chagrin, but he did have the outstanding opportunity to clerk for the Honorable Lawrence Kahn in the Northern District of New York. Following his clerkship, and after substantial experience as a practicing lawyer, Michael found his calling in 2016, bringing his experience as an attorney to the School of Business at Mount Saint Mary College in Newburgh, New York, where he serves as an Assistant Professor of Business Law and as the College's Pre-Law Advisor. Among several of Michael's areas of expertise is the subject of this book, the handling of electronically stored information, which happens to be a particular interest of mine, as well. In recent years I have had the honor of sharing space on continuing legal education panels with Michael on the topic of e-discovery.

I have only developed an interest in the world of electronically stored information as a result of being confronted with e-discovery in my work. For many years I was a committed Luddite, determined not to have a cellphone or a home computer. Eventually, though,

Contributed by Hon. Lisa Margaret Smith. © Kendall Hunt Publishing Company.

the electronic revolution won me over, with a home desktop, laptop, tablet, and smartphone in my life. The need for this technology has been highlighted for many of us over the past few months, as we have worked from home, honoring stay-at-home orders from our various governors. We have learned to use a variety of video and audio platforms for meetings and conferences, for presentations and other gatherings, and have become aware of just a few of the attendant risks of this use. As I changed my own habits over time, the presence of e-discovery as a component of the cases I supervised led to my having to learn about discovery of electronically stored information. The learning curve for my e-discovery education was significant, but not unmanageable, and it is my belief that since I was able to learn enough to navigate the topic, then others could, too. This leads me to focus on the value of the 2nd Edition of Primer for an Evolving eWorld. In my experience many attorneys, as well as their clients, do not have a comprehensive grasp of e-discovery in all its facets, or even of the need to manage electronically stored information in a business setting, in the absence of litigation. As all of us have complied with stay-at-home orders, this lack of knowledge has presented acute issues at many levels, and especially has hampered the ability of many attorneys to advise their clients effectively, and to comply with Rule 1 of the Federal Rules of Civil Procedure, which requires that counsel, as well as the court, interpret and apply the rules of civil procedure in a way that "secure[s] the just, speedy, and inexpensive determination of every action and proceeding." FED. R. CIV. P. 1.

Primer for an Evolving eWorld, 2nd Edition, is the perfect resource for attorneys (and clients) who have not yet fully embraced knowledge about electronically stored information as a part of their practice, and who are now dealing on a daily basis with the requirements of communicating and doing business via technology. Chapter 1 provides a detailed and comprehensive overview of the components and methods of e-discovery. This is a resource that can be used to refresh knowledge for someone who only occasionally deals with e-discovery, or to introduce a neophyte to this important component of discovery. Counsel have an ethical obligation to have and maintain sufficient competence in any field in which they practice, and in the field of e-discovery this book is an essential resource to assure such competence.

Primer for an Evolving eWorld, 2nd Edition, is a remarkable updated compendium of information that should be on the bookshelf of every attorney, and of every client. The information does not pretend to be exhaustive, but it does give a thorough overview of the issues that are raised by the use of electronically stored information, each and every day. Such use has become sufficiently ubiquitous, especially during the coronavirus lockdown, that each of us will benefit from the knowledge imparted by this work. Whether this book is used as an introduction to the vast subject of e-discovery, or as an occasional reference guide, it will provide a remarkable benefit to its readers. It is my honor and privilege to recommend this book to all those who are touched, or are potentially touched, by litigation in our courts, which will inevitably involve discussion of electronically stored information.

June 2020

Lisa Margaret Smith
United States Magistrate Judge
Southern District of New York

Table of Contents

	Foreword	*v*
	Introduction – Still An Ever-Changing World	*xi*
Chapter 1	The Ground Rules for Electronic Discovery	1
	1.1 The Rules and Governing Authorities	1
	1.2 Alternative Sources of Information	6
	1.3 Attorney and Litigant Preservation Obligations, Reasonableness of Accessibility	7
	1.4 More Specific Information–Preservation Orders	20
	1.5 Briefly of FOIL and FOIA in a Realm of e-Discovery	21
	1.6 Costs and Cost-Shifting	22
	1.7 Rule 26(b)(5)(B), "Claw-Back" and "Sneak-Peek" Procedures	26
	1.8 Impact of FRE 502(d)	28
	1.9 Technology and Computer-Assisted Review (TAR & CAR), Computer-Assisted Coding, Predictive Coding	30
	1.10 Sanctions, FRCP 37(e), Changing Seas, No More Safe Harbor	37
	1.11 Brief Mention of Privacy Laws: The European Union's GDPR, Brazil's LGPD, New York's S.H.I.E.L.D. Act, California's CPA, and Illinois' BIPA	50
	1.12 Service of Process by Electronic Means?	54
	1.13 Conclusion	58
Chapter 2	The Cloud, Its Usage, and Ethical Concerns	61
	2.1 What is the Cloud	61
	2.2 Versions of the Cloud	63
	2.3 Cloud Computing and Unintentional Storage of Information	63
	2.4 Ethical Considerations for Attorneys Utilizing Cloud Storage	64
	2.5 Survey of Several Other States' Ethical Guidance	69
	2.6 Conclusion	71

Chapter 3   Social Media and eMail Discovery   75

      3.1   Introduction   75

      3.2   Background–No Expectation of Privacy in Social Media   76

      3.3   The Law Pertaining to Discovery of Social Media Accounts   77

      3.4   The Law Pertaining to Discovery of eMail and Text Messages   84

      3.5   Use of eMail to Communicate with Attorneys   86

      3.6   Issues with Discovery Demands and Objections   88

      3.7   Reminder Concerning the Importance of Preservation   89

      3.8   Law Enforcement and Social Media Searches   89

      3.9   Electronic Materials and the United States' Border   95

      3.10 Caution! User Agreements May Address Release
of Information   96

      3.11 Conclusion   96

Chapter 4   Ethical and Legal Considerations for Social Media Discovery:
Attorneys   99

      4.1   Introduction – Attorney Ethical Obligations   99

      4.2   Relevant Ethical Rules at Issue–Attorney Activities
and Conduct   100

      4.3   Attorney Contact/Research of Parties or Witnesses Using
Social Media   102

      4.4   Attorney Advice to Clients Concerning Their Social Media   105

      4.5   Disbarment Is a Potential Penalty for Egregious Violations   107

      4.6   Conclusion   108

Chapter 5   Ethical and Legal Considerations for Social Media: Jurors and Juries   111

      5.1   Introduction   111

      5.2   The Background and Rules/Instructions for Jurors   112

      5.3   Some Jurors Ignore the Rules/Instructions, Creating
Legal Issues   113

      5.4   Jurors Who Are Social Media "Friends" with Litigants/
Parties or Others   119

      5.5   Preserving Challenge to a Juror   121

      5.6   What Contact May Attorneys Have with Potential Jurors?
What Information May Attorneys Gather in Voir Dire (Jury
Selection) and Beyond?   123

      5.7   Conclusion   125

Chapter 6   Ethical and Legal Considerations for Social Media: Judges   129

    6.1   Introduction   129

    6.2   Can a Judge Have a Social Media Account?   130

    6.3   Potential Impact of a Judge's Active Use of Social Media   132

    6.4   Conclusion   142

Chapter 7   Examples of Societal Concerns and Social Media   145

    7.1   Teenagers and Driving While Texting   145

    7.2   Teenagers and "Sexting"   147

    7.3   Teenagers and Cyberbullying   149

    7.4   Revenge Pornography, and the Dark Underbelly of
Social Media   154

    7.5   Wrongful Social Media Posts Can Have Associated "Costs"   156

    7.6   Conclusion   166

Chapter 8   A Specific Societal Concern and Social Media: Texting to Drivers   169

    8.1   Introduction   169

    8.2   In the State of New Jersey, Senders Must Be Wary   170

    8.3   In the State of New York, It Is Drivers Who Must Beware   172

    8.4   A Concluding Word on the Matter   174

Chapter 9   Authentication of Electronic Evidence for Trial   177

    9.1   Introduction   177

    9.2   Authentication versus Admissibility, Judicial Notice   177

    9.3   State Standards for Authentication of Electronic Evidence   179

    9.4   Federal Standards for Authentication of Electronic Evidence   187

    9.5   Conclusion   192

Chapter 10   Electronics in Times of Crisis: COVID-19 and Beyond   195

    10.1   Introduction   195

    10.2   Select Laws and Orders in the Face of Pandemic   198

    10.3   COVID-19 and Law Practice: How Different Things
May Look   201

    10.4   Conclusion   204

Chapter 11   The "Final Chapter" – What Happens to Electronic Assets
After Death?   207

    11.1   Introduction   207

    11.2   Scenarios to Consider   208

11.3   Relevant Provisions of Law   209

11.4   Review of the Scenarios Posed Earlier to Apply the Law   216

11.5   Conclusion   217

Appendix 1   Model Federal Rule of Evidence 502(d) Order   221

Appendix 2   July 19, 2018, Administrative Order, New York State Courts, re. New Subdivision (f) of Rule 11-e of Subdivision (g) of Section 202.70 (Rules of the Commercial Division of the Supreme Court)   223

Appendix 3   Model Internal Preservation/Litigation Hold Letter to Client   225

Appendix 4   Model Joint Electronic Discovery Submission and Order   227

Appendix 5   Model External Preservation/Litigation Hold Letter to Adversary or Third Party   235

Appendix 6   Several Gubernatorial Executive Orders Related to COVID-19 and Remote/Virtual World   237

Appendix 7   Several New York Judicial Administrative Orders and Messages Related to COVID-19 and Remote/Virtual World   247

Appendix 8   Order of the Chief Judge for the United States Court of Appeals for the Second Circuit Related to COVID-19 and Remote/Virtual World   259

Answers   *261*

About the Author   *267*

Please note that in each chapter, all case citations, ethics opinions and rule cites are provided for reference and information only, and should always be cite-checked before any further use.

Introduction – Still An Ever-Changing World

Since the first edition of this book, much has changed in our world. For instance, New York State has instituted the new Stop Hacks and Improve Electronic Data Security (S.H.I.E.L.D.) Act, and Brazil the Lei Geral de Proteção de Dados (LGPD – or General Law for the Protection of Privacy); European General Data Protection Regulation (GDPR) provisions have been impacting the operations of some American firms; states have been passing legislation to address revenge pornography; and of course, no book on electronics, law and society is complete without discussion of the myriad effects COVID-19 has had on our real world and virtual world. Within this new edition, all of the above and more will be addressed.

As I often state at the beginning of presentations, unless one handwrites information on a piece of paper, and thereafter shreds, burns or otherwise discards the paper without it coming near a computer, scanner or smartphone camera, almost nothing in our world exists without some analog in electronic storage. Such is our ever-evolving, ever-more-technological world. The numbers of documents, e-mails, text messages, websites, photographs, social media posts, and blog articles created daily, in both personal and professional lives, are beyond legion. Of course, in the legal world, as a microcosm of larger society, there is heavy focus these days on the discovery of electronically stored materials as information supportive of claims or defenses in a case, or otherwise informative during investigations.

That said, the goal of this text is to provide the reader with a condensed and to-the-point reference guide on major issues in the world of electronics and law – eDiscovery, social media, authentication, ethical guidance, death – and more. We will look at an introduction to the ground rules and laws for electronic discovery, together with preservation obligations, sanctions, TAR and CAR, some provisions of privacy and cybersecurity laws, and more in Chapter 1. In Chapter 2, we have a discussion of the Cloud, electronic storage, further mention of New York's S.H.I.E.L.D. Act, and concerns attorneys and all users should have if utilizing electronic or Cloud storage for client, business or personal information. Chapter 3 specifically concerns social media and the legal implications of online activity. The text next looks at ethical considerations when it comes to social media and attorneys, jurors and judges in Chapters 4, 5 and 6, respectively. We will briefly consider examples of societal concerns that arise in law when we look at teenagers and social media, driving, cyberbullying and sexting, as well as issues of adults and revenge pornography; and then text messages sent to drivers of vehicles in Chapters 7 and 8, respectively. Chapter 9 addresses the authentication of electronic evidence if introduced at trial. A new Chapter 10 provides insights into the ways COVID-19 created a novel and expanded online world, with new uses of social media and electronic resources, together with new and continuing security concerns. Finally, Chapter 11 concludes the book with consideration of what happens to electronic accounts when the owner passes away.

While some of the text explores New York laws, rules and case decisions in particular, a significant amount of space is dedicated to addressing federal rules and case decisions from federal trial and appellate courts across the nation, as well as rules and case decisions from a number of states across our Nation. The appendices at the end of the book provide the reader with model preservation/litigation hold letters, and a model Federal Rule of Evidence 502(d) Order, as utilized in the Southern District of New York, among other resources and reference materials, for potential guidance and use in practice.

Something that this text cannot do, however, is address every issue, case decision or rule related to the topics in each chapter, or related to computers and electronic information in law as a whole. One reason is that to do so would require a multi-volume treatise, not a hand-held reference guide. For instance, we will not address the matter of election law, security and electronic interference with elections and voters – hot-button issues in the news at the times both the first and second editions of this book were written, but tangential overall to the materials contained herein. We will not delve deeply into the GDPR of the European Union, the California CPA, or Brazil's LGPD – although brief mention is made in Chapters 1 and 7. We also will not address much in the way of cybersecurity, Blockchain or Bitcoin – although, again, brief mention of cybersecurity concerns is made in Chapter 1, and during the Cloud discussion in Chapter 2; emerging uses of Blockchain for authentication of evidence is mentioned in Chapter 9; and the security of online working and concerns in the COVID-19 and post-COVID-19 world are briefly mentioned in Chapter 10.

A second reason for this book not being exhaustive is that publication schedules, and the time required for printing and distribution, prevent an up-to-the-minute text of all current cases and rules discussions. Virtually weekly, if not daily, new court decisions are rendered, and new rules proposed, in the areas addressed in this text. Thus, my goal is to provide general information, and a place to begin research and case cite-checking between textual updates. It is not otherwise possible to have a printed and published book in this field.

Therefore, the aim of this book – whether it serves as an attorney desk reference, litigation bag supplement, corporate in-house counsel library resource, judicial bench book, paralegal research guide, informational text for business managers or clients, textbook for educators (high school, college, business or law school)[1] or as a resource for self-represented litigants or the general public – is to leave the reader educated on the major components of eDiscovery, social media, electronic resources and ethical considerations, and conversant on these important topics. All with case and rule cites in key legal areas available at one's fingertips.

---

[1] *See, e.g.,* S. Crawford, *Why Universities Need "Public Interest Technology" Courses*, Wired (Aug. 22, 2018), *available at*: https://www.wired.com/story/universities-public-interest-technology-courses-programs/.

CHAPTER 1

The Ground Rules for Electronic Discovery

1.1 – The Rules and Governing Authorities

As we begin our studies in eWorld, let us discuss some things to file away. First, the key and most cited applicable rules:

Federal—Federal Rules of Civil Procedure (FRCP) and Federal Rules of Evidence (FRE):

Rules	Applications
FRCP 16	Pretrial Conferences; Scheduling; Management
FRCP 26(b)(2)(B)	Duty to Disclose; Specific Limitations on Electronically Stored Information ("ESI")
FRCP 33	Interrogatories to Parties
FRCP 34(b)(2)(D)&(E)	Producing Documents; ESI
FRCP 37(e)	Failure to Make Disclosures; Failure to Preserve ESI; Sanctions
FRCP 45(a) 45(c)(2)(A) 45(d)(2)(A)&(B) 45(e)(1), 45(g)	Subpoenas; Provisions Concerning Commands for and Production of ESI; Contempt
FRE 502(d)	**"Controlling Effect of a Court Order**. A federal court may order that the privilege or protection is not waived by disclosure connected with the litigation pending before the court—in which event the disclosure is also not a waiver in any other federal or state proceeding."[1]
FRE 902(13), (14)	**"Certified Records Generated by an Electronic Process or System"; "Certified Data Copied from an Electronic Device, Storage Medium, or File."** For authentication of evidence.

---

[1] *See* Model FRE 502(d) Order, U.S. District Court, Southern District of New York. **Appendix 1.**

> *Discovery* – In both state and federal courts, it is the exchange of documents, witness names & identities, and other information related to the case at issue; it takes place both between parties, and sometimes between parties and non-parties.

> *e-Discovery* – Electronic Discovery; discovery involving electronically stored information (ESI) and materials.

> *Subpoena* – A formal document, issued by an attorney or court, directing someone to produce documents, other material or information, or appear at a place and time to give oral testimony.

Among the most important updates in the realm of discovery, affecting both electronic and paper sources, is the 2015 amendment to Federal Rule of Civil Procedure 26(b)(1). That amendment returned proportionality to the Rule and removed "reasonably calculated to lead to the discovery of admissible evidence" from the Rule. Rule 26(b)(1) currently provides the following:

> *Scope in General.* Unless otherwise limited by court order, the scope of discovery is as follows: Parties may obtain discovery regarding any non-privileged matter that is relevant to any party's claim or defense and proportional to the needs of the case, considering the importance of the issues at stake in the action, the amount in controversy, the parties' relative access to relevant information, the parties' resources, the importance of the discovery in resolving the issues, and whether the burden or expense of the proposed discovery outweighs its likely benefit. Information within this scope of discovery need not be admissible in evidence to be discoverable.[2]

No longer can broad, wide-ranging demands be issued in the first instance, under the guise that they "may lead to the discovery of admissible evidence." As the 2015 Committee Notes to the Rule Amendment make clear:

> Rule 26(b)(1) is changed in several ways.
>
> Information is discoverable under revised Rule 26(b)(1) if it is relevant to any party's claim or defense and is proportional to the needs of the case. The considerations that bear on proportionality are moved from present Rule 26(b)(2)(C)(iii), slightly rearranged and with one addition.
>
> . . . .
>
> The present amendment restores the proportionality factors to their original place in defining the scope of discovery. This change reinforces the Rule 26(g) obligation of the parties to consider these factors in making discovery requests, responses, or objections.
>
> Restoring the proportionality calculation to Rule 26(b)(1) does not change the existing responsibilities of the court and the parties to consider proportionality, and the change does not place on the party seeking discovery the burden of addressing all proportionality considerations.

---

[2]FED. R. CIV. P. 26(b)(1).

Nor is the change intended to permit the opposing party to refuse discovery simply by making a boilerplate objection that it is not proportional. The parties and the court have a collective responsibility to consider the proportionality of all discovery and consider it in resolving discovery disputes.

The parties may begin discovery without a full appreciation of the factors that bear on proportionality. A party requesting discovery, for example, may have little information about the burden or expense of responding. A party requested to provide discovery may have little information about the importance of the discovery in resolving the issues as understood by the requesting party. Many of these uncertainties should be addressed and reduced in the parties' Rule 26(f) conference and in scheduling and pretrial conferences with the court. But if the parties continue to disagree, the discovery dispute could be brought before the court and the parties' responsibilities would remain as they have been since 1983. A party claiming undue burden or expense ordinarily has far better information—perhaps the only information—with respect to that part of the determination. A party claiming that a request is important to resolve the issues should be able to explain the ways in which the underlying information bears on the issues as that party understands them. The court's responsibility, using all the information provided by the parties, is to consider these and all the other factors in reaching a case-specific determination of the appropriate scope of discovery.

The direction to consider the parties' relative access to relevant information adds new text to provide explicit focus on considerations already implicit in present Rule 26(b)(2)(C)(iii). Some cases involve what often is called "information asymmetry." One party—often an individual plaintiff—may have very little discoverable information. The other party may have vast amounts of information, including information that can be readily retrieved and information that is more difficult to retrieve. In practice these circumstances often mean that the burden of responding to discovery lies heavier on the party who has more information, and properly so.

. . . .

The burden or expense of proposed discovery should be determined in a realistic way. This includes the burden or expense of producing electronically stored information. Computer-based methods of searching such information continue to develop, particularly for cases involving large volumes of electronically stored information. Courts and parties should be willing to consider the opportunities for reducing the burden or expense of discovery as reliable means of searching electronically stored information become available.

A portion of present Rule 26(b)(1) is omitted from the proposed revision. After allowing discovery of any matter relevant to any party's claim or defense, the present rule adds: "including the existence, description, nature, custody, condition, and location of any documents or other tangible things and the identity and location of persons who know of any discoverable matter." Discovery of such matters is so deeply entrenched in practice that it is no longer necessary to clutter the long text of Rule 26 with these examples. The discovery identified in these examples should still be permitted under the revised rule when relevant and proportional to the needs of the case. Framing intelligent requests for electronically stored information, for example, may require detailed information about another party's information systems and other information resources.

The amendment deletes the former provision authorizing the court, for good cause, to order discovery of any matter relevant to the subject matter involved in the action. The Committee has been informed that this language is rarely invoked. Proportional discovery relevant to any party's claim or defense suffices, given a proper understanding of what is relevant to a claim or defense. The distinction between matter relevant to a claim or defense and matter relevant to the subject matter was introduced in 2000. The 2000 Note offered three examples of information that, suitably focused, would be relevant to the parties' claims or defenses. The examples were "other incidents of the same type, or involving the same product"; "information about organizational arrangements or filing systems"; and "information that could be used to impeach a likely witness." Such discovery is not foreclosed by the amendments. Discovery that is relevant to the parties' claims or defenses may also support amendment of the pleadings to add a new claim or defense that affects the scope of discovery.

The former provision for discovery of relevant but inadmissible information that appears "reasonably calculated to lead to the discovery of admissible evidence" is also deleted. The phrase has been used by some, incorrectly, to define the scope of discovery. As the Committee Note to the 2000 amendments observed, use of the "reasonably calculated" phrase to define the scope of discovery "might swallow any other limitation on the scope of discovery." The 2000 amendments sought to prevent such misuse by adding the word "Relevant" at the beginning of the sentence, making clear that "'relevant' means within the scope of discovery as defined in this subdivision . . ." The "reasonably calculated" phrase has continued to create problems, however, and is removed by these amendments. It is replaced by the direct statement that "Information within this scope of discovery need not be admissible in evidence to be discoverable." Discovery of nonprivileged information not admissible in evidence remains available so long as it is otherwise within the scope of discovery.[3]

New York State—Now a leader among the states, there are a number of rules specifically applicable in the Commercial Division of the Supreme Court:

Rules	Applications
Rule 202.12(b) [22 N.Y.C.R.R. § 202.12(b)]	Preliminary Conferences. "Where a case is reasonably likely to include electronic discovery counsel shall, prior to the preliminary conference, confer with regard to any anticipated electronic discovery issues. Further, counsel for all parties who appear at the preliminary conference must be sufficiently versed in matters relating to their clients' technological systems to discuss competently all issues relating to electronic discovery: counsel may bring a client representative or outside expert to assist in such e-discovery discussions."
Rule 202.12(b)(1) [22 N.Y.C.R.R. § 202.12(b)(1)]	Non-Exhaustive List, Considerations for ESI in Case.

(continued)

---

[3] Fed. R. Civ. P. 26, 2015 Advisory Committee Notes.

(continued)

Rules	Applications
Rule 202.12(c)(3) [22 N.Y.C.R.R. § 202.12(c)(3)]	"Where the court deems appropriate, it may establish the method and scope of any electronic discovery." Non-Exhaustive list provided.
Rule 202.70(g) [22 N.Y.C.R.R. § 202.70(g)]	Uniform Rules for the Supreme Court and County Court (Rules of Practice for the Commercial Division of the Supreme Court), Rules 1-34.
Rule 202.70(g), Rule 8(b)	Requiring, *inter alia*, meet & confer similar to Federal.
Rule 202.70(g) Rule 11-c & Appendix A	Guidelines for Discovery of Electronically Stored Information (ESI) from Nonparties.
Subdivision (f) to Rule 11-e of § 202.70(g)	Effective October 1, 2018.[4] Encourages parties to utilize predictive coding/technology assisted review in meeting disclosure obligations.
CPLR 3126	"Penalties for refusal to comply with order or to disclose." Applicable in all courts, not just the Commercial Division.
CPLR 4518(a)	Business records. Specifically mentions electronically stored records.

States, of course, have their own rules and laws governing procedure in litigation. E-discovery factors into the equation quite often today. However, the robustness of the provisions varies from jurisdiction to jurisdiction, with New York State now one of the states taking the lead in establishing rules for e-discovery issues.

> *Party* – An individual, company or other entity officially named in a lawsuit/legal action.

In New York, Commercial Division Rule 202.12(c)(3) provides that electronic discovery and related issues should be discussed early, at the Preliminary Conference if possible. The Rule follows the outline of Federal Rule of Civil Procedure 26. (One should also review NY CPLR 3126, for dealing with discovery sanctions and spoliation in New York.)

> *Plaintiff* – The party who brings/files the lawsuit or claim in a civil matter (a matter other than a criminal proceeding).

Rule 202.70(g) (Rule 1(b)) states: "Consistent with the requirements of Rule 8(b), counsel for all parties who appear at the preliminary conference shall be sufficiently versed in matters relating to their clients' technological systems to discuss competently all issues relating to electronic discovery. Counsel may bring a client representative or outside expert to assist in such discussions." As a note regarding the relevant ethical guidance—the ABA Model Rules of Professional Conduct (as amended by the ABA House of Delegates

---

[4] *See* Administrative Order of the Chief Administrative Judge of the New York State Courts (July 19, 2018). **Appendix 2.**

> **Defendant** – The party against whom a civil or criminal matter is brought.

> **Prosecutor** – The government attorney who brings criminal charges against a criminal defendant in a court of law.

in August 2012 when the author of this book was a member of the House) require attorneys to stay current on technology, including the risks and benefits of utilizing same.

Recently, in 2018, an amendment was added as subdivision (f) to Rule 11-e of § 202.70(g), which became effective on October 1, 2018. The new rule provision encourages parties to utilize predictive coding and technology-assisted review (TAR) in meeting disclosure obligations under Article 31 of the New York Civil Practice Law & Rules. We will discuss TAR and predictive coding later in this chapter.

Civil Practice Law & Rules (CPLR) 3126 addresses penalties that may be imposed by a court in New York State for the failure of a party to comply with discovery and disclosure obligations—including the loss or destruction of evidence that should otherwise have been preserved for litigation purposes. CPLR 3126 includes this commentary: "Court of Appeals Addresses Penalties Imposed for Failure to Preserve Electronically Stored Information."[5]

1.2 – Alternative Sources of Information

The Sedona Conference, formed in 1997, is a 501(c)(3) organization for research and educational initiatives.[6] The Technology Resource Panel, and its member judges, professors, attorneys, and others, have become leaders in the field of e-discovery and guidelines for effective and efficient practice related to the discovery and disclosure of electronically stored information.[7]

Courts will cite to Sedona Conference working principles and guidelines, giving them strong persuasive authority. If there is a lack of other authority in an area at issue, but the Sedona Conference has issued guidance and commentary on the matter, give serious consideration to citing that guidance in briefing to the court.[8]

Practitioners might also consider looking to the Advisory Committee Notes that accompany the FRCP and FRE. They encompass the notes/reports made by the respective United States Judicial Conference Advisory Committees. Any annotated set of the FRCP and FRE contains the Advisory Committee Notes. They are categorized to correspond to the years of respective amendments. These Notes are also highly persuasive, readily utilized by the courts and attorneys, and make a good resource for briefs, especially in areas where there is a dearth of case law addressing a particular matter.

---

[5] NY CPLR 3126, Commentary C3126:8A (McKinney's) "Sanction for Spoliation of Evidence." More on spoliation and loss or destruction of evidence will be addressed in Section 1.10 of this chapter, concerning Sanctions.

[6] The Sedona Conference, https://thesedonaconference.org/ (last visited June 18, 2020).

[7] *Id.*; https://thesedonaconference.org/trp (last visited June 18, 2020).

[8] *See City of Rockford v. Mallinckrodt ARD Inc.*, 326 F.R.D. 489, 495–496 (N.D. Ill. 2018) (citing to Sedona Conference Commentary, 18 SEDONA CONF. J. 141 (2017)).

1.3 – Attorney and Litigant Preservation Obligations, Reasonableness of Accessibility

Computers and electronics have forever changed the shape and face of data processing, and, in turn, discovery in litigation. Indeed, some commentators believe that all discovery, in every case, involves electronic discovery because even letters and documents on paper may have once been created electronically.

> *Litigation Hold* – A directive that all evidence and other relevant/related material be preserved and not destroyed, deleted or thrown away.

When you are representing a client, or when someone comes to see you believing they are likely to be a defendant, tell them to institute a litigation hold immediately! We will discuss the provisions of Federal Rule 37(e) and New York State CPLR 3126 later in this chapter. But suffice it to say that courts are growing more and more impatient with excuses from parties who lose or destroy information, and who otherwise should have been both aware of their obligations and capable of following through on them.

For example, in Connecticut, one court sanctioned a plaintiff because a consultant destroyed information, samples, and other material from a consultation that took place *before the case started*. The court's opinion discusses what triggers a hold obligation (including the "should have known" or reasonable anticipation of litigation standard).[9]

New York's courts have similarly spoken often about the obligations of attorneys, and the duties of parties, in preserving ESI.[10]

At the time of the initial client intake and interview meeting, attorneys should sit with clients and discuss both the obligations of counsel and the obligations of the individual or company as a client. If the client is a company with large databases, attorneys and their paralegals should sit with the IT and HR personnel to better understand where the client's relevant information is kept. If there is physical evidence, preserve that also. Follow up with a letter to the client—what we will call an "internal" litigation preservation/hold letter because it is to the client.[11] Such a letter advises the client, in writing, of all preservation obligations and makes clear that sanctions could be issued by a court if the preservation obligation is not taken seriously. This serves both to educate the client and to document the advice provided by the attorney should there later be a dispute between attorney and client.

---

[9] *See Innis Arden Golf Club v. Pitney Bowes, Inc.*, 257 F.R.D. 334 (D. Conn. 2009). *See also* Advisory Committee Note to FRCP 37(e) (as amended Dec. 1, 2015).

[10] *See McCarthy v. Philips Electronics N. Am. Corp.*, No. 112522/03, 2005 WL 6157347 (Sup. Ct. N.Y. County June 9, 2005). *But cf. China Dev. Indus. Bank v. Morgan Stanley & Co., Inc.*, —N.Y.S. 3d —, 183 A.D. 3d 504 (App. Div. 1st Dep't 2020); *Radiation Oncology Servs. of Cent. N. Y., P.C.v. Our Lady of Lourdes Mem. Hosp., Inc.*, — N.Y.S. 3d —, 2020 WL3246747 (Sup. Ct. Cortland County June 9, 2020).

[11] *See* Model Internal Preservation/Litigation Hold Letter to Client, **Appendix 3**. For an interesting decision concerning litigation holds, *see Roberts v. Corwin*, 41 Misc.3d1210(A), 2013 WL 5575866 (Sup. Ct. N.Y. County Oct. 3, 2013) (Court discusses that the parties presented no authority for the proposition that a litigation hold needs to be in writing; the hold and need for the hold can change with circumstances; no sanction was required or issued, because although the party did not have the requested e-mails, a non-party law firm did have and produce them, and there was no history of willful non-compliance). *Roberts* does not, however, provide best practice guidance, since a written litigation hold provides clear direction and a record of the instructions given.

If the company has a preservation policy that will prevent automatic deletion of material from document storage and email accounts, activate the policy immediately. If the client has no such policy and procedure, consider developing one with the client. Certainly, if the client's IT system or email provider has an automatic deletion policy that deletes material after a certain period of time without further instruction, take steps to halt it. If the client is an individual, then it may be incumbent on the attorney to retain an IT professional—should the attorney's firm not have such an individual on staff—to advise the client concerning the above. Ultimately, the client must know where their data is, and must act to protect and preserve it.

It is vital that attorneys and paralegals remember their obligations in the process of preserving, searching, and producing ESI—just as clients have their obligations—obligations to both preserve and to seek discovery from opponents.[12] The question may come down to whether one should have known the information might be relevant, or whether there was reasonable anticipation of litigation. Be careful! Further, when parties and counsel do preserve, if the material is on back-up tapes, look "around" the information being preserved. If other things look relevant, save them. Computers do not "think" in the same way that human beings think. Related material is not always saved in contiguous spaces. Utilize information technology specialists. Courts do not look kindly on cherry-picking of information.

Additionally, when it later comes to whether material is accessible or inaccessible for discovery purposes, New York's courts, and other states' courts, have adopted standards developed by the federal line of cases stemming from *Zubulake v. UBS Warburg L.L.C.*, 220 F.R.D. 212 (S.D.N.Y.2003), and *Pension Comm. of the Univ. of Montreal Pension Plan v. Banc of Am. Sec., LLC.*, 685 F. Supp. 2d 456, 473 (S.D.N.Y. 2010), and readily address electronic discovery matters, preservation, and spoliation.[13]

---

[12] *See, generally, Ahroner v. Israel Discount Bank of New York*, Index No. 602192/2003 (Sup. Ct. NY County 2009); *Ahroner v. Israel Discount Bank of New York*, 79 A.D.3d 481 (1st Dep't 2010) (cites to and applies Federal *Zubulake* standard); *Fitzpatrick v. Toy Indus. Assoc., Inc.*, Index No. 116548/2009, 2009 WL 159123 (Sup. Ct. NY County Jan. 5, 2009); 22 NYCRR 202.12(c)(3); Rule 8(b) of the Uniform Rules of the Commercial Division (22 NYCRR 202.70(g)); Fed. R. Civ. P. 26, 34, 37, *et al.*

[13] *See VOOM HD Holdings L.L.C. v. EchoStar Satellite L.L.C.*, 93 A.D.3d 33 (1st Dep't 2012) (applying the federal framework with regard to the duties of parties, litigation holds, and sanctions). However, be aware that the Second Circuit abrogated the holding of *per se* negligence and sanctions from the *Pension Committee* opinion when it rendered its decision in *Chin v. Port Auth. of N.Y. & N.J.*, 685 F.3d 135 (2d Cir. 2012). The federal framework has been further changed since the revisions to Federal Rule 37(e), discussed later in this chapter. *Voom* was distinguished by the New York Appellate Division, Third Department's decision in *Atiles v. Golub Corp.*, 141 A.D.3d 1055, 36 N.Y.S.3d 533 (App. Div. 3d Dep't 2016). However, in 2015 the New York Court of Appeals had cited to *Voom* with approval, stating: "A party that seeks sanctions for spoliation of evidence must show that the party having control over the evidence possessed an obligation to preserve it at the time of its destruction, that the evidence was destroyed with a 'culpable state of mind,' and 'that the destroyed evidence was relevant to the party's claim or defense such that the trier of fact could find that the evidence would support that claim or defense'. . . . Where the evidence is determined to have been intentionally or wilfully destroyed, the relevancy of the destroyed documents is presumed. . . . On the other hand, if the evidence is determined to have been negligently destroyed, the party seeking spoliation sanctions must establish that the destroyed documents were relevant to the party's claim or defense." *Pegasus Aviation I, Inc. v. Varig Logistica S.A.*, 26 N.Y.3d 543, 46 N.E.3d 601, 26 N.Y.S.3d 218 (2015) (citing *Voom* and *Zubulake*). *See also Cantey v. City of N.Y.*, —N.Y.S. 3d —, 2020 WL 3067393 (App. Div. 2d Dep't June 10, 2020).

A concern, and dispute, often arises when there is a question about the reasonableness or unreasonableness of accessing particular data, or data existing in a particular format—specifically if the format is one that is not current.

The Federal Rules provide for such a situation in Rule 26(b)(2):

(B) Specific Limitations on Electronically Stored Information. A party need not provide discovery of electronically stored information from sources that the party identifies as not reasonably accessible because of undue burden or cost. On motion to compel discovery or for a protective order, the party from whom discovery is sought must show that the information is not reasonably accessible because of undue burden or cost. If that showing is made, the court may nonetheless order discovery from such sources if the requesting party shows good cause, considering the limitations of Rule 26(b)(2)(C). The court may specify conditions for the discovery.

(C) When Required. On motion or on its own, the court must limit the frequency or extent of discovery otherwise allowed by these rules or by local rule if it determines that:

    (i) the discovery sought is unreasonably cumulative or duplicative, or can be obtained from some other source that is more convenient, less burdensome, or less expensive;

    (ii) the party seeking discovery has had ample opportunity to obtain the information by discovery in the action; or

    (iii) the proposed discovery is outside the scope permitted by Rule 26(b)(1).[14]

Thus, if the parties cannot agree whether "not reasonably accessible" sources should be searched, or cannot agree upon the terms under which to search, resolution may be sought by filing a motion to compel OR a motion for a protective order (depending on which party is filing). But, the parties MUST confer and attempt to resolve issues before filing a discovery motion.[15]

Now, what is the extent of "not reasonably accessible"? What if a party's own actions lead to the loss of more accessible data? What if that loss occurs before the case is filed or after the case is filed? Consider the case of *Disability Rights Council of Greater Washington v. Washington Metropolitan Transit Authority.*[16] In that case, the defendants failed to preserve emails for more than *two years* after the complaint was filed, *and* an auto-delete policy continued to dispose of information. The defendants then sought *protection* under Amended Rule 26(b)(2), claiming that data sought by plaintiffs was inaccessible or not readily accessible! Basically,

---

[14]Fed. R. Civ. P. 26(b)(2)(B) & (C).

[15]*See* Fed. R. Civ. P. 26(c) & 37(a)(1); U.S. District Court, N.D.N.Y. Local Rule 7.1(d); U.S. District Court, District of Minnesota Local Rule 7.1(a). A good faith effort to confer can mean the Court "'requires a face-to-face meeting or a telephone conference.'" *Spurgeon v. Olympic Panel Prods., LLC,* 2008 WL 1969654 (W.D. Wash. May 5, 2008) (citing, *inter alia,* FRCP 37 and Local Civil Rule 37(a)(2)(A)). Counsel should also check the local rules of the court, and the individual rules and practices of the district judge or magistrate judge before whom they appear, as some judges require letter briefs in lieu of motions.

[16]242 F.R.D. 139 (D.D.C. 2007) (Facciola, M.J.).

the defendants argued that they would have to go to monthly tapes with "snapshots" of the network system on any given day, and start looking to back-ups. In response, Judge Facciola was less than receptive to defendants' arguments. The magistrate judge found that: "While the [then-] newly amended Federal Rules . . . initially relieve a party from producing electronically stored information that is not reasonably accessible because of undue burden and cost, I am anything but certain that I should permit a party who has failed to preserve accessible information without cause to then complain about the inaccessibility of the only electronically stored information that remains."[17] The judge then added: "It reminds me too much of Leo Kosten's definition of chutzpah: 'that quality enshrined in a man who, having killed his mother and father, throws himself on the mercy of the court because he is an orphan.'"[18]

As a result, the *Disability Rights* Court discussed Rule 26(b)(2)'s provisions, and the 2006 Advisory Notes, and ordered a search of the back-up tapes, where counsel were to agree on a protocol for discovery that was to be submitted to the magistrate judge for approval and signature. The judge did, however, set forth certain issues that the protocol was expected to address, including how to restore the back-up tapes, how to search the tapes once restored, and how privilege claims would be handled (including whether the parties would "contemplate an agreement authorized by Rule 26(b)(5)(B)"—claw-back or sneak-peek).[19]

Also be certain to understand that arguments stating ESI is "not reasonably accessible" must usually be backed-up by particularity—one must present supporting facts and arguments. Courts are not interested in bare assertions, followed by citation of the rule, any more than they are in allusions lacking support for the proffered contentions.[20] For instance, in the case of *Mikron Industries, Inc. v. Hurd Windows & Doors, Inc.*,[21] a 2008 opinion out of the

---

[17] *Id.* at 147.

[18] *Id.*

[19] *Id.* at 148.

[20] *See Great Am. Ins. Co. of New York v. TA Operating Corp.*, No. 06 Civ. 13230(WHP)(JCF), 2008 WL 754646 (S.D.N.Y. Mar. 17, 2008) (Francis, M.J.) (although difficulties in complying with discovery alluded to, defendants "provided no affidavit of anyone with personal knowledge and technical expertise who could offer an informed estimate of the time and cost that might be involved;" plaintiffs' motion to compel granted); *O'Bar v. Lowe's Home Ctrs., Inc.*, 2007 WL 1299180 (W.D.N.C. May 2, 2007); *Knifesource, LLC v. Wachovia Bank, N.A.*, 2007 WL 2326892 (D.S.C. Aug. 10, 2007); *City of Seattle v. Prof'l Basketball Club, LLC*, 2008 WL 539809 (W.D. Wash. Feb. 25, 2008) (Defendant failed to object with specificity; "[a] claim that answering discovery will require the objecting party to expend considerable time and effort to obtain the requested information is an insufficient factual basis for sustaining an objection"); *but cf. Webasto Thermo & Comfort N. Am., Inc. v. BesTop, Inc.*, — F.R.D. —, 2018 WL 3198544 (E.D. Mich. June 29, 2018) (distinguishing *City of Seattle*; "reliance on *City of Seattle*[. . .], is inapposite. In *City of Seattle*, the defendant offered no facts to support its assertion that discovery would be overly burdensome, instead "merely state[ing] that producing such emails 'would increase the email universe exponentially[.]'". . . In our case, Webasto has proffered hard numbers as to the staggering amount of ESI returned based on BesTop's search requests. Moreover, while disapproving of conclusory claims of burden, the Court in *City of Seattle* recognized that the overbreadth of some search terms would be apparent on their face: "'[U]nless it is obvious from the wording of the request itself that it is overbroad, vague, ambiguous or unduly burdensome, an objection simply stating so is not sufficiently specific'") (quoting *Boeing Co. v. Agric. Ins. Co.*, 2007 WL 4358332, at *2, 2007 U.S. Dist. LEXIS 90957, *8 (W.D.Wash. Dec. 11, 2007)).

[21] 2008 WL 1805727 (W.D. Wash. Apr. 21, 2008).

U.S. District Court for the Western District of Washington, the court initially denied defendants' motion for a protective order owing to failure to confer in good faith pursuant to Rule 26(c); however, the court then evaluated the merits of the dispute and found that the defendants had failed to justify their argument that discovery should not be ordered. "Beyond the estimated costs, defendants have not demonstrated an unusual hardship beyond that which ordinarily accompanies the discovery process."[22]

It is also vital that counsel who initially meet and confer in a proceeding be clear about the following, among other things, as suggested by the author of this text:

(a) If the parties seek production of metadata, that should be specifically included in requests and discussed. If there is a later decision on that issue by a party, a court may not permit a supplemental demand once production has taken place;

(b) Requests should specify the format in which the parties wish to have materials produced; and

(c) When discussing search terms, parties and counsel should be careful and specific, but watch for over-inclusiveness—that is, keywords should not include the client's name if it is a company, because it will likely be on many irrelevant/non-responsive documents. If the keywords are not clear and focused, the court may not relieve parties of the cost and time burden at a later date when there are thousands of pages or more to review and cull. This will be discussed later in this chapter.

During discovery, as mentioned above, it is necessary that parties request the *format* (*i.e.* paper, PDF saved to a DVD/CD, native searchable format) that disclosures should take when produced by the opposing party or a subpoenaed non-party.[23] *Why?* Because some courts have held that if the specific type of format is not requested in the discovery demands, production

> *Native Format* – The original format an electronic item is in when first created.

must be accepted in the format chosen by the producing party. It may be difficult to later obtain an order from the court for a different format to be re-produced.[24]

Additionally, if there is a dispute over the format the disclosure is to utilize, the parties will be on record and the court can then address the matter pursuant to the jurisdiction's governing rules of procedure. Consider a 2017 decision from the Texas Supreme Court for important

---

[22] *Id. See also Sung Gon Kang v. Credit Bur. Connection, Inc.,* 2020 WL 1689708, at *4–*5 (E.D. Cal. Apr. 7, 2020) (if a showing is made that there is good cause for production despite inaccessibility, the court may "consider a range of options, including cost-shifting, to alleviate the responding party's hardship").

[23] *See* Model *Joint Electronic Discovery Submission and Order* (Lehrburger, M.J., S.D.N.Y.) (**Appendix 4**), addressing the issues discussed in this chapter.

[24] *See, e.g., 150 Nassau Assoc. v. R.C. Dolner,* 96 A.D.3d 676 (1st Dep't 2012) (Court held that documents were produced in a "searchable PDF format" and plaintiff had not requested "native format" or another format; therefore plaintiff was required to accept production—especially where re-production in another format would only have been for plaintiff's convenience). *Cf. Alberta Ltd. v. Fossil Indus.,* 2014 N.Y. Misc. LEXIS 4110 (Sup. Ct. Suffolk County 2014) (Court directed defendant to produce metadata and other materials as directed at preliminary conference and commensurate with responses already provided, or answer would be stricken as sanction under N.Y. CPLR 3126).

guidance on a number of the issues already discussed in this chapter. The text of the decision is provided here at length for discussion and analysis:

In re State Farm Lloyds (Supreme Court of Texas)

Guzman, Justice

Electronic discovery plays an increasingly significant role in litigation and, often, at significant expense. Given the prevalence of discoverable electronic data, discovery disputes involving electronically stored information (ESI) are a growing litigation concern. With few occasions to enter the fray, we have an opportunity in these consolidated mandamus proceedings to provide further clarity regarding ESI discovery.

Though increasingly common, electronic discovery concerns manifest in variable shades and phases. In this dispute, the parties are at odds over the form in which ESI must be produced, presenting conflicting views regarding the proper interpretation and application of our discovery rules concerning such matters. The requesting party seeks ESI in native form while the responding party has offered to produce in searchable static form, which the responding party asserts is more convenient and accessible given its routine business practices. Agreeing with the requesting party, the trial court ordered production in native form, subject to a showing of infeasibility. The court of appeals denied mandamus relief.

Under our discovery rules, neither party may dictate the form of electronic discovery. The requesting party must specify the desired form of production, but all discovery is subject to the proportionality overlay embedded in our discovery rules and inherent in the reasonableness standard to which our electronic-discovery rule is tethered. The taproot of this discovery dispute is whether production in native format is reasonable given the circumstances of this case. Reasonableness and its bedfellow, proportionality, require a case-by-case balancing of jurisprudential considerations, which is informed by factors the discovery rules identify as limiting the scope of discovery and geared toward the ultimate objective of "obtain[ing] a just, fair, equitable and impartial adjudication" for the litigants "with as great expedition and dispatch at the least expense . . . as may be practicable."

Delay and expense strain not only the resources of the parties, but also the judicial system. Consequently, the discovery rules imbue trial courts with the authority to limit discovery based on the needs and circumstances of the case, including electronic discovery. Thus, when a party asserts that unreasonable efforts are required to produce ESI in the requested form and a "reasonably usable" alternative form is readily available, the trial court must balance any burden or expense of producing in the requested form against the relative benefits of doing so, the needs of the case, the amount in controversy, the parties' resources, the importance of the issues at stake in the litigation, and the importance of the requested format in resolving the issues. Even without quantifying differences in time and expense, evidence that a "reasonably usable" alternative form is readily available gives rise to the need for balancing, and if these factors preponderate against production in the requested form, the trial court may order production as requested only if the requesting party shows a particularized need for data in that form and "the requesting party pay[s] the reasonable expenses of any extraordinary steps required to retrieve and produce the information." Unless ordered otherwise, however, "the responding party need only produce the data reasonably available in the ordinary course of business in reasonably usable form."

Because neither the trial court nor the parties had the benefit of the guidance we seek to provide today, we deny the petitions for writ of mandamus without prejudice, affording the relator an opportunity to reurge its discovery objections to the trial court in light of this opinion.

I. Factual Procedure & Background

. . . .

State Farm, supported by several amici, characterize the lower court rulings as granting requesting parties "essentially unlimited power" to dictate how the responding party must conduct electronic discovery under the Texas Rules of Civil Procedure. We set the matter for oral argument and write to (1) clarify that neither the requesting nor the producing party has a unilateral right to specify the format of discovery under Rule 196.4 and (2) provide guidance regarding the application of Rule 192.4's proportionality factors in the electronic-discovery context.

II. Discussion

. . . .

B. Form of Electronic Discovery

The rules of civil procedure generally extend the scope of discovery to "any matter that is not privileged and is relevant to the subject matter of the pending action, whether it relates to the claim or defense of the party seeking discovery or the claim or defense of any other party." As a counterbalance rested in concerns about "unwarranted delay and expense," Rule 192.4 expressly constrains the scope of discovery as to otherwise discoverable matters:

> The discovery methods permitted by these rules should be limited by the court if it determines, on motion or on its own initiative and on reasonable notice, that:
>
> (a) the discovery sought is unreasonably cumulative or duplicative, or is obtainable from some other source that is more convenient, less burdensome, or less expensive; or
>
> (b) the burden or expense of the proposed discovery outweighs its likely benefit, taking into account the needs of the case, the amount in controversy, the parties' resources, the importance of the issues at stake in the litigation, and the importance of the proposed discovery in resolving the issues.

To put it succinctly, "the simple fact that requested information is discoverable . . . does not mean that discovery must be had." So while metadata may generally be discoverable if relevant and unprivileged, that does not mean production in a metadata-friendly format is necessarily required. Indeed, as a federal district court recently observed, "a weak presumption against the production of metadata has taken hold," which may be due to "metadata's status as 'the new black,' with parties increasingly seeking its production in every case, regardless of size or complexity."

Whether production of metadata-accessible forms is required on demand engages the interplay between the discovery limits in Rule 192.4 and production of electronic discovery under Rule 196.4, which provides:

> To obtain discovery of data or information that exists in electronic or magnetic form, the requesting party must specifically request production of electronic or magnetic data and specify the form in which the requesting party wants it produced. The responding party must produce the electronic or magnetic data that is responsive to the request and is reasonably available to the responding party in its ordinary course of business. If the responding party cannot—through reasonable efforts—retrieve the data or information requested or produce it in the form requested, the responding party must state an objection complying with these rules. If the court orders the responding party to comply with the request, the court must also order that the requesting party pay the reasonable expenses of any extraordinary steps required to retrieve and produce the information.

In *In re Weekley Homes*, we summarized the "proper procedure" under Rule 196.4, including the directive that the parties "make reasonable efforts to resolve the dispute without court intervention." Meeting and conferring to resolve e-discovery disputes without court intervention is essential because discovery of electronic data involves case-specific considerations and each side possesses unique access to information concerning reasonable and viable production methods, resources (technological or monetary, for instance), and needs. *In re Weekley Homes* did not consider the precise issues presented here, however—namely, whether the form requested controls and how proportionality factors into the analysis.

At its core, the homeowners' claim that native-form production is required presumes the requesting party can unilaterally determine the form of production. But Rule 196.4 cannot be construed so narrowly given its focus on "reasonable" efforts and "reasonabl[e]" availability. Though the term "reasonable" cannot be comprehensively defined, it naturally invokes the jurisprudential considerations articulated in Rule 192.4.

Thus, if the responding party objects that electronic data cannot be retrieved in the form requested through "reasonable efforts" and asserts that the information is readily "obtainable from some other source that is more convenient, less burdensome, or less expensive," the trial court is obliged to consider whether production in the form requested should be denied in favor of a "reasonably usable" alternative form. In line with Rule 192.4, the court must consider whether differences in utility and usability of the form requested are significant enough—in the context of the particular case—to override any enhanced burden, cost, or convenience. If the burden or cost is unreasonable compared to the countervailing factors, the trial court may order production in (1) the form the responding party proffers, (2) another form that is proportionally appropriate, or (3) the form requested if (i) there is a particularized need for otherwise unreasonable production efforts and (ii) the court orders the requesting party to "pay the reasonable expenses of any extraordinary steps required to retrieve and produce the information."

Here, State Farm contends searchable static form is not only adequate, but more cost-effective and convenient. In other cases, a party may resist static-format production, if requested, because producing in native format is easier and less expensive. When a reasonably usable form is readily available in the ordinary course of business, the trial court must assess whether any enhanced burden or expense associated with a requested form is justified when weighed against the proportional needs of the case. The proportionality inquiry requires case-by-case balancing in light of the following factors:

1. Likely benefit of the requested discovery: If the benefits of the requested form are negligible, nonexistent, or merely speculative, any enhanced efforts or expense attending the requested form of production is undue and sufficient to deny the requested discovery. In such cases, quantifying or estimating time and expenses would not be critical, as it may be when benefits clearly exist. At the opposite end of the spectrum, a particularized need for the proposed discovery will weigh heavily in favor of allowing discovery as requested but, depending on the force of other prudential concerns, may warrant cost-shifting for any "extraordinary steps" required.

   Courts should consider cumulative effects rather than viewing benefits and burdens in a vacuum. Here, for example, many similar cases arising from the same extreme weather event are currently pending against State Farm. The identification and retrieval process State Farm would have to develop for native-form production may be a ticket for one train only—exponentially increasing the burden when considered in the context of repeated litigation—or have broader utility, which could have a cumulatively reductive effect. The record does not tell us, but if there are likely uses for the identification and retrieval process beyond the instant mandamus cases, initial burden and expense may be substantially ameliorated, and if not, the burden and expense may be significantly enhanced.

2. The needs of the case: In these mandamus cases, the homeowners seek native production both for optimal search capability and to access metadata. Recognizing that metadata serves no genuinely useful purpose in many cases, "many parties, local rules and courts have [in current practice] endorsed the use of [static] image production formats, principally TIFF and [PDF] formats." But metadata may be important, even dispositive in some cases.

   Relevance of metadata and the relative significance to the case must be determined on a case-by-case basis. But metadata's relevance must be obvious or at least linked, more or less concretely, to a claim or defense. Hypothetical needs, surmise, and suspicion should be afforded no weight. As a general proposition, metadata may be necessary to the litigation when the who, what, where, when, and why ESI was generated is an actual issue in the case, not merely a helpful or theoretical issue. Take, for instance, a wrongful termination case where timing of the events leading up to and following termination or authorship of case-critical documents might be a central issue in the case.

Here, the homeowners have argued production in native format is necessary to ensure disclosure of all potentially relevant information. By way of example, the homeowners provided evidence that captions annotating some photographs of hail damage to a house—such as "north elevation with hail damage missed by [claims adjuster]"—were not captured when the photographs were converted to static format in the ECS. State Farm insists, however, that PDF production of photos from the ECS does not support the necessity for native-form production, as the homeowners claim, because the omitted photo captions were provided to the homeowners through an ECS "caption log" as well as via production from another database State Farm had identified as storing discoverable ESI. In evaluating whether a particular form of production is required, the court should consider not only the relative importance of the information to the central issues in the case, but also availability of that information from some other source that is more convenient, less burdensome, or less expensive.

3. <u>The amount in controversy</u>: Accessibility—or relative inaccessibility—of electronic data contributes to increased costs and burdens associated with electronic discovery. "While large companies are still learning to cope with e-discovery costs, e-discovery remains costly and complex for the small company, small case, and unrepresented litigant. Because e-discovery is very expensive and quite complicated, the advent of e-discovery is forcing settlements, and thus, denying litigants an opportunity to litigate the merits of the case."

When the discovery rules were adopted, an explanatory guide explained an initial impetus for discovery constraints that rings just as true in today's electronic discovery frontier:

> For four decades following adoption of the federal rules, discovery procedures were continually expanded. In the 1970's, however, it became apparent that unrestricted discovery could be used to undermine the cause of justice if litigants with resources and motive to do so could drive up the cost of litigation, effectively pricing their opponents out of court and delaying disposition. Innovations in computer word processing, facsimile transmissions, and photocopying quickly made it possible for litigants of even modest means to drive up litigation costs and by "burying" their opponents in voluminous "boilerplate" discovery requests or objections, often with little more than the touch of a button. Technological changes have greatly increased the volume of documents and things that can be discoverable in a lawsuit. These developments in discovery practice have been compounded by an unfortunate weakening of professional norms that in earlier times would have made misuse or abuse of discovery unthinkable.

For these reasons, the amount in controversy plays a pivotal role in determining whether production in a specified form is justified given the burden or expense required to meet the demand.

4. <u>The parties' resources</u>: Whether the producing party has the means to fairly and realistically produce in the requested format is a significant proportionality consideration. An expense that is a drop in the bucket to one party, may be insurmountable to another. While this factor is important to the balancing inquiry "considerations of the parties' resources does not foreclose discovery requests addressed to an impecunious party, nor justify unlimited discovery requests addressed to a wealthy party." Rather, "'the court must apply the standards in an even-handed manner that will prevent use of discovery to wage a war of attrition or as a device to coerce a party, whether financially weak or affluent.'"

But beyond financial resources, one must also consider whether the requesting party has the technological resources to make proper use of ESI in the form requested. A high-powered luxury sports car is useless to someone who lacks a license to drive it. To that point, the homeowners' discovery expert testified it would be more convenient for lawyers lacking advanced technology to use image formats, but the homeowners' counsel in this case has "invested considerably to have the tools necessary to be able to deal with advanced forms of information." This may be another way of looking at the benefit versus the burden; if the potential benefits could not be realized, any associated burden would be unwarranted.

5. <u>Importance of the issues at stake in the litigation</u>: Legal disputes are always important to those who are litigating them. For one side, however, the precedential value may be more significant than the other, justifying an outlay of time and expenses that would otherwise be unwarranted. Likewise, "'many cases in public policy spheres, such as employment practices, free speech, and other matters, may have importance far beyond the monetary amount involved.' Many other substantive areas also may involve litigation that seeks relatively small amounts of money, or no money at all, but that seeks to vindicate vitally important personal or public values."

6. <u>The importance of the proposed discovery in resolving the litigation</u>: Discovery must bear at least a reasonable expectation of obtaining information that will aid the dispute's resolution. Reasonable discovery does not countenance a "fishing expedition."

7. <u>Any other articulable factor bearing on proportionality</u>: The foregoing factors are derived directly from the discovery rules, but are certainly not exclusive. As history tells, technology is constantly evolving at rapidly increasing rates. The legal system is not nearly as agile, leaving us in a perpetually responsive posture. Trial courts have flexibility to consider any articulable factor that informs this jurisprudential inquiry.

C. Parity with the Federal Rules of Civil Procedure

Our application of proportionality principles in this context aligns electronic-discovery practice under the Texas Rules of Civil Procedure with electronic-discovery practice under the Federal Rules of Civil Procedure.

Rule 34 of the federal procedural rules, which governs electronic discovery in federal-court proceedings, states:

- A party may request "electronically stored information . . . stored in any medium from which information can be obtained either directly or, if necessary, after translation by the party into a reasonably usable form."

- The request for ESI "may specify the form or forms in which electronically stored information is to be produced."

- "The responding party may state that it will produce copies of documents or of electronically stored information instead of permitting inspection."

- "The response may state an objection to a requested form for producing electronically stored information. If the responding party objects to a requested form—or if no form was specified in the request—the party must state the form or forms it intends to use."

- Absent agreement or court order:

   "If a request does not specify a form for producing electronically stored information, a party must produce it in a form or forms in which it is ordinarily maintained or in a reasonably usable form or forms"; and

   "A party need not produce the same electronically stored information in more than one form."

Rule 34's plain language does not permit either party to unilaterally dictate the form of production for ESI. The default form is "a form or forms in which [ESI] is ordinarily maintained or in a reasonably usable form or forms," and the trial court retains discretion to order discovery in a format that is appropriate to the circumstances.

The Advisory Committee Notes to Rule 34 explain:

   If the requesting party is not satisfied with the form stated by the responding party, or if the responding party has objected to the form specified by the requesting party, the parties must meet and confer . . . in an effort to resolve the matter before the requesting party can file a motion to compel. If they cannot agree and the court resolves the dispute, the court is not limited to the forms initially chosen by the requesting party, stated by the responding party, or specified in this rule for situations in which there is no court order or party agreement.

The federal rules thus place the power to decide the form of electronic discovery not with the parties, but within the trial court's sound discretion.

In *Dizdar v. State Farm Lloyds*, a similar hail-storm lawsuit against State Farm involving a similar ESI protocol, a federal district court exercised its discretion to deny the plaintiff's request for native and near-native production. The court explained that while the federal rules give "preference to documents produced as they are ordinarily maintained absent a specific request from the other party . . . it does not follow that if the requesting party specified a form for production, the information must be produced in that form." The court determined that the plaintiffs could gain all the information necessary to litigate the claims against State Farm from the ECS claim file such that the existence of any additional burden was not justified. Finding State Farm's proffered form of production "a reasonably usable format" within the meaning of the federal discovery rules, the court granted State Farm's request to produce ESI in the searchable static image claims files archived in the ECS.

To be sure, there are differences in language between the Texas rule and the federal rule. But as we affirmed in *In re Weekley Homes*, "our rules as written are not inconsistent with the federal rules or the case law interpreting them," even though they may not "mirror the federal language."

In this regard, we observe the proportionality principles under the federal rules similarly limit discovery of otherwise discoverable information. Rule 26 delineates the scope of discovery as inherently limited by proportionality:

> Unless otherwise limited by court order, the scope of discovery is as follows: Parties may obtain discovery regarding any nonprivileged matter that is relevant to any party's claim or defense and proportional to the needs of the case, considering the importance of the issues at stake in the action, the amount in controversy, the parties' relative access to relevant information, the parties' resources, the importance of the discovery in resolving the issues, and whether the burden or expense of the proposed discovery outweighs its likely benefit.

The factors expressly identified as limitations on the scope of discovery accord with the proportionality analysis we have explicated under our discovery rules, keeping our procedures in line with their federal counterparts.

Though the proportionality factors were recently relocated within the federal rules, proportionality has long been a required constraint on the scope of discovery, enacted decades ago " 'to deal with the problem of overdiscovery.'" Proportionality, it is said, acts as a governor " 'to guard against redundant or disproportionate discovery by giving the court authority to reduce the amount of discovery that may be directed to matters that are otherwise proper subjects of inquiry.'" The 2015 amendments to the federal rules have not altered "the existing responsibilities of the court and the parties to consider proportionality," do not require the requesting party to address all proportionality considerations, and do not permit the opposing party to rely on boilerplate objections of disproportionality. Rather, the amendment is directed to changing the existing "mindset" that relevance is enough, restoring proportionality as the "collective responsibility" of the parties and the court.

Discovery is necessarily a collaborative enterprise, and particularly so with regard to electronic discovery. The opposing party must object and support proportionality complaints with evidence if the parties cannot resolve a discovery dispute without court intervention, but the party seeking discovery must comply with proportionality limits on discovery requests and "may well need to . . . make its own showing of many or all of the proportionality factors." Recent federal cases provide helpful examples of proportionality analyses in e-discovery cases. Consistent with an individualized, case-specific inquiry, all manner of outcomes are represented—denials of requested discovery on proportionality grounds, rejection of proportionality complaints, and cases taking a more graduated approach by allowing limited bell-weather discovery to inform the propriety of further discovery. Ultimately, the "court's responsibility, using all the information provided by the parties, is to consider [the proportionality] factors in reaching a case-specific determination of the appropriate scope of discovery." The same applies with regard to electronic-discovery practices under the Texas Rules of Civil Procedure.

III. Conclusion

Today, we elucidate the guiding principles informing the exercise of discretion over electronic-discovery disputes, emphasizing that proportionality is the polestar. In doing so, we further a guiding tenet of the Texas Rules of Civil Procedure: that litigants achieve a "just, fair, equitable and impartial adjudication . . . with as great expedition and dispatch and at the least expense . . . as may be practicable." Because the trial court and the parties lacked the benefit of our views on the matter, neither granting nor denying mandamus relief on the merits is appropriate. Accordingly, we deny the request for mandamus relief without prejudice to allow the relator to seek reconsideration by the trial court in light of this opinion.[25]

Finally, consider that the format chosen may help with presentation if you get to trial, especially with a jury. It has been found that people recall memories more easily if they are initially presented both visually and orally, as opposed to just being told something.

Many courtrooms today (especially in the federal courts) are becoming twenty-first-century courtrooms, with computer connections, and videoscreens/computer monitors/televisions connected for easy viewing by the judge, jury, and counsel. There are professional journal articles that speak to utilization of technology in the courtroom.

Consider that when requesting your format of production. Perhaps you want all information on disk. First, you can easily project this from a computer program later (many courts now permit laptops in the courtroom, sometimes first requiring permission of the judge). Second, with computerized programs that make it possible to search documents electronically, as discussed briefly below, such a production format will make it easier to store discovery productions on your office computer system, and search for particular documents and information.

1.4 – More Specific Information—Preservation Orders

There may be times, however, when regardless of service of a litigation hold and preservation letter/notice (warning an entity about preservation obligations and the relevant law),[26] an attorney believes there is a need to obtain a court order to prevent loss or destruction of information held by another entity or party. This preservation order can be thought of

---

[25] 520 S.W.3d 595, 598-615, 60 Tex. Sup. Ct. J. 1114 (2017) (Guzman, J.) (footnotes and citations omitted).

[26] *See* Model External Preservation/Litigation Hold Letter to Adversary or Third Party, **Appendix 5**.

much like a preliminary injunction—enjoining the deletion or destruction of ESI, and ordering the affirmative preservation and safeguarding of same. Depending on the relevant jurisdiction, however, the standard that must be met to successfully obtain a preservation order varies. Courts and commentators have determined that one of three paths is generally followed.

First Path—The litigant seeking the order must meet the standard for a preliminary injunction. Those courts following this line indeed interpret the matter as a party seeking to restrain another entity or an opposing party from destroying material.[27]

Second Path—On this path, the courts have found the preliminary injunction standard not appropriate, and instead the party seeking relief must demonstrate that the order is "necessary and not unduly burdensome."[28]

Third Path—In these jurisdictions, the courts follow a mixture of the first two paths. Courts have recognized that incorporation of a preliminary injunction standard is necessary but have then developed a three-pronged test: (1) level of concern the court has concerning preservation in absence of an order, (2) any irreparable harm likely to result to the requesting party absent the preservation order, and (3) the capability of the opposing party to maintain the evidence sought to be preserved.[29]

> *Metadata* – Data about data, for example when a document was created, the author, the version, the format (Excel, Word, e-mail), etc.

1.5 – Briefly on FOIL and FOIA in a Realm of e-Discovery

Courts in the United States have been assessing and evaluating government obligations in searching records and responding to requests under the Freedom of Information Act (FOIA) and Freedom of Information Law (FOIL) in our present eWorld. In New York, courts have

---

[27] J. Carroll, *Preservation of Documents in the Electronic Age—What Should Courts Do?*, 1 Fed. Cts. L. Rev. 3 (2006) (citing *Madden v. Wyeth*, No. 3-03-CV-0167-R, 2003 WL 21443404 (N.D. Tex. Apr. 16, 2003); *Pepsi-Cola Bottling Co. of Olean v. Cargill, Inc.*, Civ. No. 3-95-784, 1995 WL 783610 (D. Minn. Oct. 20, 1995)). Note both of these cases were disagreed with by *Pueblo of Laguna, infra*, in the Second Path. *See, generally, Weiller v. New York Life Ins. Co.*, 800 N.Y.S.2d 359 (Sup. Ct. New York County 2005) (Cahn, J.) (citing *Schwartz v. Lubin*, 6 A.D.2d 108 (1st Dep't 1958) (concerning paper, but nevertheless the same idea)). *See also Deggs v. Fives Bronx, Inc.*, 2020 WL 3100023 (M.D. La. June 11, 2020).

[28] J. Carroll, *Preservation of Documents in the Electronic Age—What Should Courts Do?*, 1 Fed. Cts. L. Rev. 3 (2006) (citing *Pueblo of Laguna v. United States*, 60 Fed. Cl. 133 (2004)).

[29] J. Carroll, *Preservation of Documents in the Electronic Age—What Should Courts Do?*, 1 Fed. Cts. L. Rev. 3 (2006) (citing *Capricorn Power Co., Inc. v. Siemens Westinghouse Power Corp.*, 220 F.R.D. 429 (W.D. Pa. 2004)). *See also Tellis v. Le Blanc*, 2018 WL 6006909 (W.D. La. Nov. 14, 2018).

held that documents and records stored in electronic formats, including "system" metadata, are subject to FOIL requests, and there are corresponding review requirements.[30]

The federal courts, likewise, have discussed e-discovery tools utilized by federal agencies and departments when conducting FOIA searches and responses.[31]

Therefore, be mindful that regardless of whether the party from whom records are sought is a private party in a litigation, or a governmental entity under FOIA or FOIL obligations, e-discovery principles, protocols, and tools can be applied.

1.6 – Costs and Cost-Shifting

This text only briefly addresses cost-shifting and other such topics. Some courts, such as the Sixth U.S. Circuit Court of Appeals, have permitted prevailing parties to recover costs of e-discovery as a recoverable cost under 28 U.S.C. § 1920—*Colosi v. Jones Lang LaSalle Am., Inc.*[32]—although the Third Circuit went in the other direction in *Race Tires Am. v. Hoosier Racing, Inc.*[33] Consider the *Colosi* Court's decision in-depth, for analysis and discussion:

> ### *Colosi v. Jones Lang LaSalle Americas, Inc.* (U.S. Court of Appeals, Sixth Circuit)
>
> *Cook, U.S. Circuit Judge*
>
> Plaintiff–Appellant Brenda C. Colosi lost a wrongful termination suit against her former employer, Defendant–Appellee Jones Lang LaSalle Americas, Inc. (JLL). As the prevailing party, JLL filed a $6,369.55 bill of costs that the court clerk approved without modification. See Fed.R.Civ.P. 54(d)(1). Colosi objected to most of the charges and moved the district court to

---

[30] *See Irwin v. Onondaga County Res. Recovery Ag.*, 72 A.D.3d 314, 895 N.Y.S.2d 262 (App. Div. 4th Dep't 2010) (citing *Matter of Data Tree, LLC v. Romaine*, 9 N.Y.3d 454 (2007); N.Y. PUBLIC OFFICERS LAW § 86[4]) (court did not reach issue of "substantive" and "embedded" metadata); *Kirsch v. Board of Educ. of Williamsville Cent. School Dist.*, 152 A.D.3d 1218, 57 N.Y.S.3d 870 (App. Div. 4th Dep't 2017)). However, when it comes to the *format* or *form* of *production* in response to a FOIL or FOIA request, *see Leyton v. City University of New York*, 25 Misc.3d 1214(A), 901 N.Y.S.2d 907 (Sup. Ct. N.Y. County 2009) (Table) ("FOIL provides that '[w]hen an agency has the ability to retrieve or extract a record or data maintained in a computer storage system with reasonable effort, it shall be required to do so. When doing so requires less employee time than engaging in manual retrieval or redactions from non-electronic records, the agency shall be required to retrieve or extract such record or data electronically' . . . . Here, contrary to petitioners' contention, FOIL does not 'require[ ] agencies to produce documents electronically when requested by an applicant'") (citing N.Y. PUBLIC OFFICERS LAW § 89[3]).

[31] *See W. Values Project v. U.S. Dep't of Justice*, 317 F.Supp.3d 427 (D.D.C. 2018); *Leopold v. Nat'l Secur. Ag.*, 196 F.Supp.3d 67 (D.D.C. 2016); *Families for Freedom v. U.S. Customs & Border Protection*, 837 F.Supp.2d 331 (S.D.N.Y. 2011). *See also NAACP Legal Def. & Educ., Fund, Inc. v. Dep't of Justice*, 2020 WL 2793015 (S.D.N.Y. May 29, 2020).

[32] 781 F.3d 293 (6th Cir. 2015).

[33] 674 F.3d 158 (3d Cir. 2012). *See also U.S. v. Halliburton Co.*, 954 F.3d 307, 310-312 (D.C. Cir. 2020) (citing, *inter alia, Colosi* and *Race Tires*, and discussing what qualifies for/as costs under § 1920).

reduce the bill to $253.50. The district court denied the motion, finding each cost reasonable, necessary to the litigation, and properly taxable under statute. See 28 U.S.C. § 1920. Colosi renews her objections on appeal. We AFFIRM the district court's judgment.

I.

Section 1920 circumscribes the types of costs district courts may tax against the losing party. *Crawford Fitting Co. v. J.T. Gibbons, Inc.*, 482 U.S. 437, 445, 107 S.Ct. 2494, 96 L.Ed.2d 385 (1987). We review de novo whether taxed expenses fall within § 1920's list of allowable costs. *BDT Prods., Inc. v. Lexmark Int'l, Inc.*, 405 F.3d 415, 417 (6th Cir. 2005), abrogated on other grounds by *Taniguchi v. Kan Pac. Saipan, Ltd.*, ——U.S. ——, 132 S.Ct. 1997, 182 L.Ed.2d 903 (2012). But "[a]s long as statutory authority exists for a particular item to be taxed as a cost, we do not overturn a district court's determination that the cost is reasonable and necessary, absent a clear abuse of discretion." *Id.* (quoting *Baker v. First Tenn. Bank Nat'l Ass'n*, No. 96–6740, 1998 WL 136560, at *2 (6th Cir. Mar. 19, 1998) (per curiam) (internal punctuation omitted)).

. . . .

III.

Colosi also challenges the district court's decision to tax the cost of imaging her personal computer's hard drive. She argues that, as a matter of law, "most electronic discovery costs such as the imaging of hard drives are not recoverable as taxable costs." (Appellant Br. at 13.) Yet the statute includes no categorical bar to taxing electronic discovery costs. Rather, it authorizes courts to tax "the costs of making copies of any materials where the copies are necessarily obtained for use in the case." 28 U.S.C. § 1920(4). Thus, we first ask whether imaging a hard drive, or other physical storage device, falls within the ordinary meaning of "making copies." *See Taniguchi*, 132 S.Ct. at 2002.

The Oxford English Dictionary generally defines "copy" as a "transcript or reproduction of an original." 3 Oxford English Dictionary 915 (2d ed.1989). Although initially used to describe one "writing transcribed from . . . another," speakers long ago began to use the word figuratively to mean "[s]omething made or formed, or regarded as made or formed, in imitation of something else; a reproduction, image, or imitation." Id. (emphasis added). Because Congress last amended the statute in 2008 to change "papers" to "materials," the figurative use seems the more appropriate. *See* Judicial Administration and Technical Amendments Act of 2008, Pub.L. No. 110–406, § 6, 122 Stat. 4291, 4292 (2008). Moreover, courts have long understood that the phrase "making copies" fairly includes the production of imitations in a medium or format different than the original. *See CBT Flint Partners, LLC v. Return Path, Inc.*, 737 F.3d 1320, 1329 (Fed.Cir.2013) (explaining that one can "copy" a document from paper to digital format and vice versa); *Race Tires Am., Inc. v. Hoosier Racing Tire Corp.*, 674 F.3d 158, 166 (3d Cir.2012) (explaining that a parchment replica of the stone tablets containing the Ten Commandments falls within the ordinary meaning of "copy"); *BDT Prods.*, 405 F.3d at 419–20 (upholding cost award for scanning and creating electronic images of paper documents under § 1920(4)).

Imaging a hard drive falls squarely within the definition of "copy," which tellingly lists "image" as a synonym. And the name "imaging" describes the process itself. Imaging creates "an identical copy of the hard drive, including empty sectors." *CBT Flint*, 737 F.3d at 1328 (quoting The Sedona Conference, *The Sedona Conference Glossary: E–Discovery & Digital Information Management* 27 (Sherry B. Harris et al. eds., 3d ed. 2010)). The image serves as a functional reproduction of the physical storage disk. From the image file, one can access any application file or electronic document on the hard drive with all that document's original properties and metadata intact. *Id.* If not actually made or formed in the image of the hard drive, we certainly regard it as such. *See* 3 Oxford English Dictionary 915. Thus, a plain reading of the statute authorizes courts to tax the reasonable cost of imaging, provided the image file was necessarily obtained for use in the case. *See CBT Flint*, 737 F.3d at 1329–30; id. at 1334 n. 1 (O'Malley, J., concurring in part and dissenting in part).

In urging the opposite interpretation, Colosi relies on the Third Circuit's decision in *Race Tires America, Inc. v. Hoosier Racing Tire Corp.* In that case, the prevailing party sought to recover the entire cost of its electronic discovery, including imaging hard drives, deduplicating image files, populating a database, reviewing the files for discoverable information, redacting privileged information, converting the responsive documents to an agreed-upon format, and burning these document files onto a DVD for production. *See Race Tires*, 674 F.3d at 161–62, 166–67, 169, 171 n. 11. The Third Circuit construed the phrase "making copies" in § 1920(4) to exclude most of these processes in light of historical context and the Supreme Court's traditionally narrow reading of § 1920. *Id.* at 166–72 (citing *Crawford*, 482 U.S. at 441–42, 107 S.Ct. 2494). It compared many of these processes to untaxable discovery procedures from the pre-digital era like visiting a client's records room, searching for responsive documents, copying the relevant papers, and bringing them back to the law firm for review and redaction. *Id.* at 169. It concluded that only converting responsive documents to an agreed-upon format and burning those files onto a DVD were similar enough to the pre-digital act of photocopying to be "the functional equivalent of 'making copies.'" *Id.* at 171 & n. 11; *accord Country Vintner of N.C., LLC v. E. & J. Gallo Winery, Inc.*, 718 F.3d 249, 260–61 (4th Cir.2013) (adopting *Race Tires*'s rationale in a case with similar facts).

We find this construction overly restrictive. In attempting to do justice to the historically limited role of taxing costs in the American system, the Race Tires court ignored § 1920's text. We need not ask whether imaging is the "functional equivalent" of making photocopies in the era before electronic discovery because—consistent with the 2008 amendments—the procedure comes within the ordinary meaning of "making copies of any materials." *See Taniguchi*, 132 S.Ct. at 2002–06 (beginning its interpretation of § 1920 with the ordinary meaning of words before examining statutory context). While the Third Circuit rightly worried over expanding the scope of § 1920 to include expensive electronic discovery procedures not contemplated by Congress, this concern more appropriately pertains to the context-dependent question of whether the prevailing party necessarily obtained its copies for use in the case.

Generally, trial courts have the discretion to tax the cost of "copies attributable to discovery" as necessarily obtained for use in the case, even if neither party uses the copy at trial. *Jordan v. Vercoe*, No. 91–1671, 1992 WL 96348, at *1 (6th Cir. May 7, 1992) (order); accord Country Vintner, 718 F.3d at 257 & n. 9 (listing cases from other circuits). Courts often contrast copies necessarily produced to meet discovery obligations, which are recoverable, with copies produced solely for internal use or the convenience of counsel in conducting discovery, which are not. *See, e.g., In re Ricoh Co., Ltd. Patent Litig.*, 661 F.3d 1361, 1365 (Fed.Cir.2011) (applying Ninth Circuit law); *EEOC v. W & O, Inc.*, 213 F.3d 600, 623 (11th Cir.2000); *M.T. Bonk Co. v. Milton Bradley Co.*, 945 F.2d 1404, 1410 (7th Cir.1991). Even in *Race Tires*, the prevailing party's reason for imaging its hard drives—to facilitate counsel's review of discoverable documents rather than to create the actual production—steered the Third Circuit's analysis. *See Race Tires*, 674 F.3d at 169, 171 n. 11 ("It is all the other activity, such as searching, culling, and deduplication, that are not taxable.").

Here, we perceive no abuse of discretion in ruling imaging costs reasonable and necessary. Rather than produce relevant computer files in response to JLL's discovery requests and the district court's orders compelling production, Colosi delivered her computer to her attorney's office and demanded that JLL send a third-party vendor to image its hard drive under her attorney's supervision. Colosi's decision to tender the physical computer forced JLL to dispatch a vendor and make an image before it could search the hard drive for discoverable information, as the district court determined it had a right to do. JLL sent a vendor to image the hard drive not as an expedient; this was the sole avenue permitting review of Colosi's files. We analogize this situation to the more typical—and taxable—cost of a party delivering an image file in response to an opponent's production request. *See CBT Flint*, 737 F.3d at 1334 n. 1 (O'Malley, J., concurring in part and dissenting in part) ("I do not question that the cost of imaging source media would fall under section 1920(4) if it were directly imaged and provided to the opposing party as part of discovery."). The vendor's invoice excludes the cost of deduplication, indexing, and the other non-copying electronic discovery services. *See CBT Flint*, 737 F.3d at 1331–32; Race Tires, 674 F.3d at 169–70. In fact, Colosi concedes that "JLL never went beyond the mere electronic copying of all of the Colosi family's personal computer files." (Appellant Reply at 6.) And she points to nothing in the record showing that the district court abused its discretion in finding the invoice amount reasonable.

IV.

We therefore AFFIRM the district court's judgment awarding $6,369.55 in costs to JLL as the prevailing party.[34]

Also note that a general difference between Federal and New York State practice that once existed is no longer a great divide. In the federal system the producer pays (with concerns

---

[34] 781 F.3d 293, 295–299 (footnotes omitted).

potentially later shaping consideration of cost-shifting). In states such as New York, the producer now also pays, although some state courts had previously held that the requestor paid in order to self-regulate costs of litigation (with generally no cost-shifting).[35]

1.7 – Rule 26(b)(5)(B), "Claw-Back" and "Sneak-Peek" Procedures

Federal Rule 26(b)(5)(B)

In 2006, Rule 26 was amended with the addition of (b)(5)(B), providing that:

> *"Claw–Back"* – The ability of a party to take back documents or materials mistakenly given/disclosed to another party.

> *"Sneak – Peek"* – A discovery method by which one party gets to look at the documents or materials of another party before a formal discovery production or possibly even privilege review takes place. Occurs by agreement or Court order.

If information produced in discovery is subject to a claim of privilege or of protection as trial-preparation material [i.e. work product], the party making the claim may notify any party that received the information of the claim and the basis for it. After being notified, a party must promptly return, sequester or destroy the specified information and any copies it has; must not use or disclose the information until the claim is resolved; must take reasonable steps to retrieve the information if the party disclosed it before being notified; and may promptly present the information to the court under seal for a determination of the claim. . . .[36]

Furthermore, according to the Advisory Committee Notes, Federal Rule 26(b)(5)(A) provides the procedure for *initially withholding* information and materials from disclosure during discovery (*i.e.* privilege log requirements), while Federal Rule 26(b)(5)(B) provides the

---

[35] *See U.S. Bank Nat'l Ass'n v. GreenPoint Mortg. Funding, Inc.*, 94 A.D.3d 58 (1st Dep't 2012) (adopting the federal standard, such as that set forth in the *Zubulake* case, for costs, and holding that producing parties bear their own costs, subject to taxing by prevailing party—and of course subject to cost-shifting motion for prohibitive electronic discovery costs, or not readily accessible data) (disagreeing with *Lipco Electrical Corp. v. ASG Consulting Corp.*, 2004 N.Y. Misc. LEXIS 1337, 4 Misc. 3d 1019A (Sup. Ct. Nassau County Aug. 18, 2004)); *see also MBIA Ins. Corp. v Countrywide Home Loans, Inc.*, 895 N.Y.S.2d 643 (Sup. Ct. N.Y. County 2010) (rejecting "requestor pays" rule; citing and discussing changing authority; shift of discovery costs denied; each party bears burden of its case at start; if electronic material deleted or archived, then the court may shift costs). Consider also the case of *Adkins v. EQT Prod. Co.*, 2012 U.S. Dist. LEXIS 75133 (W.D. Va. 2012) (discussing electronic discovery generally, and then cost shifting and avoidance of production if materials not readily accessible due to undue burden or cost, and citing Federal Rule 26, The Sedona Conference, *et al.*; further the court applied a claw-back analysis, and cited to *Hopson v. Mayor & City Council of Baltimore*, 232 F.R.D. 228 (D. Md. 2005)). *See also Matter of Khagan*, 66 Misc. 3d 335, 114 N.Y.S. 3d 824 (Surr. Ct. Queens County 2019) (addressing granting of legal fees to non-party respondents to subpoenas, to be paid by objectors). This sets the stage for the next segment of this chapter.

[36] FED. R. CIV. P. 26(b)(5)(B).

framework and procedure to present issues of privilege and work product protection to a court *after* disclosure has been made. Rule 26(b)(5)(B) does not address or determine whether waiver was made, as the courts use other mechanisms and evaluations to make that determination (*i.e.*, number of documents, care taken in producing and reviewing documents, time taken before demand return, and like considerations).

The Advisory Committee Notes provide, in part:

> Rule 26(b)(5)(B) does not address whether the privilege or protection that is asserted after production was waived by the production. The courts have developed principles to determine whether, and under what circumstances, waiver results from inadvertent production of privileged or protected information. Rule 26(b)(5)(B) provides a procedure for presenting and addressing these issues. Rule 26(b)(5)(B) works in tandem with Rule 26(f), which is amended to direct the parties to discuss privilege issues in preparing their discovery plan, and which, with amended Rule 16(b), allows the parties to ask the court to include in an order any agreements the parties reach regarding issues of privilege or trial-preparation material protection. Agreements reached under Rule 26(f)(4) and orders including such agreements entered under Rule 16(b)(6) may be considered when a court determines whether a waiver has occurred. Such agreements and orders ordinarily control if they adopt procedures different from those in Rule 26(b)(5)(B).
>
> A party asserting a claim of privilege or protection after production must give notice to the receiving party. That notice should be in writing unless the circumstances preclude it. Such circumstances could include the assertion of the claim during a deposition. The notice should be as specific as possible in identifying the information and stating the basis for the claim. Because the receiving party must decide whether to challenge the claim and may sequester the information and submit it to the court for a ruling on whether the claimed privilege or protection applies and whether it has been waived, the notice should be sufficiently detailed so as to enable the receiving party and the court to understand the basis for the claim and to determine whether waiver has occurred. Courts will continue to examine whether a claim of privilege or protection was made at a reasonable time when delay is part of the waiver determination under the governing law.[37]

Federal Rule 26(f) was amended at the same time as Rule 26(b)(5)(B), and together they provide procedures for parties to discuss privilege and work-product at the start of discovery, and seek any necessary orders from the court. Because of concern in large part over the sheer volume of electronic discovery in many cases, and the pace of discovery, the rule amendments aimed to alleviate certain prevalent problems and worries. Some of the various "procedures" have been called "claw-back" and "sneak-peek" or "quick-peek," and over the intervening years they have addressed concerns over waiver.

One particular method under Federal Rule 26(b)(5)(B) is colloquially called "claw-back." In the instances where this arises, production is made and privileged materials slip through

---

[37] FED. R. CIV. P. 26, 2006 Advisory Committee Notes.

to the opposing side. Provisions may exist, either in agreement between the parties or under operation of Rule 26(b)(5)(B), allowing the producing party to reclaim, or the receiving party will alert the adversary and/or provide for return of the material. Courts have addressed waiver disputes under this framework.[38] Keep in mind, though, that a proper and detailed *privilege log* is also required, per the provisions of Federal Rule 26(b)(5)(A) and numerous state rules. Failure to do so may result in sanctions and the loss of a claim of privilege—as discussed further in this chapter under "Sanctions."[39]

Overall, the procedures and rules discussed in this portion of the chapter are of great importance in today's fast-paced world of discovery, where potentially hundreds of thousands of documents could change hands. Despite best efforts to utilize proper review protocols, and keyword searches, it is possible for errors to occur, and for disclosure of privileged material to take place.[40] Having agreements among the parties, supported by the rules, could very well save the day.

1.8 – Impact of FRE 502(d)

In the past, while "claw-back" provisions and agreements might have been honored by particular federal courts in which litigation was pending under Rule 26(b)(5)(B) and 26(f) (and sometimes related agreements or scheduling orders under Rule 16), state courts or even other federal courts might have held that a waiver of privilege occurred when documents were exchanged and related litigation was pending elsewhere.

Therefore, in 2008 Congress acted to abrogate or modify that problem with the passage of Federal Rule of Evidence 502 (as recommended by the United States Judicial Conference), and the provision was signed into law by President George W. Bush on September 19, 2008.

FRE 502 provides, *inter alia*, that production of privileged or work-product protected documents, even if inadvertent or applying claw-back and related procedures, will not waive

---

[38] *See U.S. ex rel. Rembert v. Bozeman Health Deaconess Hosp.*, 2018 WL 1610959 (D. Mont. Apr. 3, 2018); *U.S. ex rel. Schutte v. Supervalu, Inc.*, 2018 WL 2416588 (C.D. Ill. May 29, 2018); *In re Delphi Corp.*, 2007 WL 518626 (E.D. Mich. Feb. 15, 2007); *see also In re Adoption of Virgin Islands Rules of Civil Proc.*, 2017 WL 1293844, at *56 (V.I. Apr. 3, 2017). Note, however, that privilege does not apply to e-mails between non-attorneys discussing whether to retain an attorney/seek legal advice. *See The Solutions Team, Inc. v. Oak St. Health, MSO, LLC*, 2020 WL 30602 (N.D. Ill. Jan. 2, 2020).

[39] Related to this issue of privilege and logs, the parties may litigate over something called a "quick peek"—a procedure whereby parties are allowed to look quickly at the documents of another party, including potentially privileged documents, without those documents losing their privileged status. Thereafter, the parties can either agree or dispute the privileged nature of materials. The process can also be used per Federal Rule 26(f) by agreement of the parties to determine whether there is need to review a selection of documents in detail if the quick peek reveals nothing relevant to the litigation (thus speeding discovery without loss of privilege). For more on compelled "quick peek," *see Winfield v. City of New York*, 2018 WL 2148435 (S.D.N.Y. May 10, 2018).

[40] *Supervalu, Inc.*, 2018 WL 2416588, at *5. One should also be certain to exercise a "claw-back" correctly, such as to return material that is privileged, or risk an adverse court ruling. *See In re Aenergy, S.A.*, — F. Supp. 3d —, 2020 WL 1659834 (S.D.N.Y. Apr. 3, 2020); Karen L. Stevenson, *Inadvertent Production Can You Claw It Back? Maybe, Maybe Not*, ABA Litigation, vol. 44, no. 3, at 18 (Spring 2019).

privilege or discovery protection, and decisions by the federal district judge or magistrate judge in a particular case utilizing a Rule 502 Order will apply in or extend to state actions or other federal actions. Rule 502 applies in full only to cases filed on or after September 19, 2008. For cases previously filed, application is at the discretion of the district judge or magistrate judge, and the prior risk and concerns may remain (however, given the fact that the Rule was enacted more than ten years ago, the risk likely exists for only a short time longer—if it exists at all anymore).

As recently explained by a federal court in New York:

> Federal Rule of Evidence 502 . . . was amended . . . for two main purposes: (1) to resolve disputes involving inadvertent disclosure of information protected by the attorney-client or work product privilege and subject matter waiver due to the disclosure; and (2) to provide parties with a "uniform set of standards" pursuant to which they can determine the consequences of a disclosure of information protected by the attorney-client or work product privilege. . . . Under the amended evidence rule, a disclosure in a federal proceeding of information protected by the attorney-client or work product privilege does not operate as a waiver if the disclosure was inadvertent, the producing party took reasonable steps to prevent disclosure, and the producing party promptly took reasonable steps to rectify the error, including following Rule 26(b)(5)(B). . . . Federal Rule of Evidence 502 also empower a federal court to enter an order that "privilege . . . is not waived by disclosure connected with the litigation pending before the court," and, if such an order is entered, the disclosure will not constitute a waiver in any other federal or state proceeding. . . .
>
> The Advisory Committee Notes discussing Federal Rule of Evidence 502(d) recognize the high costs of electronic discovery, and the rule contemplates enforcement of voluntary agreements reached by parties in their initial meeting pursuant to Rule 26(f) such as "claw-back" and "quick peek" arrangements as a way "to avoid the excessive costs of pre-production [privilege] review.". . . . The rule incentivizes parties to voluntarily agree to procedures that will alleviate the burdens of pre-production privilege reviews by offering protection from waiver of privilege to the producing party. "Rule 502(d) does not authorize a court to require parties to engage in 'quick peek' and 'make available' productions and should not be used directly or indirectly to do so.". . .
>
> Importantly, amended Federal Rule of Evidence 502 "governs only certain waivers by disclosure.". . . It does not alter federal or state law on privilege nor "supplant applicable waiver doctrine generally.". . . It likewise does not address privileges other than attorney-client and work product. . . .[41]

The risks of waiver that previously existed would make some parties reticent or unwilling to consider engaging in agreements in federal court if there was a possibility or reality that other

---

[41] *Winfield*, 2018 WL 2148435, at *4 (citations and footnotes omitted; citing, *inter alia*, The Sedona Conference). *See also, generally, Clarke v. J.P. Morgan Chase & Co.*, No. 08 Civ. 02400(CM)(DF), 2009 WL 970940 (S.D.N.Y. Apr. 10, 2009) (Freeman, M.J.) (discussing waiver and privilege generally under F.R.E. 502); *U.S. v. Treacy*, S2 08 CR 366(JSR), 2009 WL 812033 (S.D.N.Y. Mar. 24, 2009) (Rakoff, D.J.) (same).

federal or state courts would be involved in related litigation. *Remember that if waiver was held, not only the document but possibly the subject matter addressed would be open for discovery.*

Thus, for a short time more, until all filed cases fall squarely under the auspices of Rule 502, there are three possible solutions suggested here by this text's author should one be stuck litigating a case that was filed prior to September 19, 2008:

1. Some judges in federal court utilize language in orders to the effect of: "The information may be utilized solely by counsel for the purpose of representing this(these) client(s), in this lawsuit, filed in this Court, under this docket number, and for no other purpose. . . . This order is backed by the full contempt authority of this Court." If experts are needed, an option is to identify them and have them sign a written acknowledgment to be bound by a confidentiality order and specific protocols or the judge's order containing the language above. If your judge does not do these things automatically, consider requesting it.

2. Enter into a stipulation with opposing counsel in the state court action using language tracking Federal Rules 26(b)(5)(B), 26(f), and 502(d), and request that the state court judge "so order" it. Then, you have a court order, and law of the case. (Potential appellate reversal is a slight risk, but if that happens, you have recourse against opposing counsel for having agreed and then challenged.)

3. Request to have a joint conference with all counsel, the state judge, and the federal judge. Determine all issues to be addressed concerning discovery, confidentiality, claw-back, and related matters. Then request each judge issue an order in the case they are overseeing. These are rare occurrences but have taken place.

Note that these three suggestions may be particularly helpful if the parties *and* counsel are the same in *both* the state and the federal court actions.

1.9 – Technology and Computer-Assisted Review (TAR & CAR), Computer-Assisted Coding, Predictive Coding

Are you familiar with the procedures mentioned in the heading for this part of the chapter? If not, you may wish to become familiar in short order. To many, these are the way of the future. But they require cautious application, and the parties must remain in communication as to all aspects applied throughout the discovery process. The case excerpt at the end of this chapter section will make that clear.

Recently, given that electronic discovery—and the volume of documents and sources subject thereto—is growing exponentially, courts have turned their attention first to keyword searches and then coding and data analytics as methods for both the finding and production of relevant documents and for privilege reviews.

Keyword searches can be employed either when a party on their own (in searching materials for internal use) or in consultation with the other party (parties) and the court (when utilized for discovery) decides on a list of words (often mixed with Boolean search methods) that have been selected and identified to narrow the scope of "hits" when a database or email system is searched, leading to relevant documents that are more likely responsive in discovery. Of course, keyword searches are not easy to create and can be perilous if relied on without a clear idea of what the search target is. "In too many cases,. . . the way lawyers choose keywords is the equivalent of the child's game of 'Go Fish.' The requesting party guesses which keywords might produce evidence to support its case without having much, if any, knowledge of the responding party's 'cards' (i.e., the terminology used by the responding party's custodians). Indeed, the responding party's counsel often does not know what is in its own client's 'cards.'"[42]

Thus, Courts and litigants are turning to TAR, CAR, and Predictive Coding (and its ability for Continuous Active Learning (CAL)) to be of significant assistance in expediting discovery, including the review and identification of relevant electronic documents and materials. Commentators have written extensively on the topic, and their articles are recommended for a deeper study.[43]

> *TAR/CAR/Predictive Coding* – Utilizing technology, software and even artificial intelligence to assist with searches of large volumes of electronic materials in discovery; coding and phrases can be used, but systems can now also "learn" as they go; these methods reduce human hours and costs.

---

[42] *Da Silva Moore v. Publicis Groupe*, 287 F.R.D. 182, 191 & n.13 (S.D.N.Y. 2012) (Peck, M.J.).

[43] *See* M. Grossman & G. Cormack, *Technology-Assisted Review in e Discovery Can Be More Effective and More Efficient Than Exhaustive Manual Review*, 17 RICH. J. L. & TECH. 1 (Spring 2011), available online at: http://jolt.richmond.edu/jolt-archive/v17i3/article11.pdf; B. Dimm, *TAR, Proportionality, and Bad Algorithms (1-NN)*, BLOG, *available at:* https://blog.cluster-text.com/2018/08/13/tar-proportionality-and-bad-algorithms-1-nn/ (blog article asking, and exploring, "[s]hould proportionality arguments allow producing parties to get away with poor productions simply because they wasted a lot of effort due to an extremely bad algorithm?"). Furthermore, "[i]t is well settled that TAR 'is an acceptable way to search for relevant ESI in appropriate cases.'… TAR is defined as '[a] process for prioritizing or coding a collection of [ESI] using a computerized system that harnesses human judgments of subject matter expert(s) on a smaller set of documents and then extrapolates those judgments to the remaining documents in the collection.'… In other words, human reviewers 'code a "seed set" of documents. The computer [then] identifies properties of those documents that it uses to code other documents.'… With TAR processes like the Predict tool… which use continuous active learning (sometimes referred to as 'TAR 2.0'),

> the software continuously analyzes the entire document collection and ranks the population based on relevancy. Human coding decisions are submitted to the software, the software re-ranks the documents, and then presents back to the human additional documents for review that it predicts as most likely relevant. This process continues until the TAR team determines that the predictive model is reasonably accurate in identifying relevant and nonrelevant documents, and that the team has identified a reasonable number of relevant documents for production."

Lawson v. Spirit AeroSystems, Inc., 2020 WL 1813395, at *6-*7 (D. Kan. Apr. 9, 2020) (citing, *inter alia, Da Silva Moore*, 287 F.R.D. at 183, 184; *Youngevity Int'l, Corp. v. Smith*, 2019 WL 1542300, at *11 (S.D. Cal. Apr. 9, 2019) ("Predictive coding or TAR has emerged as a far more accurate means of producing responsive ESI in discovery than manual human review of keyword searches."); *Entrata, Inc. v. Yardi Sys., Inc.*, 2018 WL 5470454, at *7 (D. Utah Oct. 29, 2018) (stating that "it is 'black letter law' that courts will permit a producing party to utilize TAR"); The Sedona Conference Glossary: E-Discovery & Digital Information Management (Fourth Edition), 15 Sedona Conf. J. 305, 357 (2014); Bolch Judicial Inst. & Duke Law, Technology Assisted Review (TAR) Guidelines 4 (Jan.

Several court decisions have addressed keyword searches and predictive coding, providing valuable guidance. One is a discovery order out of the U.S. District Court for the Northern District of Illinois, by Special Master Maura Grossman, and it provides a terrific discussion of search methodology and its application to the litigation.[44] Two other decisions were penned by now-former federal judges in the Southern District of New York, who were well-regarded for their writings in e-discovery while serving on the federal bench.

In *Da Silva Moore v. Publicis Groupe*,[45] Judge Peck discussed predictive coding and TAR in detail. This is believed to be the first decision approving the use of computer-assisted predictive coding searches in discovery proceedings. Predictive coding, while not completely replacing human review, is a method of "teaching" computers to recognize relevant documents from a manually coded "seed set" and search terms. A random sample of documents is identified, searched, and evaluated for "recall" and "precision," and the machines are taught to search. In essence, it is built from spam filter programs that have saved many from endless junk email while greatly reducing "false positives" and "mis-hits." To utilize CAR, TAR, and coding, attorneys and computer/IT experts with case familiarity will need to work closely together.

In *National Day Laborer Organizing Network v. U.S. Immigration & Customs Enforcement Agency*,[46] Judge Scheindlin addressed a massive FOIA litigation and ordered the government agency to conduct more responsive searches, identify documents, and produce affidavits that contained "reasonable specificity of detail" and not "conclusory statements." The judge also held that predictive coding could be utilized in aspects of the discovery and production protocols. This case further advanced the consideration and use of such protocols in modern litigation and discovery practice—which, as the next case excerpt illustrates, has found its way into litigation across the globe.

It is of the utmost importance to remember that the parties must agree both as to the use of TAR,[47] and on the coding utilized in the TAR/CAR, as well as on whether the yield from

---

2019)). In fact, some argue that TAR is "as good ... if not better" than review by human eyes. *See* Zach Warren, *There's Still a Misconception That Eyes-On Review Is the 'Gold Standard'*, N.Y. L.J. at 5 (Jan. 17, 2020).

[44] *See In re Broiler Chicken Antitrust Litig.*, 2018 WL 1146371 (N.D. Ill. Jan. 3, 2018) (Grossman, Special Master) (Gilbert, M.J.).

[45] 287 F.R.D. 182.

[46] 877 F. Supp. 2d 87 (S.D.N.Y. 2012) (Scheindlin, D.J.). *But cf. Heffernan v. Azar*, 417 F. Supp. 3d 1 (D.D.C. 2019).

[47] "While 'the case law has developed to the point that it is now black letter law that where the producing party wants to utilize TAR for document review, courts will permit it'... no court has ordered a party to engage in TAR over the objection of that party. The few courts that have considered this issue have all declined to compel predictive coding ... Despite the fact that it is widely recognized that 'TAR is cheaper, more efficient and superior to keyword searching' ..., courts also recognize that responding parties are best situated to evaluate the procedures, methodologies, and technologies appropriate for producing their own electronically stored information...." *In re Mercedes-Benz Emissions Litig.*, 2020 WL 103975, at *1 (D.N.J. Jan. 9, 2020) (citing, *inter alia*, *City of Rockford v. Mallinckrodt ARD Inc.*, 326 F.R.D. 489, 493 (N.D. Ill. 2018) (discussing the advantages of TAR but deferring to the parties' choice to use search terms); *Hyles v. N.Y.C.*, 2016 WL 4077114, at *3 (S.D.N.Y. Aug. 1, 2016) (refusing to order a party to use TAR and stating that a party is free to decide how to search so long as its process is reasonable); The Sedona Principles: Second Edition, Best Practices Recommendations & Principles for Addressing

the search is of sufficiently low percentage to warrant no further review of the pool of documents. These are not decisions for parties to make unilaterally, without consultation. In 2018, a decision was reported out of the Queen's Bench Division of the Business and Property Courts of England and Wales, Technology and Construction Court, addressing that very issue.[48] Claimants (plaintiffs) and defendants had agreed to a search that would be applied to a pool of electronic documents in the case. Any documents that were returned from that search, hitting the agreed-upon key words, were to be reviewed manually, and apparently the parties did not agree to any CAR. Following several rounds of agreements on keywords to reduce the number of responsive materials, 450,000 documents were returned.[49] Thereafter, however, claimants disclosed a list of 12,476 documents, followed by a supplemental list of 4,163 out of 230,000 reviewed. Although claimants had stated that the documents were "manually reviewed," it turned out that the 230,000 documents had been reviewed manually *with the aid of CAR*.[50] Additionally, the remaining 220,000 documents that had not yet been searched were never searched, the argument being that CAR "de-prioritized" them, because according to claimants, a 1% sample size yielded only a 0.38% return that remaining documents would be relevant, thus meaning any further cost and time were unwarranted.[51]

The *Triumph* Court addressed the relevant English *Practice Directions* and issued its holding. That holding is provided in great detail below, for assessment, analysis, and discussion:

Electronic Document Production, Principle 6 (available at www.TheSedonaConference.org)). *See also, generally,* Max Mitchell, *'A Tidal Wave of Change': Plaintiffs Firms Are Tapping Data Analytics, but Some Are Still Reluctant,* N.Y. L.J. at 5 (Jan. 31, 2020). *But cf. Kaye v. N.Y.C. Health & Hosp. Corp.,* 2020 WL 283702, at *2 (S.D.N.Y. Jan. 21, 2020) ("When documents are produced in discovery, whether they be produced electronically or otherwise, the Court does not believe that, in the first instance, the receiving party has a right to examine and evaluate the way the production was made or require collaboration in the review protocol and validation process") (citation omitted). But note that in the case of *Lightsquared* (which was cited to by the court in *Kaye*), that court held: "If the defendants unilaterally develop and implement a search protocol, rather than negotiate a stipulated search protocol with the plaintiffs, they run the risk that [the magistrate judge] will later deem their search strategy insufficient and require them to conduct additional searches." *Lightsquared Inc. v. Deere & Co.,* 2015 WL 8675377, at *8 (S.D.N.Y. Dec. 10, 2015). Separately, and of great interest, since the emergence of COVID-19, Continuous Active Learning (CAL) Technology Assisted Review (TAR) has moved from law firms and courthouses to research labs and hospitals, as it is put to work in the review of clinical studies for the fight against COVID-19—both in safety of care and a search for a cure. *See Using Technology-Assisted Review to Find Effective Treatments and Procedures to Mitigate COVID-19,* University of Waterloo Cheriton School of Computer Science (May 7, 2020), *available at* https://cs.uwaterloo.ca/news/using-technology-assisted-review-find-treatments-procedures-to-mitigate-covid-19; Doug Astin, Maura Grossman and Gordon Cormack, *Put TAR to Work to Help with the COVID-19 Fight: eDiscovery Trends,* eDiscoveryToday (May 7, 2020), *available at* https://ediscoverytoday.com/2020/05/07/maura-grossman-and-gordon-cormack-put-tar-to-work-to-help-with-the-covid-19-fight-ediscovery-trends/.

[48] *Triumph Controls UK, Ltd. v. Primus Int'l Holding Co.,* [2018] EWHC 176 (TCC), Case No: HT-2016-000104 (Coulson, J.).

[49] *Id.*

[50] *Id.*

[51] *Id.*

Triumph Controls UK, Ltd. v. Primus Int'l Holding Co.
(Queen's Bench Division of the Business and Property Courts of England and Wales)

Coulson, Justice

27. . . . I have much greater concerns about the claimants' approach. First, what they did is not what they said they would do in the EDQ [Electronic Documents Questionnaire], which promised a manual review of all documents responsive to the keyword searches. Neither is what they did at all clear from their Disclosure List.

28. Although the broad outline of the searches undertaken by the claimants was subsequently explained in September 2016, it was not a detailed exposition. At no time have the claimants provided relevant details as to how the CAR was set up or how it was operated. In circumstances where the decision to use the CAR was unilateral, and where the defendants had no input into it at all, that is unsatisfactory. . . .

29. This problem has been compounded by the lack of information as to the sampling exercise. All that the defendants, and the court, have been told is that there was a sampling exercise which produced a predictive figure of 0.38%. But there is no information as to precisely how that sampling exercise was conducted. There are, for example, no stated tolerances and no explanation of how many rounds of sampling were undertaken. That again is unsatisfactory. It is not in accordance with the TCC ediscolsure protocol. I am bound to say that, given that both the CAR and the sampling were unilateral, I am slightly surprised that there is not now better evidence as to what actually happened. There is always a risk,. . . that a unilateral decision will be carefully scrutinised by the court at a later date, and a different course may be ordered. That will be more likely if the evidence as to what was done remains vague.

30. There is a further point about the number of people involved in the CAR process. The evidence suggested that there were perhaps ten paralegals and four associates involved in the searches. It is not apparent that there was any overseeing senior lawyer,. . . I think that [defendants' counsel] is right to say that the sheer volume of those involved with the CAR system in this case may mean that it has not been 'educated' as well as it might have been, particularly in respect of the criteria for relevance.

31. In all those circumstances, therefore, I agree that both the CAR exercise, and the sampling exercise that it produced, cannot be described as transparent, and cannot be said to be independently verifiable.

32. But does that matter? Notwithstanding the flaws in the system adopted, is it reasonable or proportionate to require the claimants to do anything more in relation to disclosure of the balance of 220,000 documents? I have considered this question carefully, and I have concluded that it is. There are a number of reasons for that.

33. First, the total number of documents disclosed (16,500 originally, and 3,000 since, or about 19,500 in total) appears very modest in circumstances where 450,000 documents have been identified as being responsive to the agreed keywords. In my experience, even in an ordinary case, the percentage of disclosable documents, compared to the total of responsive documents, would tend to be larger than the 4.3% or so which the 19,500 documents represent. That potential discrepancy is even starker in the present case, which is based on a wide-ranging series of allegations as to non-disclosure in respect of many different elements of an international business. Such allegations will almost always lead to a document-heavy trial.

34. Secondly, as I have noted above, around 2,000 of the documents disclosed by the claimants subsequently (i.e. after the first list and the first supplemental list) were responsive to the original keywords. That strongly suggests that they should have been disclosed originally. That raises a further question-mark about the adequacy of the claimants' disclosure process.

35. Thirdly, the evidence makes plain that at least some of the total of 3,000 further documents have been disclosed because they were documents that the claimants' own witnesses and experts have referred to and/or wish to rely on. Something has potentially gone wrong with disclosure when, for example, a witness of fact can remember a particular document when preparing his or her witness statement which, on the one hand, is so important that he/she wants to refer to it in their evidence but which, on the other hand, has not even been disclosed.

36. Finally, the only sampling exercise which has been done on the 220,000 documents is that which produced the 0.38% prediction. For the reasons that I have given, that figure may well not be reliable. Indeed, I consider that all the other evidence demonstrates that the 0.38% is likely to be an underestimate.

37. For all these reasons, I conclude that the steps taken by the claimants in relation to the balance of the 220,000 documents have not been adequate. What therefore should be done to correct that, considering all the circumstances of the case?

38. First, I consider that some form of manual search is required. Whilst I agree with [claimants' counsel] that, as he put it, there is no magic in a manual review, the absence of any real explanation of the CAR process in this case means that it is difficult to contemplate any alternative which would not be potentially controversial. That is one of the dangers of operating a system without agreement or explanation. Since time is of the essence, only a manual review will do.

. . . .

40. As to the costs, whilst of course the court is anxious to keep them to a minimum, that figure must be set against the claim for US$65 million. As to the estimated time, I consider that two months is very much a 'worst case scenario' because, whatever the arguments might be as to their status as proper comparators, there was other evidence in the documents which indicated that it would not take anything like as long as two months.

41. However, I recognise the importance of proportionality. I also recognise that the trial is less than six months away.

42. Thus, at the hearing on 31 January 2018, having given my reasons for arriving at the conclusion summarised in paragraph 37 above, I ordered that the parties were forthwith to agree a methodology by which a sample of 25% of the 220,000 documents was to be manually searched. That search was to take no longer than three weeks. The results are to be put into an agreed letter. I can then be shown the letter and told about the results of that search at the hearing on 22 February 2018, which is the date on which the parties are back before this court to argue about another aspect of this case.[52]

> **Spoliation** – Destroying, altering, losing or failing to preserve evidence.

Therefore, be aware of and familiar with the concepts of TAR, CAR, and Predictive Coding. They are tools in the e-discovery toolbox that could be of great assistance when large amounts of electronic documents must be marshaled and reviewed. There are pitfalls and technological complications that must be addressed, and the parties together with the court overseeing the case must agree on protocols. If handled properly, though, the value should not be underestimated.

---

[52] *Id.* (citations omitted).

1.10 – Sanctions, FRCP 37(e), Changing Seas, No More Safe Harbor

You are working one day, and all of a sudden virtual warning bells go off—because sometimes deletion policies or accidents happen—spoliation[53] (loss/destruction) happens—and ESI is gone! *Now what*?![54]

---

[53] "Spoliation is the destruction or significant alteration of evidence, or failure to preserve property for another's use as evidence in pending or reasonably foreseeable litigation." *Tchatat v. O'Hara*, 249 F.Supp.3d 701, 706 (S.D.N.Y. 2017) (citing and quoting *In re Terrorist Bombings of U.S. Embassies in E. Afr.*, 552 F.3d 93, 148 (2d Cir. 2008)). "'A party seeking sanctions for spoliation has the burden of establishing the elements of a spoliation claim.'… 'These elements are "(1) that the party having control over the evidence had an obligation to preserve it at the time it was destroyed; (2) that the evidence was destroyed with a culpable state of mind; and (3) that the destroyed evidence was relevant to the party's claim or defense such that a reasonable trier of fact could find that it would support that claim or defense."'… In other words, 'for sanctions to be appropriate, it is a necessary, but insufficient, condition that the sought-after evidence actually existed and was destroyed.'… The party seeking sanctions need not produce direct evidence of spoliation; rather, 'circumstantial evidence may be accorded equal weight with direct evidence and standing alone may be sufficient.'… If that party proves 'negligent' spoliation, it 'must adduce sufficient evidence from which a reasonable trier of fact could infer that the destroyed or unavailable evidence would have been of the nature alleged by that party.'… However, if that party proves 'bad faith'—that is, intentional spoliation—'relevance may be presumed from the fact of the evidence's destruction.'… 'If the requisite showing is made, sanctions may be proper under Fed. R. Civ. P. 37(b) when a party spoliates evidence in violation of a court order.'… 'Determining the proper sanction to impose for spoliation is confined to the sound discretion of the trial judge and is assessed on a case-by-case basis.'… At all times, however, a sanction should '(1) deter parties from engaging in spoliation; (2) place the risk of an erroneous judgment on the party who wrongfully created the risk; and (3) restore the prejudiced party to the same position [it] would have been in absent the wrongful destruction of evidence by the opposing party.'… '[A] court should always impose the least harsh sanction that can provide an adequate remedy.'" *Rivera v. Hudson Valley Hospitality Group, Inc.*, 2019 WL 3955539, at *3 (S.D.N.Y. Aug. 22, 2019) (citing, *inter alia, Tchatat*, 249 F.Supp.3d at 706-707; *CAT3, LLC v. Black Lineage, Inc.*, 164 F. Supp. 3d 488, 500 (S.D.N.Y. 2016); *Res. Funding Corp. v. DeGeorge Fin. Corp.*, 306 F.3d 99, 109 (2d Cir. 2002)). *And see Nida v. Allcom*, 2020 WL 2405251, at *3 (C.D. Cal. Mar. 11, 2020). *See also Eisenband v. Pine Belt Automotive, Inc.*, 2020 WL 1486045, at *7 (D.N.J. Mar. 27, 2020) ("Before reaching this analysis, however, the Court must determine whether spoliation has, in fact, occurred …. Spoliation occurs where 'the evidence was in the party's control; the evidence is relevant to the claims or defenses in the case; there has been actual suppression or withholding of evidence; and, the duty to preserve the evidence was reasonably foreseeable to the party.'"). This standard also results in another question to be settled by the courts—can a party be sanctioned for loss of ESI that is possessed by a third-party/non-party. Some courts have read FRCP 34's "control" language into FRCP 37—and it is a jurisdiction-specific and fact-intensive consideration, which attorneys should undertake if ESI is possessed by others not their client. Have a plan. *See* Brian A. Zemil, *Navigating "Control" in a Matrix of ESI Discovery*, ABA Litigation, vol. 45, no. 2, at 20 (Winter 2020); *Rosehoff, Ltd. v. Truscott Terrace Holdings LLC*, 2016 WL 2640351, at *5 (W.D.N.Y. May 10, 2016) (cited and discussed by B. Zemil, *Navigating "Control", supra*).

[54] We are not addressing intentional loss, deletion, or destruction here. For more on that instance, *see* the discussion of the *Hausman* case in Chapter 3. In addition, *see Paisley Park Enterprises, Inc. v. Boxill*, 330 F.R.D. 226 (D. Minn. Mar. 5, 2019) (sanctions awarded for intentional destruction and wiping of cellular phones and loss/destruction of other ESI, pursuant to FRCP 37(b) and 37(e)). *See also* Christopher Boehning & Daniel J. Toal, *Intentional, Bad Faith Spoliation From Use of Ephemeral Messaging*, N.Y. L.J. at 5 (Dec. 3, 2019) (discussing the use of social media platforms that utilize "ephemeral messaging"—such as Signal, WeChat, Wire, or Wickr—where messages can be set to disappear soon after receipt and reading, sometimes almost instantaneously, and whether that subjects parties in a litigation to potential sanctions); and see *Herzig v. Ark. Found. for Med. Care, Inc.*, 2019 WL 2870106

Federal Rule 37(e) (and previously 37(f)) of the FRCP used to be a "safe harbor," *but not anymore*. The rule has since been amended, as of December 1, 2015.

Former Rule 37(e) provided: "Absent exceptional circumstances, a court may not **impose sanctions** *under these rules* on a party for failing to provide electronically stored information lost as a result of the routine, good-faith operation of an electronic information system" (emphasis added). Now, however, Rule 37 reads, in pertinent part, as follows—which is provided in full because of the great importance it holds:

[Rule 37](e) Failure to Preserve Electronically Stored Information. If electronically stored information that should have been preserved in the anticipation or conduct of litigation is lost because a party failed to take reasonable steps to preserve it, and it cannot be restored or replaced through additional discovery, the court:

(1) upon finding prejudice to another party from loss of the information, may order measures no greater than necessary to cure the prejudice; or

(2) only upon finding that the party acted with the intent to deprive another party of the information's use in the litigation may:

(A) presume that the lost information was unfavorable to the party;

(B) instruct the jury that it may or must presume the information was unfavorable to the party; or

(C) dismiss the action or enter a default judgment.[55]

The Advisory Committee Notes to the 2015 Amendments make clear that a new discovery standard has not been created:

> *Sanction* – Punishment for failure to meet obligations under rules or laws, or pursuant to an order or directive of a court/judge (such as a discovery order); sanctions can include a monetary fine or dismissal of a case.

The new rule applies only if the lost information should have been preserved in the anticipation or conduct of litigation and the party failed to take reasonable steps to preserve it. Many court decisions hold that potential litigants have a duty to preserve relevant information when litigation is reasonably foreseeable. Rule 37(e) is based on this common-law duty; it does not attempt to create a new duty to preserve. The rule does not apply when information is lost before a duty to preserve arises.[56]

(W.D. Ark. July 3, 2019) (cited and discussed in C. Boehning & D. Toal, *Intentional, Bad Faith Spoliation, supra*), possibly the first case on this issue of ephemeral messages, which granted a spoliation motion, but cited no authority because the court also granted a companion motion for summary judgment dismissing the case on the merits, and the spoliation decision and sanction appears to have been subsumed therein—thus there was no mention or application of FRCP 37(e).

[55] FED. R. CIV. P. 37(e).

[56] Advisory Committee Notes to Federal Rule 37(e) (Dec. 1, 2015). *But cf.* Will Young, *How Corporate Lawyers Made It Harder to Punish Companies That Destroy Electronic Evidence*, ABA Journal (online Feb. 7, 2020), *available at* https://www.abajournal.com/news/article/how-corporate-lawyers-made-it-harder-to-punish-companies-that-destroy-electronic-evidence (discussing the background and drafting of FRCP 37(e), and how it removed some of the fear that negligence in spoliation could result in severe sanctions). One should also remember that

The new Rule, and the Advisory Committee, took issue with the line of cases that had discussed sanctions for material lost because of simple negligence. The amendment to Rule 37 was a fundamental shift in the world of e-discovery sanctions—a world that has seen much tumult in just the last ten years, with changing pronouncements of what constitutes sanctionable conduct and the severity of sanctions available based on the level of culpable conduct by parties. Prior to the Rule 37(e) amendment, federal courts across the country were engaged in disputes concerning the level of culpable conduct warranting sanctions, as well as whether pre-litigation conduct could result in sanctions.[57] It is indispensable to have a working knowledge of the rules governing sanctions at the time one has a pending litigation with discovery and preservation obligations.

Note that sanctions, if imposed, can vary at the discretion of the court, on a case-by-case basis per the provisions of Rule 37(e) and potentially the court's inherent authority. It should be understood, though, that there is a controversy over application of inherent authority in an area of sanctions for discovery abuses otherwise governed by Rule 37(e). The Advisory Committee Notes specifically state that Rule 37(e) "forecloses reliance on inherent authority or state law to determine when certain measures should be used", and some courts have read the rule to remove inherent authority from sanctions decisions for ESI spoliation.[58] However, the U.S. District Court for the Southern District of New York, in *CAT3, LLC v. Black Lineage, Inc.*, held that the court maintained its inherent authority to issue sanctions for spoliation, even if sanctions under the rules would not be applicable.[59]

In any event, the potential sanctions range across a spectrum. On the less severe end is an award of costs and fees to the party suffering expense owing to the improper action of the opposing party. Contempt is also an available sanction. On the most severe end of the spectrum

---

perfection in the process is neither required, nor likely possible, but attempts at something called "discovery on discovery" (seeking to gather information from opposing parties as to their storage, collection, review, and production of material) appears to be on the rise. Robert Lindholm, *et al.*, *Successfully Defend Against **Discovery on Discovery** Requests*, N.Y. L.J. at 11 (Feb. 3, 2020) (discussing how to avoid or defend against discovery on discovery demands; citing, *inter alia*, FRCP 37(e) and Advisory Committee Notes 2015 ("perfection… often impossible"); *Orbit One Commc'ns v. Numerex*, 271 F.R.D. 429, 441 (S.D.N.Y. 2010)).

[57] *See, e.g., inter alia, Pension Committee of the Univ. of Montreal Pension Plan v. Banc of Am. Sec., LLC*, 685 F.Supp.2d 456 (S.D.N.Y. 2010) (Scheindlin, D.J.) (abrogated by *Chin v. Port Auth. of N.Y. & N.J.*, 685 F.3d 135 (2d Cir. 2012) (Court held that mere failure to institute litigation hold did not constitute gross negligence *per se*)); *Surowiec v. Cap. Title Ag. Inc.*, 790 F. Supp. 2d 997, 1007 (D. Ariz. 2011) (Campbell, D.J.) (Court disagreed with *Pension Committee* holding that failure to issue a litigation hold after litigation is anticipated is gross negligence *per se* in sanctions analysis for spoliation); *Merck Eprova AG v. Gnosis S.P.A.*, 2010 WL 1631519 (S.D.N.Y. Apr. 20, 2010) (Sullivan, D.J.); *Rimkus Consulting Group, Inc. v. Cammarata*, 688 F.Supp.2d 598 (S.D. Tex. 2010) (Rosenthal, D.J.); *Spanish Peaks Lodge, LLC v. Keybank Nat'l Ass'n*, 2012 WL 895465, at *3 & n.5 (W.D. Pa. Mar. 15, 2012) (Ambrose, D.J.); *GenOn Mid-Atl, LLC v. Stone & Webster, Inc.*, 282 F.R.D. 346 (S.D.N.Y. 2012) (Maas, M.J.); *Victor Stanley, Inc. v. Creative Pipe, Inc.*, 269 F.R.D. 497, 516 (D. Md. 2010) (Grimm, Chief M.J.).

[58] FED. R. CIV. P. 37 Advisory Committee Notes, 2015 Amendment. *See also Bistrian v. Levi*, --- F. Supp. 3d ---, 2020 WL 1443735 (E.D. Pa. Mar. 24, 2020).

[59] 164 F.Supp.3d 488 (S.D.N.Y. 2016) (Francis, M.J.). *See also In re Gorsoan Ltd.*, 2020 WL 3172777, at *10-*11 (S.D.N.Y. June 15, 2020).

exist adverse inference sanctions (where the court advises a jury that it may consider the lost material to adversely affect the party that lost it) and default of the defendant/dismissal of the plaintiff in the case, basically directing a verdict in favor of the party suffering prejudice and against the party committing the misconduct.[60] Under the current Rule 37(e), though, the most severe sanctions are reserved for those situations where parties act with intent to deprive others of ESI that should have been preserved and disclosed.[61]

Remember, further, that preservation obligations—warranting a "litigation hold"—may arise from a number of sources, including statutes, common law, or court orders. Consider the rules, obligations, and potential sanctions given recent extreme weather events, natural disasters, and malicious computer attacks and viruses. Ensure that your law office or business document storage systems are sufficiently backed-up, so you do not have to argue extenuating circumstances and convince a court that the loss was not due to deliberate action to avoid severe sanctions.

Additionally, keep in mind that willful or negligent disobedience of a court order or directive is also sanctionable under Federal Rule 37(b)(2).[62] Rule 37 in the federal courts, for discovery abuses or failure to abide by discovery rules, is only one source of sanctions available to trial judges. Again, as mentioned earlier, some judges cite sources maintaining they are empowered to sanction parties' discovery abuses per inherent authority to manage cases and move dockets. As stated by the U.S. District Court in Nevada, in a 2017 decision post-Rule 37 amendment:

> Rule 37 of the Federal Rules of Civil Procedure provides the court with a wide range of sanctions for a party's failure to adequately engage in discovery. "The Rule provides a panoply of sanctions, from the imposition of costs to entry of default.". . . "Discovery sanctions serve the objectives of discovery by correcting for the adverse effects of discovery violations and deterring future discovery violations from occurring.". . . The court may levy sanctions based on its inherent authority to respond to abusive litigation practices or on a party's failure to obey a discovery order. . . . The court may apply "dispositive sanctions" and dismiss a party's claim or defense. However, the court should impose dispositive sanctions only where "a party has engaged deliberately in deceptive practices that undermine the integrity of judicial proceedings.". . .[63]

---

[60] See Desert Palace, Inc. v. Michael; 2018 WL 401534, at *1 (D. Nev. Jan. 11, 2018); Rockman Co. (USA), Inc. v. Nong Shim Co., Ltd., 229 F.Supp.3d 1109, 1121 (N.D. Cal. 2017); Garvin v. Arcturus Venture Partners Inc., 2017 WL 4296273 (N.D. Tex. Sept. 28, 2017); see also U.S. v. 5124 Gumwood Ave., McAllen, Hidalgo Cty., Tex., 2018 WL 6737410 (E.D. Tex. Nov. 1, 2018). See Christopher Boehning & Daniel J. Toal, "Staggering" Spoliation Leads To Case Terminating Sanctions, N.Y. L.J. at 5 (June 2, 2020) (citing and discussing WeRide Corp. v. Huang, 2020 WL 1967209 (N.D. Cal. Apr. 24, 2020)).

[61] Nuvasive, Inc. v. Madsen Med., Inc., 2016 WL 305096 (S.D. Cal. Jan. 26, 2016).

[62] See Comlab Corp. v. Kal Tire, 2018 WL 4333987 (S.D.N.Y. Sept. 11, 2018).

[63] Dinkins v. Schinzel, 2017 WL 4183115, at *2 (D. Nev. Sept. 19, 2017). See also Brian A. Zemil, Tug of War Over Authority for ESI Spoliation Sanctions, ABA Litigation, vol. 45, no.1, at 20 (Fall 2019) (addressing Rule 37(e), inherent authority, and the split amongst courts around the nation).

Now, a quick inquiry—*if* your case was filed *prior* to December 1, 2015, which version of Rule 37 applies? This is an important question since Rule 37(e) was only amended at the end of 2015, and litigation often takes several years to resolve from filing through conclusion if a trial is required. The courts will apply a test of whether application of the amended rule to the facts and circumstances of a case is "just and practicable," to determine whether the pre-2015 or post-2015 Rule 37(e) will govern your matter.[64] One court held:

> [i]t would be unjust to apply a new rule retroactively when that rule governs a party's conduct. But Rule 37(e) does not govern conduct; a party has the same duty to preserve evidence for use in litigation today as before the amendments. It is not unjust to apply the new rule when it merely limits the Court's discretion to impose particular sanctions, especially when the motion seeking sanctions was filed after the amendment took effect.[65]

Next consider that under Federal Rule 26 other instances of failures, such as with regard to privilege logs, can warrant sanctions.[66] In *Blackrock*, out of the U.S. District Court for the Southern District of New York, the court addressed failures of the defendant to properly produce a privilege log. Among the documents at issue were email strings. Defendant's failures caused a loss of privilege for a number of the documents for which there were deficient entries. Below is the court's opinion, in detail, for analysis and discussion.

BlackRock Balanced Capital Portfolio (FI) v. Deutsche Bank Nat'l Trust Co. (U.S. District Court, Southern District of New York)

Netburn, U.S. Magistrate Judge

On February 14, 2018, the Court ruled on BlackRock's initial motion. As relevant here, the Court ruled that the use of metadata to import information for a privilege log, while not inherently problematic, did not "absolve[ ]" Deutsche Bank "of its obligations to review, supplement and correct a metadata privilege log to ensure that it satisfies the rigorous standards for a compliant log.". . . The Court directed Deutsche Bank to review and revise its log to make sure that the information provided is sufficient to allow BlackRock to understand the privilege assertions and make informed challenge decisions. Thereafter, the Court ordered a "substantive and detailed" meet and confer. And if disputes remained, the plaintiffs would select 10 exemplars from the log for the Court's *in camera* review ("Deficient Log Exemplars"). In addition, BlackRock challenged Deutsche Bank's common interest privilege assertions, arguing that Deutsche

---

[64] *See Distefano v. Law Offices of Barbara H. Katsos*, 2017 WL 1968278 (E.D.N.Y. May 11, 2017) (citing, *inter alia*, *Cat3, LLC v. Black Lineage, Inc.*, 164 F. Supp. 3d 488; 28 U.S.C. § 2074(a)). *See also Fashion Exchange LLC v. Hybrid Promotions, LLC*, 2019 WL 6838672, at*3 (S.D.N.Y. Dec. 16, 2019).

[65] *Security Alarm Fin. Enter., L.P. v. Alarm Protection Tech., LLC*, 2016 WL 7115911 (D. Alaska Dec. 6, 2016) (footnotes omitted); *see also Nuvasive, Inc.*, 2016 WL 305096 (denying most severe sanctions; although conduct took place before rule change, the sanctions would be applied at trial after rule change).

[66] *See BlackRock Balanced Capital Portfolio (FI) v. Deutsche Bank Nat'l Trust Co.*, 2018 WL 3584020 (S.D.N.Y. July 23, 2018).

Bank had improperly withheld 18,000 documents under this doctrine. Here, the Court directed BlackRock to identify 30 exemplars, and ordered Deutsche Bank to submit a sworn affidavit of a "party representative" explaining why the communications are protected under the common interest doctrine. Thereafter, the parties were directed to meet and confer in good faith, and if resolution was not possible, BlackRock was to select 10 exemplars for the Court's *in camera* review ("Common Interest Exemplars").

The parties were unable to resolve these issues, and BlackRock timely renewed its motion, identified 10 Deficient Log Exemplars and 10 Common Interest Exemplars, and moved the Court for an order finding that Deutsche Bank has waived its privilege as to all claims.

I. Deficient Log Exemplars

BlackRock alleges that Deutsche Bank's privilege log remains woefully inadequate.... Specifically, BlackRock contends that the log does not provide the general subject matter for 73,364 documents, does not identify the author for at least 15,211 documents, and fails to identify a lawyer for 651 documents. On this record, BlackRock argues that Deutsche Bank should be found to have waived its privilege assertions as to all claims.

Of the 10 Deficient Log Exemplars, Deutsche Bank withdrew its privilege assertion with respect to five documents. Most of these documents were examples where the privilege log failed to provide enough information to allow BlackRock to understand the privilege assertion. For example, descriptions such as "Fw:15Ga-1 Reporting [I]," "Missing Exception Reports—All Trustees-06-20-11.xls," "Trade Updated Investment Policy I," and "[s]preadsheet reflecting legal advice from unspecified in-house counsel regarding repurchase obligations," all proved not to be privileged upon a challenge from BlackRock. In addition, two exemplars were previously produced in a less redacted form, and Deutsche Bank has agreed to reissue these exemplars in a similar manner following BlackRock's challenge.

Deutsche Bank's rolling corrections leaves the Court with only three documents out of the 10 exemplars to examine on the merits. Document 3 (**ICN ROK 003 1 00000111-2522**) discusses how to respond to a letter from Gibbs & Bruns regarding Wells Fargo's role as master servicer for certain loans. Document 9 (**ICN PAS 001 1 00000012-1614**) discusses attorney advice as to the appropriate course of action if a breach of R&Ws occurred with respect to a specific loan. Both are privileged discussions between outside counsel and Deutsche Bank. Furthermore, these entries adequately describe the contents of the e-mails in either the subject or description fields. The entry for Document 3 contains sufficient information to show that the document relates to a response to a Gibbs & Bruns letter, while the entry for Document 9 contains sufficient information to show that the document relates to repurchase obligations for a JP Morgan Chase matter.

Finally, Document 6 (**ICN CAB 002 1 00000080-10243-1**) appears to be a balance sheet. Deutsche Bank seeks to redact the portions of the sheet that describe activities related to its counsel, but fails to show how those portions are for the purpose of seeking legal advice. The redacted portion of this document is not privileged and Deutsche Bank is ORDERED to produce it in full.

II. Common Interest Exemplars

As with the Deficient Log Exemplars, Deutsche Bank appears to have used the process of BlackRock's selection of Common Interest Exemplars as a way to make rolling corrections to its privilege log. As ordered by the Court, BlackRock selected 10 exemplars. Deutsche Bank has withdrawn its privilege assertion for two documents. . . .

Before even reviewing the documents *in camera*, it is clear that Deutsche Bank's revised privilege log still lacks adequate information for a majority of the documents submitted for review. Only two documents actually have a description in the "DESCRIPTION" field (**PIJ_002_1_00000025-0855** and **RER_002_1_00000178-0733**). Log rows for the other seven documents do not provide BlackRock with information sufficient to decide whether to challenge Deutsche Bank's privilege assertions. For example, **CAB_006_1_00000036-13219**'s "DESCRIPTION" field is left blank, and its "SUBJECT" is "FW: Deutsche Bank v. Landeros Reyes 202 Park St., Gypsum, CO." Similarly, **COD_002_1_00000006-0564**'s "SUBJECT" is "City of Chicago Debts on Unpaid Cases Violations Bank of America Updated Administrative Hearing Debts" and also does not provide a description. Further, **STA_001_1_00000013-0950**'s "TO/RECIPIENT" field neglects a clear name and only provides: "cn=amystoddard/ou=newyork/ou=dbna/o=deuba@dbamericas."

Additionally, **VIR 003 1 00000126-00196** and six other exemplars do not identify an attorney in the "IDENTIFY COUNSEL" field despite Deutsche Bank's assertion that these documents are attorney-client privileged or protected work product. Deutsche Bank admits that they fail to identify an attorney for 670 documents for which they assert attorney-client or work product protection. Yet, Deutsche Bank did not edit their privilege log to identify an attorney or modify privilege assertions.

Upon review of the remaining eight Common Interest Exemplars, nearly all are found to be non-privileged. The common interest doctrine is not a separate privilege, but an extension of the work product or attorney client privilege. United States v. Schwimmer, 892 F.2d 237, 243 (2d Cir. 1989). Thus, the communication in question must be attorney-client privileged or protected work product. Allied Irish Bank v. Bank of Am., 252 F.R.D. 163, 171 (S.D.N.Y. 2008). The party asserting the common interest privilege must show: (1) all clients and attorneys with access to the communication had in fact agreed upon a joint approach to the matter communicated, and (2) the information was shared with the intent to further that common purpose. S.E.C. v. Wyly, No. 10 Civ. 5760 (SAS), 2011 WL 3055396, at *2 (S.D.N.Y. July 19, 2011). In reviewing the communications, "the key question is whether the parties are collaborating on a legal effort that is dependent on the disclosure of otherwise privileged information between the parties or their counsel." AU New Haven, LLC v. YKK Corporation, No. 15 Civ. 03411 (GHW)(SN), 2016 WL 6820383, at *3 (S.D.N.Y. Nov. 18, 2016). The common interest doctrine only shields communications between codefendants, coplaintiffs, or persons who reasonably anticipate that they will become colitigants. Ambac Assur. Corp. v. Countrywide Home Loans, Inc., 27 N.Y.3d 616, 628 (2016).

. . . .

Deutsche Bank must partially produce **ICN RER 002 1 00000178-0733**. This communication is an email string between a paralegal with Residential Credit Solutions ("RCS") and Deutsche Bank concerning pending litigation involving the Aames Mortgage Investment Trust 2005-2, to which Deutsche Bank was the indenture trustee and RCS was the servicer. The bottom half of the 09/22/2011 9:21 AM email reveals an RCS attorney's legal advice regarding RCS's legal strategy in the litigation. This portion of the email is protected by the common interest privilege because Deutsche Bank and RCS share a common interest as trustee and servicer. However, the remaining emails in this string are cover emails that attach copies of the 2005-2 trust. These emails do not seek or reflect legal advice and are therefore not attorney-client privileged or protected work product. Accordingly, the 09/26/2011 9:22 AM, 09/23/2011 6:34 PM, and redacted version of the 09/22/2011 9:21 AM emails should be produced. . . .

. . . .

Deutsche Bank must produce **ICN COD 002 1 00000006-0564**. This communication is an email string between a Deutsche Bank employee and various employees in Bank of America's BAC Home Loans Servicing group concerning violations the City of Chicago is asserting against properties titled to Deutsche Bank. Although the communication discusses impending litigation in which Deutsche Bank and the servicer would be colitigants, it is merely administrative. An attorney is not a party to the conversation, and the emails do not reveal legal advice or counsel's impressions. The predominate purpose of this communication is to notify Bank of America, as servicer, to contact the City of Chicago to negotiate or resolve Deutsche Bank's liabilities. Thus, this communication is not privileged and should be produced.

. . . .

Deutsche Bank must produce **ICN WOE 002 1 00000021-0343**. This communication is an email string from a Deutsche Bank employee to a paralegal in the Wells Fargo Law Department. Although a paralegal is a party to the communication, the emails do not seek or reflect legal advice. Rather, the paralegal is notifying the Deutsche Bank employee that it contacted the wrong servicer. This communication is not privileged and should be produced.

. . . .

Deutsche Bank properly withheld **ICN CAB 006 1 00000036-13219-1**. This communication is a draft affidavit prepared by Brown & Camp, LLC, counsel for Deutsche Bank in a foreclosure action, and is therefore protected work product. Work product protection is extended where it is shared "between codefendants, coplaintiffs, or persons who reasonably anticipate that they will become colitigants." Ambac Assur. Corp. v. Countrywide Home Loans, Inc., 27 N.Y.3d 616, 628 (2016). This document was shared with Deutsche Bank's Servicer, Title Insurer, and the Title Insurer's counsel, who are all colitigants and share a common interest with Deutsche Bank. Accordingly, this document is protected by the common interest privilege and should not be produced.

III. Waiver

Rule 26 of the Federal Rules of Civil Procedure provides that when a party withholds documents on the grounds of privilege, it must both "expressly make the claim" and "describe the nature of the documents, communications, or tangible things not produced or disclosed—and do so in a manner that, without revealing information itself privileged or protected, will enable other parties to assess the claim." Fed. R. Civ. P. 26(b)(5)(A). When a party submits a privilege log that is deficient, the claim of privilege may be denied. United States v. Constr. Prods. Research, 73 F.3d 464, 473 (2d Cir. 1996); see also Fed. R. Civ. P. 26(b)(5), Advisory Committee Notes ("To withhold materials without such notice is contrary to the rule, subjects the party to sanctions under Rule 37(b)(2), and may be viewed as a waiver of the privilege or protection.").

BlackRock contends that Deutsche Bank's revised privilege log shows little improvement from the original, such that a finding of waiver is appropriate. It claims that many of the documents still lack meaningful descriptions or intelligible identifiers, making it impossible to assess whether the assertion of privilege was appropriate. It provides several pages of the log that exemplify these alleged deficiencies and cites case law that it claims supports its position. Deutsche Bank opposes waiver, arguing that BlackRock failed to engage in the Court-ordered meet-and-confer process and that the exemplar pages are not deficient on the merits.

The Court finds that Deutsche Bank has failed to provide an adequate privilege log notwithstanding multiple opportunities to do so. While some log entries are better than others—for example, where descriptions make clear that documents are draft discovery responses or attorney bills—Deutsche Bank's obligation is to provide sufficient information so that BlackRock is able to assess the privilege fully. Moreover, as explained in the Court's review of the exemplars, these deficient log entries often result in overly broad privilege assertions. Indeed, Deutsche Bank itself withdrew its privilege claims as to seven of the 20 documents that BlackRock selected as exemplars, and the Court found that an additional ten were improperly withheld.

A privilege log is not a mere administrative exercise. Its purpose is to ensure that a withholding party can justify a privilege designation. By submitting a deficient log, Deutsche Bank attempted to bypass this requirement, resulting in vastly overinclusive privilege designations. In an attempt to avoid this very problem, the Court ordered Deutsche Bank to provide a sworn affidavit by a party representative explaining why 30 of the Common Interest Exemplars were protected under the common interest doctrine. Instead, Deutsche Bank offered the sworn statement of a Morgan, Lewis & Bockius partner. The Court requested a party representative to force the party to defend its designations in connection with these loan-level litigations. Counsel's views were not requested.

The failure to provide a party affidavit is demonstrative of Deutsche Bank's broader failings with respect to its privilege log. The Court has given Deutsche Bank multiple opportunities to correct these deficiencies and it has failed to do so. Instead, Deutsche Bank waits until a document is challenged to review whether its privilege designation is correct. And as discussed earlier, when challenged, Deutsche Bank frequently realizes that the privilege was improperly asserted. This stance inappropriately shifts the burden to BlackRock to challenge a privilege assertion when Deutsche Bank should have established why a document was protected in the first place. A privilege log is not an iterative process and the Court will not offer Deutsche Bank another opportunity to follow the rules established in this Circuit.

Accordingly, Deutsche Bank has waived its privilege with respect to all documents listed on its privilege log (except as otherwise ruled in this Order) unless it can make a particularized showing as to individual documents that it believes are (1) adequately described on its log and, (2) in fact, privileged. Only documents listed on the privilege log with complete information— that is, the name of the author of the document, the name of any attorney, a clear description of the document, etc.—could qualify for this safety valve. Absent an application to the Court within 30 days on a document-by-document basis, all documents on the privilege log must be produced. The parties are ordered to file their letters on the docket.[67]

The several states of the United States, of course, have their own rules and sanctions frameworks. New York's and Michigan's are highlighted here as examples, although one must, of course, be sure to understand the framework in his or her particular jurisdiction.

Recognize that while Rule 37(e) clarified and standardized, to an extent, the framework in the federal world—one that eased the severity of when and how sanctions are applied—the framework in the New York world, which previously adopted federal standards, has not yet been similarly affected. Recall the case of *Pegasus*, discussed earlier in this chapter. In that case, New York's highest court began its analysis with the following holding:

A party that seeks sanctions for spoliation of evidence must show that the party having control over the evidence possessed an obligation to preserve it at the time of its destruction, that the evidence was destroyed with a "culpable state of mind," and "that the destroyed evidence was relevant to the party's claim or defense such that the trier of fact could find that the evidence would support that claim or defense". . . . Where the evidence is determined to have been intentionally or wilfully destroyed, the relevancy of the destroyed documents is presumed. . . . On the other hand, if the evidence is determined to have been negligently destroyed, the party seeking spoliation sanctions must establish that the destroyed documents were relevant to the party's claim or defense.

On this appeal, we are asked to decide whether the Appellate Division erred in reversing an order of Supreme Court that imposed a spoliation sanction on the defendants. We hold that it did, and remand the matter to the trial court for a determination as to whether the evidence, which

---

[67] *Id.*

the Appellate Division found to be negligently destroyed, was relevant to the claims asserted against defendants and for the imposition of an appropriate sanction, should the trial court deem, in its discretion, that a sanction is warranted.[68]

In addition, the states, of course, each have separate rules governing discovery failures, and the resultant penalties. For instance, courts and counsel can look to CPLR 3126 for guidance in New York, can look to Court Rule 2.313 in Michigan, or can look to Code of Civil Procedure § 2023.030 in California.

CPLR 3126 in New York provides:

If any party, or a person who at the time a deposition is taken or an examination or inspection is made is an officer, director, member, employee or agent of a party or otherwise under a party's control, refuses to obey an order for disclosure or wilfully fails to disclose information which the court finds ought to have been disclosed pursuant to this article, the court may make such orders with regard to the failure or refusal as are just, among them:

(1) an order that the issues to which the information is relevant shall be deemed resolved for purposes of the action in accordance with the claims of the party obtaining the order; or

(2) an order prohibiting the disobedient party from supporting or opposing designated claims or defenses, from producing in evidence designated things or items of testimony, or from introducing any evidence of the physical, mental or blood condition sought to be determined, or from using certain witnesses; or

(3) an order striking out pleadings or parts thereof, or staying further proceedings until the order is obeyed, or dismissing the action or any part thereof, or rendering a judgment by default against the disobedient party.[69]

Although New York has not adopted the Federal Rule 37(e) framework *per se*, the New York courts do apply a searching analysis to the facts when considering whether a certain level of sanctions is appropriate given behavior by a party.[70] A violation standing alone, without consideration of the level of seriousness, will not result in application of sanctions on the

---

[68] *Pegasus Aviation I, Inc. v. Varig Logistica S.A.*, 26 N.Y.3d 543, 547–548, 46 N.E.3d 601, 602 (2015) (citing *Zubulake v. UBS Warburg LLC*, 220 F.R.D. 212, 220 (S.D.N.Y. 2003); *VOOM HD Holdings LLC v. EchoStar Satellite L.L.C.*, 93 A.D.3d 33, 45, 939 N.Y.S.2d 321 (1st Dep't 2012)).

[69] N.Y. CPLR 3126. New York courts are also empowered to award sanctions against frivolous conduct. *See* 22 N.Y. Comp. Code R. & Regs. § 130-1.1.

[70] *See Lantigua v. Goldstein*, 149 A.D.3d 1057, 53 N.Y.S.3d 163 (App. Div. 2d Dep't 2017) ("Supreme Court providently exercised its discretion in denying that branch of the plaintiff's motion which was pursuant to CPLR 3126 to strike the . . . defendants' answer. The drastic remedy of striking an answer is inappropriate absent a clear showing that a defendant's failure to comply with discovery demands is willful and contumacious. . . . Under the circumstances, the plaintiff did not make a clear showing that the . . . defendants' failure to timely comply with certain discovery demands was willful and contumacious such that the drastic remedy of striking their answer was warranted") (citations omitted).

most severe end of the spectrum.[71] Consider the following majority opinion of the Appellate Division, Fourth Department, evaluating sanctions issued by a trial court under CPLR 3126:

Sarach v. M & T Bank Corp. (N.Y. Supreme Court, Appellate Division, Fourth Department)

Memorandum Opinion

Plaintiff commenced this action on March 1, 2012, for injuries he allegedly sustained when he slipped and fell on ice on March 23, 2009, as he was walking into defendant's bank in Buffalo, New York. On August 10, 2010, prior to the commencement of the action, plaintiff sought an order pursuant to CPLR 3102(c) for pre-action disclosure and preservation of evidence. Defendant opposed plaintiff's request for any pre-action disclosure, but represented to Supreme Court that it had voluntarily undertaken preservation of certain evidence, including accident reports, photographs, and surveillance videotapes, and ultimately "consent[ed] to an order of preservation." On October 29, 2010, the court granted plaintiff's application and ordered defendant to preserve, inter alia, all "photographs [and] video tapes, including but not limited to security and surveillance video related to the subject accident." During discovery after the action was commenced, plaintiff requested, inter alia, surveillance films related to the subject accident, and defendant responded that those materials had not been preserved. Thereafter, on July 30, 2014, plaintiff brought a motion pursuant to CPLR 3126 to strike defendant's answer on the ground that defendant had violated the court's 2010 order of preservation. The court granted plaintiff's motion and struck defendant's answer and affirmative defenses. Defendant appeals.

Initially, we agree with plaintiff that a sanction was warranted inasmuch as defendant "wilfully fail[ed] to disclose information" that the court had ordered to be preserved (CPLR 3126). Nevertheless, we conclude that the court abused its discretion in striking defendant's answer and affirmative defenses. It is well established that "a less drastic sanction than dismissal of the responsible party's pleading may be imposed where[, as here,] the loss does not deprive the nonresponsible party of the means of establishing his or her claim or defense". . . . Indeed, we note that the record does not demonstrate that the plaintiff has been "'prejudicially bereft'" of the means of prosecuting his action. . . . Thus, we conclude that an appropriate sanction is that an adverse inference charge be given at trial with respect to the unavailable surveillance footage . . . and we therefore modify the order accordingly.

Our dissenting colleague agrees that a "remedy is necessary," but disagrees with the sanction we have imposed, our analysis in reaching that sanction, and ultimately our directive to the court

---

[71] *See, generally, Brandsway Hospitality, LLC v Delshah Capital LLC*, 173 A.D.3d 457, 103 N.Y.S.3d 42 (App. Div. 1st Dep't 2019). *See also* Mark A. Berman, *Email Spoliation Sanctions: Timing Is Everything*, N.Y. L.J. at 5 (Jan. 7, 2020) (citing and discussing, *inter alia, Kohl v. Trans High Corp.*, No. 655200/2016, 2019 N.Y. Misc. LEXIS 5613 (Sup. Ct. N.Y. County Oct. 15, 2019); *ABL Advisor, LLC v. Patriot Credit Co.*, LLC, No. 651985/2015, 2019 N.Y. Misc LEXIS 4826 (Sup. Ct. N.Y. County Sept. 3, 2019); *Dantzig v. Orix Am Holdings, LLC*, No. 653368/2016, 2019 N.Y. Misc. LEXIS 5464 (Sup. Ct. N.Y. County Oct. 4, 2019)).

on how to effectuate the sanction. In our view, our resolution of this case requires us simply to determine whether defendant violated an order and whether such violation requires a sanction pursuant to CPLR 3126. The dissent refers to the "minimal prejudice suffered by plaintiff in not having been able to inspect the surveillance video in question." That reference overlooks the undisputed fact that plaintiff sought an order pursuant to CPLR 3102(c) for pre-action disclosure, and counsel for defendant not only volunteered to preserve certain items, including surveillance video related to the subject accident, but "consent[ed] to an order of preservation." Naturally, the court then granted the relief requested by plaintiff, and defendant never challenged the resulting order. Under those circumstances, we are unable to conclude that defendant's failure to comply with the order was anything but wilful. As for our dissenting colleague's concern with respect to the form of the adverse inference charge, we anticipate that the court will follow the Pattern Jury Instructions.[72]

Another New York court, the Appellate Division, First Department, did find a basis for an adverse inference sanction following a party's violation of discovery obligations. In *Douglas Elliman LLC v. Tal*,[73] the First Department held that an adverse inference sanction, imposed under CPLR 3126, was appropriate owing to plaintiff's gross negligence in destroying electronically stored information:

The record demonstrat[ed] that plaintiff acted with gross negligence in destroying ESI not only after commencement of the action triggered a duty to preserve, but after defendant['s] [ ] deposition, in which she referenced an email exchange in which she allegedly advised plaintiff that she had started working at Itzhaki Properties, and requested dual licensure, which plaintiff approved. . . . Accordingly, the court properly exercised its discretion in presuming the relevance of the email exchange and imposing spoliation sanctions. . . . Further, the court engaged in "an appropriate balancing under the circumstances" by ordering a tailored adverse inference charge limited to the alleged contents of the email exchange regarding defendant's [ ] work at Itzhaki Properties, and precluding plaintiff from presenting contrary evidence.

In Michigan, as another example, the state rules permit sanctions that mirror those available in the federal courts. In 2018, the Michigan Court of Appeals held:

Trial courts have the inherent authority to enforce their orders. . . . The court rules further provide that a trial court may "order such sanctions as are just" when a party "fails to obey an order to provide or permit discovery." MCR 2.313(B)(2). A trial court has a range of sanctions available to it to punish violations of its discovery orders: it may treat a fact as established, prohibit the disobedient party from supporting or opposing a designated claim or defense, prohibit the disobedient party from introducing particular evidence, strike pleadings in whole or in part, stay proceedings, dismiss an action or enter a default, order the disobedient party to compensate the

---

[72] 140 A.D.3d 1721, 1721–1723, 34 N.Y.S.3d 303, 303–305 (App. Div. 4th Dep't 2016) (citations omitted).

[73] 156 A.D.3d 583, 65 N.Y.S.3d 697 (Mem) (1st Dep't 2017). *See also Sanchez v. City of N.Y.*, 181 A.D. 3d 522, 119 N.Y.S. 3d 54 (mem) (App. Div. 1st Dep't 2020).

opposing party for his or her costs and fees, hold the disobedient party in contempt, or fashion any other remedy that is just under the circumstances. MCR 2.313(B)(2);. . . Trial courts must carefully review all the factors involved in the case and consider the full range of possible sanctions and then select the sanction that is "just and proper in the context of the case before it.". . .

When selecting the appropriate sanction from the wide range of available sanctions, the trial court should review several factors. It should consider

> (1) whether the violation was willful or accidental; (2) the party's history of refusing to comply with discovery requests (or refusal to disclose witnesses); (3) the prejudice to the defendant; (4) actual notice to the defendant of the witness and the length of time prior to trial that the defendant received such actual notice; (5) whether there exists a history of plaintiff's engaging in deliberate delay; (6) the degree of compliance by the plaintiff with other provisions of the court's order; (7) an attempt by the plaintiff to timely cure the defect; and (8) whether a lesser sanction would better serve the interests of justice. . . .

The sanction selected by the trial court must be proportionate and just in light of the violation at issue.[74]

In conclusion, with regard to sanctions, although they exist to provide an assurance that parties will obey their obligations in discovery at the risk of penalty, one should really abide by the obligations contained in the laws and rules of discovery to ensure a proper functioning of the justice system and a fair hearing for all parties involved.

1.11 – Brief Mention of Privacy Laws: The European Union's GDPR, Brazil's LGPD, New York's S.H.I.E.L.D. Act, California's CPA, and Illinois' BIPA

In this portion of the chapter, very brief mention is made of several data privacy and security laws/regulations. The European Union's General Data Protection Regulation (GDPR) is one such provision.[75] The reader's attention is directed to the GDPR if one has business or legal matters involving electronic data and communications with persons in Europe.

Briefly, the GDPR was created in 2016, but it was not enforced until May 25, 2018.[76] The European Commission's official website makes clear that the GDPR regulates personal

---

[74] *Blackburn v. Fabi*, 2018 WL 1832081, at *1-*2 (Mich. Ct. Apps. Apr. 17, 2018) (unpublished) (other citations omitted) (note that although the case concerns non-electronic issues, the same principles can be applied in the electronic world).

[75] Regulation (EU) 2016/679 of the European Parliament and of the Council of 27 April 2016 (repealing Directive 95/46/EC).

[76] European Commission COM(2018) 43 Final (Communication from the Commission to the European Parliament and the Council, *Stronger Protection, New Opportunities—Commission Guidance on the Direct Application of the General Data Protection Regulation as of 25 May 2018*), available at: https://eur-lex.europa.eu/legal-content/EN/TXT/?qid=1517578296944&uri=CELEX%3A52018DC0043.

data of individuals—but not legal entities or those who have passed away. Furthermore, the GDPR applies to professional and commercial activity, not to activities that take place at home or for personal, non-business activities. So personal data that enters the business world, or that is used for economic purposes, is governed and protected by the European GDPR.[77]

Thus, be very careful if you are transacting business cross-border, particularly with people located in Europe. If those individuals, for example, provide personal information to a company providing services—say banking, travel, insurance—the GDPR provisions apply to the information provided to the company by the individuals.[78] But in a circumstance where individual people contact other individual people and exchange information on a private matter—such as for a gathering of friends—then, even if they are within Europe, GDPR does not apply.[79]

An entire text could be written addressing, in great detail, the provisions and application of the GDPR. That, however, is not the purpose of this text. Therefore, if one has anything other than personal, non-business interactions with European persons, one must educate himself or herself on the regulations and nuances of the GDPR.

Now, as is clear from the heading to this subchapter section, a number of other jurisdictions have instituted laws and rules governing data privacy and security. For instance, in New York State the Stop Hacks and Improve Electronic Data Security (SHIELD) Act was codified at New York General Business Law §§ 899-aa & 899-bb (and effective October 23, 2019 and March 21, 2020, respectively).[80] It should be noted that, in actuality, when it comes to data breaches and loss or potential loss of personal identifying information, all 50 states as well as the District of Columbia, Guam, Puerto Rico, and the U.S. Virgin Islands require notification of affected persons if there is a breach of data security involving personal or private information, and legislation is always being introduced or evolving.[81] Furthermore, it has been reported that "at least 25 states … have laws that address data security practices" when it comes

---

[77] *See* European Commission, *What Does the General Data Protection Regulation (GDPR) Govern?*, available at: https://ec.europa.eu/info/law/law-topic/data-protection/reform/what-does-general-data-protection-regulation-gdpr-govern_en.

[78] *See id.*

[79] *Id.* However, as with everything else in life, there are few if any absolutes, and generalities run the risk of being too broad. In late spring 2020, a court in The Netherlands ruled on a matter where a daughter sued her own mother, after the mother refused to remove photos of her grandchildren from Pinterest and Facebook. Normally this would seem to be a purely private matter, not subject to the GDPR. However, the judge held that when the grandmother posted the photos to the social media platform, they became public, and then subject to the European GDPR. The grandmother was to be fined €50 per day, up to a maximum fine of approximately €1,000, until the photos were removed. *See* Matt Binder, *Grandma Ordered to Delete Facebook Photos of Grandkids or Face Fine*, Mashable (online May 22, 2020), *available at* https://mashable.com/article/grandma-ordered-to-delete-facebook-photos/.

[80] *See* N.Y. Gen. Bus. L. §§ 899-aa, 899-bb.

[81] *See* Mark Krotoski & Martin Hirschprung, *Preparing for New Requirements Under the N.Y. SHIELD Act*, N.Y. L.J. at 4 (Oct. 2, 2019); Shari Claire Lewis, *FTC Acts on Online Privacy, With State Laws Looming*, N.Y. L.J. at 5 (Dec. 17, 2019).

to firms and organizations operating in the private sector.[82] New York's SHIELD Act adds additional protections, including a broader definition of "breach," expansion of the elements that trigger a breach notification, and creation of a new: "Reasonable security requirement. (a) Any person or business that owns or licenses computerized data which includes private information of a resident of New York shall develop, implement, and maintain reasonable safeguards to protect the security, confidentiality, and integrity of the private information including, but not limited to, disposal of data."[83] New York's law specifically governs the entire life-cycle of data, from creation through and including deletion/destruction. The entity possessing the information must protect it from unauthorized access.[84]

With the COVID-19 pandemic resulting in much more online and remote work activity, in the United States and around the world, simultaneous growing threats emerged—one a threat to human health and well-being; the other a threat to data security and privacy. Thus, New York's actions, although they pre-dated the COVID-19 pandemic, were in line with California's new Consumer Privacy Act (CCPA) (effective January 1, 2020, and scheduled to begin enforcement July 1, 2020), the EU General Data Protection Regulation (GDPR), and the laws of myriad other jurisdictions in this field.[85]

California's Consumer Protection Act[86] resembles the European Union's right to be forgotten (or de-referenced), in that California's law allows consumers to ask businesses to delete personal information.[87] Meanwhile, Illinois' Biometric Information Privacy Act (BIPA)[88] regulates the collection, use, and storage of biometric information, governs how the information

---

[82] Daniel S. Wittenberg, *Data Privacy: More than Just "The New Black,"* ABA Litigation, vol. 45, no. 3, at 26-27 (Spring 2020).

[83] N.Y. Gen. Bus. L. § 899-bb(2)(a); *see also* M. Krotoski & M. Hirschprung, *Preparing for New Requirements, supra.* In 2020 there was also additional legislation pending in the New York State Legislature to create an even more robust law, the New York Privacy Act (NYPA). *See* Joseph V. DeMarco, *Implications of the "Data Fiduciary" Provision in the Proposed New York Privacy Act*, N.Y. L.J. Cybersecurity Special Report, at S6 (Mar. 2, 2020).

[84] *See* Gail Gottehrer, *Defensible Data Disposal: Once a Risk Mitigation Strategy, Now a Compliance Requirement*, 91 N.Y. St. Bar J. 8 (Sept./Oct. 2019). Louisiana has a similar statute—codified at La. Database Security Breach Notification Law, R.S. 51:3074. *Id.* New York's law also has a disclosure exemption, found in Gen. Bus. L. § 899-aa(2)(a), however its application may be narrow, and care should be taken if a firm plans to assert the exemption. *See* Keara Gordon, *et al., New York's SHIELD Act: How Much Will Your Inadvertence Cost You?*, N.Y. L.J. at 3 (May 7, 2020).

[85] *See* David Jacoby & Linda Priebe, *SHIELD and Sword: NY's Far Reaching Statute Governing Data Breaches*, N.Y. L.J. at 3 (May 11, 2020). *See also* Eric B. Stern & Andrew A. Lipkowitz, *Cyber Coverage for Penalties Imposed Under the SHIELD Act*, N.Y. L.J. at 5 (May 12, 2020); Christopher A. Iacono & Gabrielle I. Weiss, *New SHIELD Act Provisions Take Effect In March, Additional Legislation Pending*, N.Y. L.J. at 4 (Feb. 27, 2020); Peter Brown, *Trends in 2020 for Data And Personal Privacy*, N.Y. L.J. at 5 (Jan. 28, 2020); Francis Serbaroli, *The SHIELD Act: NY's New Data Protection Requirements Take Effect*, N.Y. L.J. at 3 (Nov. 26, 2019).

[86] Cal. Civ. Code § 1798, *et seq.*

[87] *See* G. Gottehrer, *Defensible Data, supra. See also* F. Paul Greene, *Experimentation in Privacy Law Leads to Increased Complexity*, N.Y. L.J. at 4 (Nov. 26, 2019) (discussing and comparing, among others, California, Nevada, New York and European Union laws and regulations).

[88] 740 ILCS 14/1, *et seq.*

can be disclosed, and requires firms and organizations that collect biometric information (fingerprints, iris scans, voiceprints, facial geometry, and the like—but not biological samples for testing, eye color, hair color, tattoo descriptions, or other such information) to destroy that information permanently and securely when the purpose of collection has been met—or within three (3) years of the human being's last interaction with the collector.[89] While, as mentioned above, numerous jurisdictions have addressed data privacy and actions to be taken in the case of breach, as of the fall of 2019 it was reported that only Illinois and three other states (California, Texas, and Washington) had taken action to protect *biometric information*. And, further, only a few local governments (places such as Oakland, California; San Francisco, California; and Somerville, Massachusetts) had acted to ban local government use of facial recognition technology.[90] Furthermore, as of summer 2019, Illinois was the only state whose law provided for a private right of action where aggrieved individuals could sue the violating company directly.[91]

There is clear import as to why biometric data should have its own specific protection under the law—a reason raised right in the language of the Illinois statutes: a victim of data breach could apply to change their social security number but they would not be able to change their biometric information (their facial geometry, for example)[92]—at least not without very great difficulty and much additional expense.

Briefly, in Brazil, the Lei Geral de Proteção de Dados (LGPD) (or General Law for the Protection of Privacy) is very similar to the GDPR and CCPA, and regulates personal data. It applies to all persons and entities, including the government—and also applies to those collecting information within the nation from outside the borders of Brazil. However, much like the GDPR it does not apply to purely private and personal interactions that are not based in business/economic transactions; and it also does not apply to journalistic or artistic purposes, or national defense or state security activities, among other carve-outs.[93]

To operate in business today, and to advise businesses concerning provisions of law, one must be knowledgeable when it comes to storage, protection, use, and destruction of personal identifying and biometric data. If you own/manage a business (including law firms), or you are an attorney advising a business, you must be aware of all jurisdictions within which the enterprise in question reaches individuals and obtains private, personal identifying and/or biometric data (even if the business organization is not located in said jurisdiction), because

---

[89] *See* G. Gottehrer, *Defensible Data, supra;* Ross Todd, *Google Hit With Class Action Under Illinois Biometric Privacy Law Over Facial Recognition*, N.Y. L.J. at 5 (Feb. 11, 2020); Kristen L. Burge, *Growing Patchwork of Biometric Privacy Laws*, ABA Litigation, vol. 44, no. 4, at 10 (Summer 2019).

[90] *See* Victoria Hudgins, *Why Facial Recognition Tech Got Banned By 4 Local Governments*, N.Y. L.J. at 4 (Nov. 26, 2019).

[91] *See* K. Burge, *Growing Patchwork, supra.*

[92] *See* 740 ILCS 14/5.

[93] *See* John Isaza & Hannah Katshir, *Brazil Passes Landmark Privacy Law: The General Law for the Protection of Privacy*, ABA Business Law Today (online Apr. 23, 2020), *available at* https://businesslawtoday. org/2020/04/brazil-passes-landmark-privacy-law-general-law-protection-privacy/?utm_source=newsletter&utm_medium=email&utm_campaign=may20_articles.

the provisions of the rules and laws addressed in this subchapter section, or others specific to a jurisdiction, may apply to the activities. One commentator has reported that there are over 120 jurisdictions around the world with pending or passed privacy laws or regulations, but apparently 79% of companies have been found failing to comply or keep current—which failure cost Facebook, Inc., for instance, a record *$5 billion* fine.[94]

Finally, in addition to all of the foregoing, attorneys and companies should be familiar with the fact that discovery, including eDiscovery, taking place both within a jurisdiction and cross-border, may involve activities implicating application of data protection and privacy laws and regulations, which the parties would have to address accordingly.[95]

Indeed, in this part of eWorld, being forewarned is definitely being forearmed.

1.12 – Service of Process by Electronic Means?

> *Service of Process* – Procedure where a plaintiff gives notice of initiation of a legal proceeding to a defendant.

Have you ever thought about service of process using Facebook or email? Well, counsel and the courts have been addressing just such action for years.

To begin our analysis, there is the case of *Fortunato v. Chase Bank USA*.[96] In that case, the defendant filed a motion seeking, in part, leave of the court to serve a third-party complaint via methods including Facebook private message and email to the address on the Facebook profile. The judge denied that portion of the motion, ordering substituted service by other means pursuant to Federal Rule of Civil Procedure 4(e) (requiring service pursuant to the laws of the state—N.Y. CPLR 308 & 316). It was held that service must be reasonably calculated to provide notice/ apprise parties of the pendency of litigation—and there was no showing of that in *Fortunato* with regard to service via Facebook.

However, that was just the start, not the end, of the story. Thereafter, another request for service by electronic means was considered in *F.T.C. v. PCCare247, Inc.*[97] In that case, foreign service of the summons and complaint was accomplished by submitting them to the

---

[94] *See* D. Wittenberg, *Data Privacy, supra*; Jacqueline Thomsen, *Judge Warns Facebook in Approving Record $5B Fine for Alleged Privacy Violations*, N.Y. L.J. at 2 (Apr. 27, 2020). *See also* Stefan Ducich & Jordan L. Fischer, *The General Data Protection Regulation: What U.S.-Based Companies Need to Know*, 74 The Bus. Law. 205 (Winter 2018-2019).

[95] *See, e.g.*, Christopher Boehning & Daniel J. Toal, *Confidentiality Order Sufficiently Protects EU Data in U.S. Discovery*, N.Y. L.J. at 5 (Apr. 7, 2020); William Schwartz, *How the CLOUD Act Is Likely To Trigger Legal Challenges*, N.Y. L.J. at 5 (Mar. 31, 2020) (discussing the federal Clarifying Lawful Overseas Use of Data (CLOUD) Act, a subsequent data-sharing agreement between the United States and United Kingdom, and how the process changed from letters rogatory and Mutual Legal Assistance Treaties to foreign law enforcement officials serving demands directly on U.S.-based companies overseas, and those companies having to appear in foreign courts to challenge the demands instead of U.S. courts).

[96] 2012 WL 2086950, 2012 U.S. Dist. LEXIS 80594 (S.D.N.Y. June 7, 2012) (Keenan, D.J.).

[97] 2013 WL 841037 (S.D.N.Y. Mar. 7, 2013). Be aware, though, that commentators and critics have criticized the result in *PCCare247*.

Indian Central Authority for service on the defendants, in accordance with Federal Rule of Civil Procedure 4(f)(1) and the Hague Convention on the Service Abroad of Judicial and Extrajudicial Documents in Civil or Commercial Matters, as well as by three alternative means. FTC then, however, sought leave to serve other documents by Facebook and email. The court granted the application, evaluating the jurisprudence surrounding Federal Rule 4(f)(3) and finding that the proposed means of service were not barred by international agreement, and did comport with notions of due process.[98]

Furthermore, in 2013, the U.S. District Court for the Eastern District of Missouri in *Joe Hand Promotions, Inc. v. Shepard*[99] summarized the state of the law, and held that under Federal Rule 4(f)(3), service of process by electronic means (*i.e.*, email) was permitted only in the case of an international party—one outside of the United States. The *Joe Hand* court at that time found no basis to utilize the federal rules for authorization of electronic service on a party inside of the United States. However, that was with regard to Federal Rule 4(f)—concerning service on individuals in foreign jurisdictions. When it comes to Federal Rule 4(e), service on individuals within the jurisdiction of a district, federal courts, such as the Northern District of California in *Pacific Bell Telephone Company v. 88 Connection Corporation*,[100] have indeed held electronic means to be one of the methods acceptable for service.

Since the "early" beginnings in this area of law, questions surrounding service of process utilizing social media and email have arisen in both state and federal actions, with court decisions coming down on both sides of the issue.[101]

In New York, CPLR 308(5) provides that the state courts may devise a method of appropriate service, if the case warrants, on *ex parte* application. In *Safadjou v. Mohammadi*,[102] the Appellate Division, Fourth Department, approved additional service by email when other forms of service were not possible. The defendant was located in Iran. In *Safadjou*, the

---

[98] *See also St. Francis Assisi v. Kuwait Fin. House*, 2016 WL 5725002, at *2 (N.D. Cal. Sept. 30, 2016) (Beeler, M.J.) ("As in . . . *PCCare*, service by the social-media platform, Twitter, is reasonably calculated to give notice to and is the 'method of service most likely to reach' al-Ajmi. . . . Al-Ajmi has an active Twitter account and continues to use it to communicate with his audience. Service by Twitter is not prohibited by international agreement with Kuwait"). *But cf. Asiacell Communications PJSC v. Doe*, 2018 WL 3496105, at *4 (N.D. Cal. July 20, 2018) ("the court finds that plaintiff's request to effect service via email and Facebook does not comply with the requirements of Rule 4(f)(3), nor is it reasonably calculated to apprise defendants of the pendency of the action and afford them an opportunity to present their objections; therefore, it does not comport with the requirements of due process"); *Entrepreneur Media, Inc. v. Doe*, 2019 WL 8638802 (C.D. Cal. Dec. 5, 2019).

[99] 2013 WL 4058745 (E.D. Mo. Aug. 12, 2013).

[100] 2016 WL 946132 (N.D. Cal. Mar. 14, 2016).

[101] *See, e.g., K.A. v. J.L.*, 450 N.J.Super. 247, 161 A.3d 154 (N.J. Super. Ct. 2016) ("There are only a handful of unpublished decisions, mostly from Federal District Courts, that have addressed the issue of service of process being accomplished through social media, with there being an almost even split between those decisions approving it and those rejecting it. The cases permitting such service have done so only on condition that the papers commencing the lawsuit be served on the defendant by another method as well"; the Court held that under the circumstances of this case, service by Facebook was acceptable).

[102] 105 A.D.3d 1423 (App. Div. 4th Dep't 2013). *But cf. Corson v. Power Moves, Inc.*, 2020 WL 3318099 (S.D.N.Y. June 18, 2020) (citing, *inter alia, Safadjou*).

Appellate Division affirmed the judgment of divorce, rejecting a challenge to the efficacy of the e-service. The Court specifically found that the parties had communicated by email for several months before service, the defendant used two email addresses that were utilized for the service, and the defendant acknowledged following the e-service that she received an email from the plaintiff's attorney, which the attorney sent to the two email addresses.

Safadjou was subsequently distinguished, though, by *Qaza v. Alshalabi*.[103] In *Qaza*, also a matrimonial action, the wife sought leave of the court, by *ex parte* order, to serve the summons on the husband via social media. The court held:

> Unlike the facts and circumstances presented in *Safadjou*, in the application before this Court plaintiff has failed to sufficiently authenticate the Facebook profile as being that of defendant and has not shows [*sic*] that, assuming arguendo that it is defendant's Facebook profile, that defendant actually uses this Facebook page for communicating. As such, plaintiff has not demonstrated that, under the facts presented here, service by Facebook is reasonably calculated to apprise defendant of the matrimonial action.
>
> The act for divorce has a multitude of ancillary affects on the rights and liabilities of parties. The Court must be scrupulous in allowing service by a methodology most likely to give notice. . . .[104]

The court made very clear in its conclusion that "[b]efore the Court could consider allowing service by Facebook pursuant to CPLR 308(5) the record must contain evidence that the Facebook profile was one that defendant actually uses for receipt of messages. The Court notes that anyone can create a Facebook profile using accurate, false or incomplete information and there is no way, under the application currently pending, for the Court to confirm whether the profile proffered by plaintiff is in fact the defendant's profile and that he accesses it."[105]

Then consider the case of *Sulzer Mixpac AG v. Medenstar Indus. Co. Ltd.*, where the court held:

> While email communications may also go astray or fail to come to the relevant individuals' attention, the Court finds that in this case, service to the email address listed on defendant's website is "reasonably calculated, under all circumstances, to apprise interested parties of the pendency of the action and afford them an opportunity to present their objections.". . .The email address in question is listed prominently on Medenstar['s]. . . Internet homepage. . . . Defendant

---

[103] 54 Misc.3d 691, 43 N.Y.S.3d 713 (Sup. Ct. Kings County Dec. 5, 2016).

[104] *Id.* at 695, 716 (citations omitted).

[105] *Id.* at 696, 717 (citing *Fortunato*). *See also NYKCool A.B. v. Pacific Intern. Serv., Inc.*, 66 F.Supp.3d 391 (S.D.N.Y. 2014) (Court acknowledged service via e-mail under some circumstances, pursuant to F.R.C.P. 4(f)(3); "in those cases where service by email has been judicially approved, the movant supplied the Court with some facts indicating that the person to be served would be likely to receive the summons and complaint at the given email address"; but in this particular case the Court did not permit service via an email address that was only on a humanitarian website linked to defendant because it was not sufficient); *Zanghi v. Ritella*, 2020 WL 589409, at *6 -*7 (S.D.N.Y. Feb. 5, 2020) (one e-mail address acceptable for service, but another one not).

Medenstar presumably relies at least partially on contact through export@medenstar.com to conduct overseas business, and it is reasonable to expect Medenstar to learn of the suit against it through this email address. Service to this email address . . ."comports with constitutional notions of due process."[106]

In the state courts of New York, a similar analysis is performed, and in the case of *Baidoo v. Blood-Dzraku*, the Court held:

> plaintiff has shown that it would be an exercise in futility to attempt the two alternative service methods provided for by CPLR 308. Both "substitute service" and "nail and mail" service require knowledge of the defendant's "actual place of business, dwelling or usual place of abode". . . The record establishes that plaintiff has been unsuccessful in obtaining either a business or home address for defendant, even though she has diligently sought that information. . . . Having demonstrated a sound basis for seeking alternative service pursuant to CPLR 308(5), plaintiff must now show that the method she proposes is one that the court can endorse as being reasonably calculated to apprise defendant that he is being sued for divorce.[107]

In Maryland, the U.S. District Court for the District of Maryland held, on this question of service of process by electronic means:

> [t]he Court first finds that service by electronic mail is appropriate where Defendant is not represented by counsel, where Defendant has left the country-apparently having moved to Thailand, where Plaintiff has searched diligently yet unsuccessfully for Defendant's mailing address, and where Defendant has exhibited a willingness to communicate with Plaintiff by electronic mail as discussed supra. . . . Further, the Court finds no evidence that alternative service by electronic mail is prohibited by any international agreement including Thailand.[108]

Finally, there is the 2017 decision in *Baez v. City of Schenectady*, out of the U.S. District Court for the Northern District of New York. In the action, two witnesses were directed to appear for depositions via subpoena, but failed to appear. The magistrate judge discussed contempt, but in the meantime ordered the witnesses to appear for deposition on a specific date. The judge further directed "[d]efendants' counsel to prepare new subpoenas, to be signed by the Court, for a deposition to occur on November 28, 2017, at Defendants' counsel's office located in Albany, New York. . . . *In addition to the requirements of Rule 4, the Court is also requiring that a copy of the subpoena together with a copy of this Order also be transmitted to the witnesses via any known social media (i.e. Facebook), email, or text address.*"[109]

---

[106]312 F.R.D. 329, 332 (S.D.N.Y. 2015) (citations omitted); *see also Terrestrial Comms LLC v. NEC Corp.*, 2020 WL 3270832 (W.D. Tex. June 17, 2020).

[107]5 N.Y.S.3d 709, 713 (Sup. Ct. N.Y. County 2015) (service of divorce summons via Facebook private message permitted).

[108]*Enovative Tech., LLC v. Leor*, 2014 WL 7409534, *1 (D. Md. Dec. 24, 2014).

[109]*Baez v. City of Schenectady*, 2017 WL 4990646 (N.D.N.Y. Nov. 1, 2017) (Stewart, M.J.) (emphasis added).

Ultimately, continuing the trend of social media having a great impact on the world of litigation, these decisions make clear that courts are slowly becoming more comfortable with service of process via Facebook, email, or other social media when more traditional means of service are unavailable or unavailing.

1.13 – Conclusion

As is clear from this chapter, there is much to be aware of in our ever-changing eWorld. It is important to know the rules and recognize where more information and understanding must be sought. Only then can one truly walk forth without trepidation and with comfort in knowing that the next step taken will not be head-long into quicksand.[110]

Chapter End Questions

Note: Answers may be found on page 261.

1. Which of the following Federal Rules has nothing to do directly with e-discovery?
   a. Rule 16
   b. Rule 26(b)
   c. Rule 37(e)
   d. Rule 72

2. Which of the following NYS Rules governs penalties and sanctions?
   a. Rule 202.12(c)(3)
   b. Rule 202.12(b)(1)
   c. CPLR Rule 3212
   d. CPLR Rule 3126

3. What does "ESI" stand for?
   a. Electronically Stored Information
   b. Eronneously Strained Information
   c. Everyone Stays Informed
   d. Evaluate Strong Intelligence

4. In most circumstances, concerning discovery costs, which party usually pays?
   a. The Court
   b. Requesting party
   c. Producing party
   d. None of these

---

[110]For a valuable case, addressing a number of major issues discussed in this chapter, *see Stinson v. City of New York*, 2016 WL 54684 (S.D.N.Y. Jan. 5, 2016) (Sweet, D.J.). In *Stinson*, the Court addressed, *inter alia*, (a) preservation of text messages; (b) preservation of other ESI; (c) ineffectiveness of litigation hold; (d) adverse inference sanction; and (e) application and timing for a new rule of procedure.

5. Which of the following court decisions is considered a key decision in the area of Technology Assisted Review (TAR)?

   a. *Smith v. Jones*
   b. *Marbury v. Madison*
   c. *Da Silva Moore v. Publicis Groupe*
   d. None of these

6. Under Federal Rule 37(e), which of the following does not warrant the most severe sanctions?

   a. Intent to deprive the other party of relevant information
   b. Destruction of specific material at the direction of upper level managers in a company
   c. Negligent loss of information through accidental destruction
   d. All of the above receive the most severe sanction

7. Given current caselaw, which of the following ways might be used to serve process in a legal matter electronically?

   a. Facebook
   b. E-mail
   c. Text message
   d. All of the above

8. True or False: Attorneys are responsible, together with clients, for the preservation of information.

   a. TRUE
   b. FALSE

9. True or False: If a party does not request the specific format of the discovery sought (i.e. native format electronic, PDF electronic, paper), then once the responding party provides discovery, the receiving party cannot later demand re-production in a different specified form.

   a. TRUE
   b. FALSE

10. True or False: When requesting information from a government source under FOIL or FOIA, electronic materials are not subject to disclosure.

    a. TRUE
    b. FALSE

CHAPTER 2

The Cloud, Its Usage, and Ethical Concerns

2.1 – What Is the Cloud?

You have doubtless heard of the term "the Cloud." You may even utilize the Cloud to store information from your office files or from your personal devices, such as an iPhone. But, you may still ask, "Just what is the Cloud or cloud computing?" Well, to

> ***Cloud*** – Storage not on the local hard drive of a computer or device; using the Internet and network of storage capacity to access material and information from anywhere at any time.

start with, the Cloud we are talking about is not a fluffy cotton ball floating in the sky from which rain falls. However, believe it or not, that is actually the influence for the name—from the cloud symbol used in flow charts as far back as the 1960's.[1]

Cloud computing is not about local storage, though. Anything stored on the hard drive of your desktop or laptop computer is not utilizing the Cloud. "The Cloud" is also not the term used for network-attached storage (NAS) hardware—that is, for data stored on a network dedicated only to your business or home computing needs. According to *PC Mag*:

> [f]or it to be considered "cloud computing," you need to access your data or your programs over the Internet, or at the very least, have that data synced with other information over the Web. . . . The end result is the same: with an online connection, cloud computing can be done anywhere, anytime.[2]

---

[1] *See* Mark Koba, *Cloud Computing: CNBC Explains*, CNBC (online June 29, 2011), *available at* cnbc.com/id/43483060.

[2] E. Griffith, *What Is Cloud Computing?*, PC MAG (online), available at: http://www.pcmag.com/article2/0,2817,2372163,00.asp (May 2016). *See also ProExpress Distrib. LLC v. Grand Elec., Inc.*, 2017 WL 2264750, at *1 n.1 (Md. Ct. Spec. Apps. May 24, 2017) (citing *PC Mag*). In other words, "[t]he 'cloud' is 'merely "a fancy way of saying stuff's not on your [own] computer."'" Ohio State Bar Assoc. Informal Advisory Op. 2013-03 (citing Formal Op. 2011-200, 1 (Pa. Bar Assoc. Comm. on Legal Ethics & Prof'l Resp. 2011)).

Utilizing the Cloud, large amounts of information and documentation can be stored without the need to "download 'cumbersome software . . . onto office desktops or laptops.'"[3] Thus, the Cloud is basically centralized storage of data—just not on your own device or hard drive. A series of networks is formed, utilizing the Internet, to store material off-site, in secure locations.[4] Following September 11, 2001, and the loss or threat of loss of information faced by companies storing only on local drives, the Cloud and network storage became attractive as it developed. Thus, "a cloud computing environment consists of resources, such as datacenters, that are operated by a cloud provider. . . . Those resources are used to provide users with access to applications or services, such as 'web portal[s] with email functionality, database programs, word processing programs, accounting programs, inventory management programs, [or] numerical analysis programs[.]'"[5]

Further explanation has been provided by the courts as follows. Basically, the Cloud

leverages "[v]irtualization technology." . . . For example,. . . cloud computing "facilitates the operation of multiple virtual servers within a single physical server system[ ]" and [there are]. . . benefits that stem from the use of such technology (e.g., that users can run software "without impacting users of other virtual servers operating within the same physical server system" or that cloud providers can offer users access to their physical resources even though the users are remote to the cloud provider).[6]

You may have heard of "Dropbox," or you may even have utilized Dropbox. That is an example of Cloud computing/storage and file sharing. "Those apps live in your file system so that you can easily move files from your computer to the cloud and vice versa by dragging and dropping them into your Dropbox folder. The service automatically and quickly syncs your files across all of your devices, so you can access everything, everywhere. There is no size limit on files you upload to Dropbox with the desktop or mobile apps, but larger files can take several hours to upload, depending on your connection speed."[7]

With the growth of online events, availability of virtual resources and sites, and in light of the COVID-19 pandemic's impact on off-site and work-from-home activities, the Cloud is ever-expanding, and attorneys and law firms are turning to Cloud computing more and more in their practices and their everyday lives—meaning that understanding the ethical uses, security threats, risks, benefits and options is vital.[8]

---

[3] *See In re Jobdiva, Inc.*, 843 F.3d 936 (Fed. Cir. Dec. 12, 2016) (citing and quoting DICTIONARY OF COMPUTER AND INTERNET TERMS 434 (11th ed. 2013)).

[4] Ohio State Bar Assoc. Informal Advisory Op. 2013-03 (citing Andrew L. Askew, *iEthics: How Cloud Computing Has Impacted the Rules of Professional Conduct*, 88 N. DAK. L. REV. 453, 457 (2012)).

[5] *Kaavo Inc. v. Cognizant Tech. Solutions Corp.*, 2016 WL 476730, at *1 (D. Del. Feb. 5, 2016) (*Kaavo I*).

[6] *Kaavo Inc. v. Amazon.com Inc.*, 2016 WL 6562038, at *8 (D. Del. Nov. 3, 2016) (*Kaavo II*).

[7] *ProExpress Distrib. LLC v. Grand Elec., Inc.*, 2017 WL 2264750, at *1 n.3 (Md. Ct. Spec. Apps. May 24, 2017) (citing CNET.com).

[8] *See* Jack Newton, *The Legal Industry's Rapid Transition to the Cloud*, 92 N.Y. St. B.J. 30 (May 2020); Victoria Hudgins, *Pandemic Places Greater Emphasis on the Cloud*, N.Y. L.J. at 5 (Mar. 20, 2020).

2.2 – Versions of the Cloud

There are multiple versions of cloud computing available to consumers: private, public, community, and hybrid.

When we talk about the *Public Cloud*, we mean: "[t]he cloud infrastructure is made available to the general public or a large industry group and is owned by an organization selling cloud services."[9] This would be akin to the cloud storage used by iPhone owners, where many users utilize storage on a cloud operated by a major service provider.

The *Private Cloud*, on the other hand, "delivers similar advantages to public cloud, including scalability and self-service, but through a proprietary architecture. Unlike public clouds, which deliver services to multiple organizations, a private cloud is dedicated to a single organization."[10] As opposed to a public cloud, the infrastructure of a private cloud "is operated solely for an organization. It may be managed by the organization or a third party and may exist on premise or off premise."[11] Nevertheless, it is the cloud storage for a single entity (although, again remember, not on local servers or hard drives).

With regard to the *Community Cloud*, "[t]he cloud infrastructure is shared by several organizations and supports a specific community that has shared concerns (e.g., mission, security requirements, policy, and compliance considerations). It may be managed by the organizations or a third party and may exist on premise or off premise."[12]

Finally, when we look at a *Hybrid Cloud*, "[t]he cloud infrastructure is a composition of two or more clouds (private, community, or public) that remain unique but are bound together by standardized or proprietary technology that enables data and application portability (e.g., cloud bursting for load-balancing between clouds)."[13]

> ***Internet of Things (IoT)*** – Everything connected to the Internet, and using data & networks to speak to each other or convey information.

2.3 – Cloud Computing and Unintentional Storage of Information

Cloud-based services are a significant part of the make-up and function of the "Internet of Things" (IoT). The IoT includes mobile devices and computers, as well as smart TVs, smart cars, smart watches, home personal assistant devices (such as Amazon "Alexa" or Google "Home"), and other devices.[14] "It's about networks, it's about devices, and it's about data."[15]

---

[9] *Google, Inc. v. U.S.*, 95 Fed.Cl. 661, 667 (Fed. Cl. 2011).

[10] TechTarget (online), available at: https://searchcloudcomputing.techtarget.com/definition/private-cloud.

[11] *Google, Inc.*, 95 Fed.Cl. at 667.

[12] *Id.*

[13] *Id.*

[14] *See* Matt Burgess, *What is the Internet of Things? WIRED Explains*, WIRED (online Jan. 2017), *available at* wired.co.uk/article/internet-of-things-what-is-explained-iot; *see also Weller v. Scout Analytics, Inc.*, 230 F. Supp. 3d 1085, 1089 (N.D. Cal. 2017).

[15] *See* M. Burgess, *supra.*

As technology expands, so will the efforts of law enforcement and civil attorneys to obtain the information contained thereon. Searches of devices and phones by law enforcement, and the protocols for the same, may also involve or implicate data that is not stored on the device itself, but that law enforcement personnel could access through cloud computing services. That does not, however, necessarily invalidate the searches,[16] and therefore the public should be aware of new developments affecting legal rights.

You may have read or heard about efforts by law enforcement to serve a search warrant on Amazon.com for access to information that may have been "overheard" and "stored" by an Echo device (answering to the name "Alexa," as seen on TV).[17] It is believed the device may have been activated during a struggle that resulted in an alleged murder. The device sent information to the Cloud but also stored it on the device itself. Similarly, a Fitbit's recorded data, together with separate computer and e-mail data, was the subject of evaluation as potential evidence in a separate murder investigation in 2017.[18]

2.4 – Ethical Considerations for Attorneys Utilizing Cloud Storage

As with almost everything else in life, aside from legal concerns, separate ethical guidance affects one's activities. In New York and other jurisdictions, there are governing Rules of Professional Conduct, and they impact an attorney's use of cloud storage.

For instance, New York Rule of Professional Conduct (NYRPC) 1.6(a) provides:

A lawyer shall not knowingly reveal confidential information, as defined in this Rule, or use such information to the disadvantage of a client or for the advantage of the lawyer or a third person, unless: (1) the client gives informed consent, as defined in Rule 1.0(j); 2) the disclosure is impliedly authorized to advance the best interests of the client and is either reasonable under the circumstances or customary in the professional community; or (3) the disclosure is permitted by paragraph (b).

---

[16] *See People v. Lopez*, 2018 WL 328745 (Ca. Ct. Apps. 6th Dist. Jan. 9, 2018) (not published); *cause transferred, rehearing granted, opinion after transfer, People v. Lopez*, 2019 WL 7037476 (Cal. Ct. Apps. 6th Dist. Dec. 19, 2019) (not published).

[17] *See* D. Weiss, *Can Amazon's Alexa Provide Murder Clues? Digital Assistants Could Aid Prosecutions*, ABAJournal.com (Jan. 4, 2017).

[18] *See* D. Weiss, *Murdered Woman's Fitbit Data Inconsistent with Husband's Story, Police Say*, ABAJournal.com (Apr. 25, 2017). Some companies are seizing on the public's concerns as these topics become more widely known, with one selling a "bracelet of silence" that emits ultrasonic signals so Alexa, smart watches, *et al.*, cannot hear you. *See* Kashmir Hill, *Don't Want Alexa to Listen? Wear This*, The N.Y. Times *Sunday Business* at 3 (Feb. 16, 2020).

As of January 1, 2017, New York Rule of Professional Conduct 1.6(c) provides:

A lawyer shall make reasonable efforts to prevent the inadvertent or unauthorized disclosure or use of, or unauthorized access to, information protected by Rules 1.6, 1.9(c), or 1.18(b).

New York Rule of Professional Conduct 5.3 governs the work of non-lawyers, and provides that it must be adequately supervised by attorneys.

Furthermore, New York RPC 1.6(b) provides:

A lawyer may reveal or use confidential information to the extent that the lawyer reasonably believes necessary:

(1) to prevent reasonably certain death or substantial bodily harm; (2) to prevent the client from committing a crime; (3) to withdraw a written or oral opinion or representation previously given by the lawyer and reasonably believed by the lawyer still to be relied upon by a third person, where the lawyer has discovered that the opinion or representation was based on materially inaccurate information or is being used to further a crime or fraud; (4) to secure legal advice about compliance with these Rules or other law by the lawyer, another lawyer associated with the lawyer's firm or the law firm; (5) (i) to defend the lawyer or the lawyer's employees and associates against an accusation of wrongful conduct; or (ii) to establish or collect a fee; or (6) when permitted or required under these Rules or to comply with other law or court order.

We look to these rules because the ethics authorities in the various jurisdictions that have addressed this area speak to cloud computing and storage of information as implicating "confidential information" of a client. Indeed, that makes intuitive sense—when most or all information of a client is stored electronically on a given file, no doubt confidential information is included therewith. That, therefore, implicates, at the least, NYRPC 1.6.

The New York State Bar Association issued a formal ethics opinion on this very matter in 2010, stating:

A lawyer may use an online data storage system to store and back up client confidential information provided that the lawyer takes reasonable care to ensure that confidentiality is maintained in a manner consistent with the lawyer's obligations under Rule 1.6. A lawyer using an online storage provider should take reasonable care to protect confidential information, and should exercise reasonable care to prevent others whose services are utilized by the lawyer from disclosing or using confidential information of a client.[19]

The New York County Lawyers Association (NYCLA) also issued ethical guidance concerning shared computing services. NYCLA addressed inquiries by a solo practitioner who wished to share office space, computers, and personnel with other professionals (an accountant and an investment adviser), but with each professional having their own, separate business entities. The NYCLA Committee determined that regardless of whether the other professionals

---

[19] New York State Bar Association Ethics Opinion 842 (2010).

were deemed designated professionals by the Appellate Division, computer services and access should not be shared. As specifically stated in the Opinion:

> The inquirer asked whether the arrangement could involve sharing computer services. Without knowing what particular computer service the inquirer is referring to, and without having particular computer expertise of its own, the Committee believes that any computer services which involve communications with or contain client information and which are shared with or accessible to non-attorneys could run afoul of the attorney's ethical obligations to preserve client confidences. The attorney must diligently preserve the client's confidences, whether reduced to digital format, paper, or otherwise. The same considerations would also apply to electronic mail and websites to the extent they would be used as vehicles for communications with the attorney's clients.[20]

In comments to Rule 1.6, the New York State Bar Association has further stated that when it comes to the communication or transmission of information:

> that includes information relating to the representation of a client, the lawyer must take reasonable precautions to prevent the information from coming into the hands of unintended recipients. Paragraph (c) does not ordinarily require that the lawyer use special security measures if the method of communication affords a reasonable expectation of confidentiality. However, a lawyer may be required to take specific steps to safeguard a client's information to comply with a court order (such as a protective order) or to comply with other law (such as state and federal laws or court rules that govern data privacy or that impose notification requirements upon the loss of, or unauthorized access to, electronic information). For example, a protective order may extend a high level of protection to documents marked "Confidential" or "Confidential – Attorneys' Eyes Only"; the Health Insurance Portability and Accountability Act of 1996 ("HIPAA") may require a lawyer to take specific precautions with respect to a client's or adversary's medical records; and court rules may require a lawyer to block out a client's Social Security number or a minor's name when electronically filing papers with the court. The specific requirements of court orders, court rules, and other laws are beyond the scope of these Rules.[21]

> **Bug** – Usually innocent error or issue in code that causes an unintentional harm or malfunction.

In 2017, the American Bar Association, issued an opinion of its own, at the same time revisiting issues addressed in prior Formal Opinion 99-413 and Formal Opinion 11-459. In its 2017 Opinion, the ABA considered not only the communication itself, but also the need for attorneys to ensure that the method selected and the storage provider utilized complied with attorney obligations to safeguard client information.

---

[20] New York County Lawyers Association Formal Opinion 733. The NYCLA Opinion also cited to NYSBA Opinion 709 (1998) (use of Internet e-mail; attorney should exercise care in preserving privilege).

[21] New York State Bar Association Comment 17 to Rule 1.6 of the New York Rules of Professional Conduct (amended as of January 1, 2017).

A lawyer generally may transmit information relating to the representation of a client over the internet without violating the Model Rules of Professional Conduct where the lawyer has undertaken reasonable efforts to prevent inadvertent or unauthorized access. However, a lawyer may be required to take special security precautions to protect against the inadvertent or unauthorized disclosure of client information when required by an agreement with the client or by law, or when the nature of the information requires a higher degree of security. . . .

. . . Comment [18] to Model Rule 1.6(c) includes nonexclusive factors to guide lawyers in making a "reasonable efforts" determination. Those factors include: •the sensitivity of the information, •the likelihood of disclosure if additional safeguards are not employed, •the cost of employing additional safeguards, •the difficulty of implementing the safeguards, and •the extent to which the safeguards adversely affect the lawyer's ability to represent clients. . . .[22]

Virus – Any of a number of malicious computer data codes that, if they gain access to a device or system, cause damage.

Malware – Computer software designed to intentionally cause damage.

Ransomware – Malware that once it has access to a system, blocks the user/owner, and threatens to either block material/stored information forever, or publish sensitive information to the world, unless a specified ransom is paid.

Attorneys always have duties of competence and confidentiality.[23] And even given the approval of cloud storage for attorney use, some recommend not using it for truly sensitive data and information – which, of course, is a judgment call by attorney and client.[24]

Furthermore, as everyone is aware, security on the Internet is always a concern, as is the ever-present existence of hackers, those who conduct Ransomware attacks, and those who introduce malware, viruses, and other evils into computer systems.[25] Therefore most, if not all, of the ethics authorities who speak to the issue of cloud storage also speak to an attorney's obligation to ensure that the cloud utilized has adequate and updated security measures

---

[22] American Bar Association Formal Op. 477R (2017) (citing, *inter alia*, MODEL RULES OF PROF'L CONDUCT R. 1.6 cmt. [18] (2013); and referencing ABA Formal Opinion 08-451 and issues attorneys should evaluate if utilizing an outsource vendor, so that the attorney ensures that she or he fulfills their duties of diligence and supervision).

[23] *See* SOCIAL MEDIA ETHICS GUIDELINES OF THE COMMERCIAL & FEDERAL LITIGATION SECTION OF THE NEW YORK STATE BAR ASSOCIATION, Guidelines 1.A & 5.E (2019).

[24] *See* Matt Reynolds, *Lawyers Should Weigh Risks and Ethics in Cloud Computing*, ABA Journal (online Feb. 28, 2020), available at https://www.abajournal.com/news/article/lawyers-should-weigh-risks-and-ethics-in-cloud-computing.

[25] *See* Debra Cassens Weiss, *Websites for Texas Courts Are Shut Down After Ransomware Attack*, ABA Journal (online May 12, 2020), *available at* https://www.abajournal.com/news/article/websites-for-texas-courts-are-shut-down-after-ransomware-attack. *See also* David Thomas, *Lady Gaga's Law Firm Got Hacked. Now What?*, N.Y. L.J. at 1, 7 (May 18, 2020) (ransomware attack on law firm demanded $21 million, and then $42 million, or else sensitive data would be released related to clients and individuals, including Lady Gaga, Robert DeNiro, the rock band AC/DC, and Donald Trump).

– indeed, something about which corporate/business clients should now also be advised.[26] Attorneys should always be certain of their firm's cybersecurity and the obligations imposed on them.[27] Courts have acknowledged that the Cloud is part of the "vast cosmos of the Internet."[28] But secure third-party cloud-based file storage has generally been acknowledged by courts as an acceptable security option to protect materials identified as confidential or protected in discovery proceedings.[29]

---

[26] *See* David J. Rosenbaum & Kevin Ricci, *Cybersecurity: An Ethical Responsibility*, N.Y. L.J. at 4 (Mar. 23, 2020) (citing, *inter alia*, RPC 1.1, 1.4, 1.15 & 1.16, & NYSBA Op. 842). Furthermore New York State enacted new legislation under the S.H.I.E.L.D. (Stop Hacks and Improve Electronic Data Security) Act, codified at General Business Law §§ 899-aa & 899-bb, and effective October 23, 2019 and March 21, 2020, respectively. All 50 states, as well as the District of Columbia, Guam, Puerto Rico and the U.S. Virgin Islands require notification of affected persons if there is a breach of data security involving personal or private information. *See* Mark Krotoski & Martin Hirschprung, *Preparing for New Requirements Under the N.Y. SHIELD Act*, N.Y. L.J. at 4 (Oct. 2, 2019). New York's recent Act adds additional protections, including a broader definition of "breach," expansion of the elements that trigger a breach notification, and creation of a new: "Reasonable security requirement (a) Any person or business that owns or licenses computerized data which includes private information of a resident of New York shall develop, implement and maintain reasonable safeguards to protect the security, confidentiality, and integrity of the private information including, but not limited to, disposal of data." N.Y. Gen. Bus. L. § 899-bb(2); *see also* M. Krotoski & M. Hirschprung, *Preparing for New Requirements, supra*. With the COVID-19 pandemic resulting in much more work-from-home activity, and remote sharing of information via electronic means, a growing threat simultaneously presented itself with regard to hacks of personal and private information. New York's actions were in line with California's new Consumer Privacy Act (CCPA), scheduled to begin enforcement July 1, 2020, and the EU General Data Protection Regulation (GDPR). *See* David Jacoby & Linda Priebe, *SHIELD and Sword: NY's Far Reaching Statute Governing Data Breaches*, N.Y. L.J. at 3 (May 11, 2020). *See also* Eric B. Stern & Andrew A. Lipkowitz, *Cyber Coverage for Penalties Imposed Under the SHIELD Act*, N.Y. L.J. at 5 (May 12, 2020); Christopher A. Iacono & Gabrielle I. Weiss, *New SHIELD Act Provisions Take Effect In March, Additional Legislation Pending*, N.Y. L.J. at 4 (Feb. 27, 2020); Francis Serbaroli, *The SHIELD Act: NY's New Data Protection Requirements Take Effect*, N.Y. L.J. at 3 (Nov. 26, 2019). New York's S.H.I.E.L.D. Act is also mentioned in Chapter 1, *supra*, in discussion along with California's CCPA, Brazil's LGPD, and the European Union's GDPR. Additionally, one other thing should be noted—do not assume that all data breaches occur remotely using the Internet. It has been reported that there is also a potential hardware weakness presenting an opportunity for data breach. If your laptop computer has a port with a thunderbolt icon over it or next to it, then it has a "Thunderbolt connection." That connection usually allows for charging, external storage, and other functionality. However, it is also the source of potential cybersecurity vulnerability because a hacker having only brief physical access to the laptop can read and copy data, and otherwise utilize attacks such as "evil maid direct memory access (DMA)." Courtney Linder, *If Your Laptop Has a Thunderbolt Port, You Could Get Hacked*, Popular Mechanics (online May 13, 2020), *available at* https://www.popularmechanics.com/technology/security/a32449230/thunderbolt-vulnerability-hacking/. Always be vigilant.

[27] *See, generally,* Medina, *Defensible Cybersecurity Tailoring an Organization's Security Posture to Applicable Legal Standards*, 88 N.Y.St. B.J., at 38 (May 2016); Treglia, *Increasing Cybersecurity Requirements for Lawyers*, N.Y. L.J., at 7 (May 30, 2017); Hudson, *21st Century Standards*, ABA Journal, at 24 (July 2017).

[28] *U.S. v. Mayo*, No. 2:13-CR-48, 2013 WL 5945802, at *8 (D. Vt. Nov. 6, 2013) (citing M. Orso, *Cellular Phones, Warrantless Searches, and the New Frontier of Fourth Amendment Jurisprudence*, 50 Santa Clara L. Rev. 183, 211 (2010)).

[29] *See LeBlanc v. Halliburton Co.*, 2018 WL 566436 (D. N.M. Jan. 25, 2018) (Fouratt, M.J.).

2.5 – Survey of Several Other States' Ethical Guidance

In a 2015 ethics opinion from the Tennessee Supreme Court Board of Professional Responsibility, it was stated that an attorney may ethically store client confidential information in the "cloud." However, in doing so, an attorney must be careful that: "(1) all such information or materials remain confidential; and (2) reasonable safeguards are employed to ensure that the information is protected from breaches, loss, and other risks. Due to rapidly changing technology, the Board doesn't attempt to establish a standard of care, but instead offers guidance from other jurisdictions."[30]

The Tennessee Board further stated that it was not opining about specific standards of care. Instead, the opinion is a compendium of citations and guidance from numerous other jurisdictions, including Maine, Ohio, Vermont, Pennsylvania, Kentucky, Florida, Alaska, Alabama, New Hampshire, and New York. The Board opined: "[a] lawyer owes the same ethical duties, obligations and protections to clients with respect to information for which they employ cloud computing as they otherwise owe clients. . . with respect to information in whatever form."[31]

Although a high burden is placed on attorneys, it is not an unreasonable standard, and perfection is not the measure applied. An attorney must simply utilize reasonable methods, ensuring competency, and protecting client confidential information with secure measures complying with Rules 1.1, 1.6, and 1.9. For instance, if an attorney has truly sensitive or confidential information—consider, perhaps, trade secrets of a business—he or she may have to reconsider use of a service provider or whether other security is necessary to safeguard material.[32]

The Tennessee Board also discussed the provisions of four specific ethics opinions, from Alabama, Maine, North Carolina, and Pennsylvania, in which the commissions or ethics authorities in those states provided particular considerations and safeguards that attorneys should heed before engaging in any use of cloud storage. For example, in North Carolina, an attorney should consider entering into an agreement with the cloud provider whereby the provider will hold itself bound to the same ethical responsibilities and safeguarding requirements as attorneys.[33] In Pennsylvania and Maine, there are internal policies and procedures that lawyers should adopt for cloud storage, so data that is lost can be obtained from appropriate back-up storage sources, and so that passwords are routinely changed to prevent unauthorized access to sensitive material.[34]

Ultimately, the Tennessee Board determined that attorneys are permitted to utilize cloud-based services, including for confidential information, so long as they follow and adhere to the appropriate requirements of the rules of professional responsibility.

---

[30] Tenn. Formal Ethics Op. 2015-F-159 (2015).
[31] *Id.*
[32] *See id.* (citing Fla. Ethics Op. 12-3 (2012)).
[33] *See id.* (citing N.C. 2011 Formal Ethics Op. 6 (2012)).
[34] *See id.* (citing Pa. Formal Ethics Op. 2011-200 (2011); Me. Ethics Op. 207 (2013)).

Separately, the Board of Professional Conduct of the Supreme Court of Ohio issued, in 2016, an *Ethics Guide on Client File Retention*. Toward the end of that *Guide*, on pages 7–8, the Board discusses digital media and cloud storage of client files. The *Guide* specifically advises that attorneys should provide information to all clients if cloud storage is utilized, as some may have valid security concerns. In addition, the *Guide* advises attorneys—as have other authorities—that cloud storage vendors are considered non-attorney assistants, and thus attorneys have obligations under RPC 5.3 and must ensure that any vendor procedures satisfy attorney ethical obligations.[35]

Ohio further warns attorneys, and their staff, that

> [i]n selecting a vendor, the lawyer must "act competently to safeguard information relating to the representation of a client against inadvertent or unauthorized disclosure by the lawyer or other persons or entities who are participating in the representation of the client or who are subject to the lawyer's supervision or monitoring." Consequently, a lawyer using the services of an outside service provider for digital "cloud" storage is required to undertake reasonable efforts to prevent the unauthorized disclosure of client information. This may require a reasonable investigation by the lawyer of the methods employed by the third party vendor.[36]

The Ohio *Guide* closes the discussion with this valuable guidance: "At a minimum, the lawyer employing 'cloud' storage methods should ensure that: 1. The vendor understands the lawyer's obligation to keep the information confidential; 2. The vendor is itself obligated to keep the information confidential; and 3. Reasonable measures are employed by the vendor to preserve the confidentiality of the files."[37]

Staying in the State of Ohio for a moment, the Ohio State Bar Association issued an informal ethics advisory opinion in 2013, in which it was stated that use of cloud storage was not prohibited by the Ohio Rules of Professional Conduct, but that the cloud "is a permutation on traditional ways of storing client data," and as mentioned by the other ethics authorities identified in this chapter, attorneys must comply with ethics rules and opinions speaking to protection of client information.[38]

According to the ABA, as well, the Model Rules of Professional Conduct will allow for outsourcing of legal and nonlegal support services, *but* the lawyer must ensure compliance with ethical rules relating to competency, confidentiality, and supervision. Therefore, stay up to date. Many courts, rules, and commentators speak about attorney obligations to provide competent representation, maintain requisite knowledge, and keep abreast of the benefits and risks of technology.[39]

---

[35] Ohio *Ethics Guide on Client File Retention*, pp. 7–8.

[36] Id.

[37] *Id.*

[38] Ohio State Bar Association Informal Ethics Advisory Op. 2013-03 (2013).

[39] *See* New York Rule of Professional Conduct 1.1 and cmt. 8; New York City Bar Association Op. 2015-3 (2015) (issue of attorneys falling victim to scams; if fail to secure client data could violate Rule 1.6; if receive funds, have to make sure they are confirmed and honored by the bank, as with all funds in the trust account, per Rule 1.15). *See also* William Turton, *Corporate email Hazards Abound*, Times Herald-Record at 13 (Feb. 24, 2020).

2.6 – Conclusion

In summary, attorneys should always be wary, and cautious, when utilizing new technology.[40] Failure to understand and properly secure new technologies could result in unintended disclosures and potential waiver of privilege or work product. Use of cloud storage is not unethical, nor is it prohibited by the Rules of Professional Conduct. However, the use of cloud storage must, at the same time, be commensurate and compliant with the laws, as well as the provisions of the Rules of Professional Conduct and the guidance contained in your jurisdiction's relevant ethics opinions. An attorney utilizing the Cloud should certainly, among other things, understand the vendor's operations; ensure that the vendor is reputable, secure, and experienced; monitor the Cloud vendor's security measures and updates; and maintain back-ups of any material and information that is stored on the Cloud.

Chapter End Questions

Note: Answers may be found on page 261.

1. A cloud infrastructure that is made available to the general public or a large industry group and is owned by an organization selling cloud services, is called:
   a. Private Cloud
   b. Public Cloud
   c. Hybrid Cloud
   d. Rain Cloud

2. A cloud infrastructure that delivers similar advantages of scalability and self-service, but through a proprietary architecture dedicated to a single organization is called:
   a. Private Cloud
   b. Public Cloud
   c. Hybrid Cloud
   d. Fluffy Cloud

3. A cloud infrastructure that is shared by several organizations and supports a specific community that has shared concerns, and may exist on premise or off premise, is called:
   a. Private Cloud
   b. Public Cloud
   c. Hybrid Cloud
   d. Community Cloud

---

[40] *See Harleysville Inc. Co. v. Holding Funeral Home, Inc.*, Case No. 1:15cv00057, 2017 WL 1041600 (W.D. Va. Feb. 9, 2017), sustained in part and overruled in part by 2017 WL 4368617 (W.D. Va. Oct. 2, 2017), for a cautionary tale.

4. A cloud infrastructure that is a composition of two or more clouds that remain unique but are bound together by standardized or proprietary technology that enables data and application portability is called:

    a. Private Cloud
    b. Public Cloud
    c. Hybrid Cloud
    d. Storm Cloud

5. Which of the following ethical rules must attorneys be concerned with if they utilize Cloud services?

    a. RPC 1.6
    b. RPC 5.3
    c. Both RPC 1.6 & 5.3
    d. Neither of the Rules cited

6. If attorneys use Cloud services for storage of client information, which of the following considerations must they evaluate under the ethical rules?

    a. The sensitivity of the information
    b. The cost of employing additional safeguards
    c. The likelihood of disclosure if additional safeguards are not employed
    d. All of these considerations

7. Which state's advisory guide recommends the following for attorneys utilizing the Cloud: "ensure that: 1. The vendor understands the lawyer's obligation to keep the information confidential; 2. The vendor is itself obligated to keep the information confidential; and 3. Reasonable measures are employed by the vendor to preserve the confidentiality of the files."

    a. New York
    b. Kansas
    c. Tennessee
    d. Ohio

8. Which of the following court decisions has held that secure third-party cloud-based file storage is acknowledged as an acceptable security option to protect materials identified as confidential or protected in discovery proceedings?

    a. *Smith v. Smith*
    b. *Zublulake v. UBS Warburg*
    c. *LeBlanc v. Halliburton Co.*
    d. *Sherwood v. Walker*

9. Which of the following ethics opinions does not concern, in some way, an attorney's duties with regard to client information stored or communicated on the Internet or in the Cloud?

   a. NYSBA Op. 842
   b. ABA Op. 477R
   c. Tenn. Op. 2015-F-159
   d. All of these opinions concern duties of attorneys and the Internet

10. True or False: Attorneys are prohibited from using Cloud services to store client information.

    a. TRUE
    b. FALSE

CHAPTER 3

Social Media and eMail Discovery

3.1 – Introduction

Text messages, electronic mail, Facebook, Twitter, and other forms of electronic media have become the fruitful subject of electronic discovery. In 2010, the United States Supreme Court, in the case of *City of Ontario, Cal. v. Quon*,[1] accepted that the jurisprudence of Fourth Amendment analysis extends to searches of text messages contained on mobile devices—although in *Quon* there was no expectation of privacy, and therefore the search did not violate the Fourth Amendment to the U.S. Constitution. Based on the Court's ruling, and pre-

> *e-mail* – Messages sent by electronic means (computer, phone, etc.) from sender's mailbox to recipient's specific mailbox address on a server; electronic mail.
>
> *Social Media* – Any of countless electronic platforms on the Internet where users have accounts, and can communicate with each other using written messages, pictures, photos, graphics or other methods.

decessor and progeny cases, text messages have been a fast-growing subject of electronic discovery preservation, especially if they contain relevant information on the subject of the litigation and are still in existence when notice of litigation exists.[2] This is, of course, in addition to any and all other electronic media

---

[1] 560 U.S. 746 (2010) (Kennedy, J.).

[2] *See also Rosen v. Evolution Holdings, LLC*, 24 Misc.3d 1205(A), 890 N.Y.S.2d 370 (Dist. Ct. Nassau County 2009) (Table) ("The lease executed by the two parties is sufficient proof to overcome the motion to dismiss and the language of the text messages further indicates the existence of a fully executed lease agreement").

evidence—social media and emails—which we will discuss in this chapter. The same principles can be applied for voicemails and any other electronic material.[3]

3.2 – Background—No Expectation of Privacy in Social Media

To begin, expanding on the above mention of the *Quon* case, it has been clearly acknowledged by the courts that: "The protections of the Fourth Amendment extend only to information over which the party has a reasonable expectation of privacy. . . . [T]he Fourth Amendment does not protect against the disclosure of information 'knowingly expose[d] to the public.'. . . Generally speaking, "[c]ourts . . . do not consider social media content as 'private.'". . . . Courts routinely have found that there is no right to privacy in internet postings that are publicly accessible. . . ."[4]

For instance: "'Twitter is a social networking and micro-blogging service' that allows users to 'send and read electronic messages known as "tweets."'. . . . 'Any Twitter user can sign up to "follow" any other Twitter user, which means that Twitter will cause the follower to receive all tweets the author publishes.'"[5]

---

[3]Indeed, use of social media for communications in our society, and its subsequent inclusion in court proceedings and litigation, has involved the highest elected office in the United States—the presidency. *See Knight First Amendment Inst. at Columbia Univ. v. Trump*, 302 F.Supp.3d 541 (S.D.N.Y. 2018). In fact, in the year 2020 the administration of Donald Trump issued an Executive Order addressing freedom of expression and ideas on social media and the Internet, and "applying the ideals of the First Amendment to modern communications technology," after one of the president's tweets on Twitter was labeled as potentially misleading. The E.O. further addressed the provisions of Section 230(c) of the Communications Decency Act (47 U.S.C. § 230(c)), and directed federal agencies to review those provisions with regard to social media companies. See U.S. Executive Order 13925 (85 FR 34079) (May 28, 2020); *see also* J. Thomsen, N. Robson & M. Scarcella, *'Unlawful and Unenforceable': Legal Experts Deride Trump's Attempt to Target Social Media Companies*, N.Y. L.J. at 2 (May 29, 2020). The E.O. was immediately questioned and challenged as soon as it was signed and issued, with challengers arguing in federal district court that the E.O. violated the free speech protections of the social media companies and their users, and with the challengers seeking a permanent injunction against the E.O. *See* Alaina Lancaster, *Mayer Brown Leads Legal Challenge to Trump's Executive Order Targeting Social Media Companies*, N.Y. L.J. at 2 (June 4, 2020).

[4]*Burke v. New Mexico*, 2018 WL 2134030 (D. N.M. May 9, 2018) (citing, *inter alia*, *Katz v. U.S.*, 389 U.S. 347, 351 (1967); *U.S. v. Meregildo*, 883 F. Supp. 2d 523, 526 (S.D.N.Y. 2012); *Rosario v. Clark Cty. Sch. Dist.*, 2013 WL 3679375, at *6 (D. Nev. July 3, 2013) (unpublished); *Tompkins v. Detroit Metro. Airport*, 278 F.R.D. 387, 388 (E.D. Mich. 2012) ("[M]aterial posted on a 'private' Facebook page, that is accessible to a selected group of recipients but not available for viewing by the general public, is generally not privileged, nor is it protected by common law or civil law notions of privacy")).

[5]*Beter v. Murdoch*, 2018 WL 3323162 at *2 n.4 (S.D.N.Y. June 22, 2018) (citing *U.S. v. Liu*, 69 F.Supp. 3d 374, 377 n.1 (S.D.N.Y. 2014); *Nunes v. Twitter, Inc.*, 194 F.Supp. 3d 959, 960 (N.D. Cal. 2016)). The same is true when it comes to platforms like Facebook. "Facebook is a website and social media platform that allows billions of users worldwide to connect with one another by posting messages and photos, responding to messages and photos shared by other users, and interacting with other Facebook users in relation to those posts .... Each Facebook user has an account, and each account corresponds to a 'profile' on the platform, to which users publish their posts. Users may also publish posts to Facebook 'pages' or 'groups,' which are administered either by individuals or sets of Facebook users...." *Wagschal v. Skoufis*, --- F.Supp.3d ---, 2020 WL 1033873, at *1 (S.D.N.Y. Mar. 3, 2020) (McMahon, C.J.). Although users do have the ability to restrict who views their pages and posts by adjusting privacy settings, or blocking other users. *Id.* The overall purpose of social media, however, is the same—to be social with others.

Furthermore, in the 2018 case of *T.C. on Behalf of S.C. v. Metropolitan Government of Nashville*, the court held:

> [t]here is a distinction between discovery of social media postings that are available to the general public and those that the user has restricted from view. "[I]nformation posted on a private individual's social media 'is generally not privileged, nor is it protected by common law or civil law notions of privacy.'". . . However, a party does not have "a generalized right to rummage at will through information that [an opposing party] has limited from public view.". . . To obtain discovery of non-public social media, a party must show that the information sought is reasonably calculated to be relevant to the claims and defenses in the litigation.[6]

Obviously, then, there is generally no expectation of privacy for social media and no protection from discovery on that basis—although, as will be discussed, that does not mean that parties have an absolute right to view the public and private social media pages of another simply because a case is filed. There are restrictions and requirements, as addressed in this chapter.

3.3 – The Law Pertaining to Discovery of Social Media Accounts

There is a seminal case out of New York State to which the beginning of much, if not all, of social media discovery can be traced. The "Patient Zero" of social media discovery, if you will—*Romano v. Steelcase, Inc.*[7] The case comes from a time before many people tended to closely monitor all of their "public" versus "private" social media page content.

In *Romano*, a woman fell off of an office chair and thereafter claimed severe injuries, suing the manufacturer. The plaintiff claimed that she was largely home-bound and experienced a loss of enjoyment of life; she also made other claims in the tort action. However, the plaintiff's public social media posts contradicted those claims. On posts accessible to defendant, there were photos and writing describing the plaintiff's active social life and travel. The defendant, of course, referring to the posts in discovery, sought access to the plaintiff's private social media pages that were believed to have additional posts contradicting the plaintiff's claims. The plaintiff challenged the demands, claiming among other things that she had a reasonable expectation of privacy in the Facebook and MySpace posts. Secondarily, Facebook also opposed disclosure demanded by the defendant, arguing that the federal Stored Communications Act[8] prevented a social media provider from disclosing the contents of communications

---

[6]2018 WL 3348728 (M.D. Tenn. July 9, 2018) (citing, *inter alia*, *Tompkins v. Detroit Metro. Airport*, 278 F.R.D. 387, 388 (E.D. Mich. 2012)).

[7]30 Misc.3d 426, 907 N.Y.S.2d 650 (Sup. Ct. Suffolk County 2010). Some might arguably attribute a "first case" label to either *Barnes v. Cus Nashville, LLC*, 2010 WL 2196591 (M.D. Tenn. May 27, 2010), or *Crispin v. Christian Audigier, Inc.*, 717 F.Supp.2d 965 (C.D. Cal. May 26, 2010). However, for our purposes, *Romano* will be considered the seminal case—although it was decided a few months after *Barnes* and *Crispin* in 2010.

[8]Stored Wire and Electronic Communications and Transactional Records Access Act (Stored Communications Act), 18 U.S.C. §§ 2701, *et seq.*

and postings if the user does not consent—and further that even a court order should not compel disclosure.[9]

The *Romano* Court determined that the posts went to a direct issue in the case. Disclosure was necessary and required.

> In light of the fact that the public portions of Plaintiff's social networking sites contain material that is contrary to her claims and deposition testimony, there is a reasonable likelihood that the private portions of her sites may contain further evidence such as information with regard to her activities and enjoyment of life, all of which are material and relevant to the defense of this action. Preventing Defendant from accessing to Plaintiff's private postings on Facebook and MySpace would be in direct contravention to the liberal disclosure policy in New York State.[10]

Additionally, to address the Stored Communications Act concern, the court simply ordered the plaintiff to provide Facebook with a signed authorization so the defendant could obtain the information sought. Although the Court could have addressed the SCA issue, the Court instead exercised its authority over the parties in the case to compel the provision of a release and authorization, thus resolving the discovery matter.

The discovery of social media raises the good question: Why should photos that are posted online be treated any differently from those saved in a paper photo album?

Again, when it comes to Facebook, and other social media platforms, the material is potentially fair game for discovery. Take the case of *Tompkins v. Detroit Metro. Airport.*[11] In *Tompkins* (which has been cited by a number of other courts, as citations in this book make clear), the plaintiff brought suit for a slip and fall at the airport, claiming injuries and impairment of her abilities to work and enjoy life. The defendant demanded, among other things, that the plaintiff sign an authorization for access to her Facebook account. As could be expected, the plaintiff objected and claimed portions of the account were private, not open or available to the public. The defendant, in turn, cited the *Romano* case and traditional FRCP 26 arguments. The plaintiff in the meantime cited to the New York case of *McCann v. Harleysville Ins. Co. of New York*, 78 A.D.3d 1524 (4th Dep't 2010), in which case the court sustained denial of a motion to compel Facebook discovery—not on privacy grounds, but rather because

---

[9] *Romano*, 30 Misc.3d 426, 907 N.Y.S.2d 650.

[10] *Id.* at 430, 654.

[11] 278 F.R.D. 387 (E.D. Mich. 2012) (Whalen, M.J.). *See also Locke v. Swift Trans. Co. of Ariz., LLC*, 2019 WL 430930, at *2 & n.2 (W.D. Ky. Feb. 4, 2019) ("To fall within the scope of discovery, SNS [social networking site] information must meet the relevance standard, and the burden of discovering the information must be proportional to the needs of the case. Put simply, social media information is treated just as any other type of information would be in the discovery process.... 'Generally, SNS content is neither privileged nor protected by any right of privacy.'... However, '[d]iscovery of SNS requires the application of basic discovery principles in a novel context.'... In particular, several courts have found that even though certain SNS content may be available for public view, the Federal Rules do not grant a requesting party 'a generalized right to rummage at will through information that [the responding party] has limited from public view' but instead require 'a threshold showing that the requested information is likely to lead to the discovery of admissible evidence.'") (citing, *inter alia, Tompkins*).

the defendant was on a "fishing expedition."[12] At least in *Romano* the plaintiff had posted contradictory information on public social media pages.

The *Tompkins* Court denied the motion to compel disclosure. The public page photographs did not contradict the plaintiff's claims; and surveillance photographs taken by the defendant also did not justify the demand for private Facebook pages since the plaintiff was engaged in mundane activities. It was determined that requesting the entire account was overbroad.

In New York, prior to 2018, the standard utilized for discovery of, and access to, social media was one requiring that a factual predicate be established—that is, specific identification of information sought on social media believed to contradict or conflict with plaintiff's claims, since not all information on the private social media pages might be related to the action.[13] One could accomplish this showing by using deposition testimony, postings on public social media pages, testimony of witnesses, or information from private investigation.

However, in 2018, New York's highest court—the Court of Appeals—reversed a decision out of the Appellate Division in *Forman v. Henkin*, and changed the standard to one mirroring that for discovery of other materials, such as paper documents. *Forman*[14] involved a personal injury action after a fall from a horse. The trial court granted the defendant's motion to compel, and directed the plaintiff to produce all photographs of the plaintiff privately posted on Facebook prior to the accident that the plaintiff intended to use at trial, together with post-accident photos and messages. The court included built-in restrictions regarding those photographs showing nudity or romantic encounters.[15] The Appellate Division majority reversed the trial court, citing to related decisions, including *McCann*, and held that the plaintiff must disclose all photos intended for use at trial—but eliminating authorization for post-accident messages and data[16]—surveillance of a personal injury plaintiff being a "far cry"

---

[12] *Id.*

[13] *See Patterson v. Turner Constr. Co.*, 88 A.D.3d 617 (1st Dep't 2011) (citing *Offenback v. L.M. Bowman, Inc.*, 2011 WL 2491371, at *2, 2011 U.S. Dist. LEXIS 66432, at *5-8 (M.D. Pa. 2011)); *Melissa "G" v. N. Babylon Union Free Sch. Dist.*, 48 Misc.3d 389, 6 N.Y.S.3d 445 (Sup. Ct. Suffolk County 2015) (producing counsel is first judge of relevance; no *in camera* review to start; Facebook posts to be reviewed and potentially produced even if behind privacy setting); *Loporcaro v. City of New York*, 35 Misc.3d 1209(A), 950 N.Y.S.2d 723 (Table) (Sup. Ct. Richmond County 2012) (access granted to part of Facebook account, and some deleted materials, after the defendant showed material on public pages conflicting with the plaintiff's claims of loss of enjoyment of life; citing *Patterson* and *Romano*); *Doyle v. Temco Serv. Indus., Inc.*, 172 A.D.3d 554, 555, 98 N.Y.S.3d 746, 747 (Mem) (App. Div. 1st Dep't 2019) ("Private social media information can be discoverable to the extent it 'contradicts or conflicts with [a] plaintiff's alleged restrictions, disabilities, and losses, and other claims' ... Here, plaintiff alleges that injuries she sustained as the result of a slip and fall at her place of work have caused her to suffer, among other things, a loss of enjoyment of life. Defendants are entitled to discovery to rebut plaintiff's claims ... however, defendants' discovery demand seeking access to all of plaintiff's post-accident social media accounts is overbroad .... In their reply brief, defendants limit their demand to seek 'only plaintiff's post-accident social media records regarding social and recreational activities that she claims have been limited by her accident.' Accordingly, the motion to compel should be granted to that extent, which is consistent with the principles set forth in *Forman*. To the extent plaintiff's social media accounts contain 'sensitive or embarrassing materials of marginal relevance,' plaintiff can seek a protective order") (citing *Patterson*, and *Forman v. Henkin*, 30 N.Y.3d 656, 663-665 (2018)).

[14] 30 N.Y.3d 656 (2018) (DiFiore, C.J.).

[15] *Id.* (citing N.Y. CPLR 3101(a)).

[16] *Id.*

from trying to uncover private social media postings absent a factual predicate. The lower courts further cited a CPLR 3101 analysis.

The New York State Court of Appeals, though, reversed the Appellate Division, also employing an analysis of the provisions of CPLR 3101, and citing *Romano v. Steelcase*. Chief Judge DiFiore, writing for a unanimous court, held that it is inherent in discovery that parties may not know if requested material actually exists. The real purpose of discovery is to find that out. Therefore, disclosure "should [turn] on whether it is 'material and necessary to the prosecution or defense of an action.'"[17]

The *Forman* Court's holding is provided here at length for analysis and discussion:

Forman v. Henkin (New York State Court of Appeals)

DiFiore, Chief Judge

[W]e agree with other courts that have rejected the notion that commencement of a personal injury action renders a party's entire Facebook account automatically discoverable. . . . Directing disclosure of a party's entire Facebook account is comparable to ordering discovery of every photograph or communication that party shared with any person on any topic prior to or since the incident giving rise to litigation—such an order would be likely to yield far more nonrelevant than relevant information. Even under our broad disclosure paradigm, litigants are protected from "unnecessarily onerous application of the discovery statutes".

Rather than applying a one-size-fits-all rule at either of these extremes, courts addressing disputes over the scope of social media discovery should employ our well-established rules—there is no need for a specialized or heightened factual predicate to avoid improper "fishing expeditions." In the event that judicial intervention becomes necessary, courts should first consider the nature of the event giving rise to the litigation and the injuries claimed, as well as any other information specific to the case, to assess whether relevant material is likely to be found on the Facebook account. Second, balancing the potential utility of the information sought against any specific 'privacy' or other concerns raised by the account holder, the court should issue an order tailored to the particular controversy that identifies the types of materials that must be disclosed while avoiding disclosure of nonrelevant materials. In a personal injury case such as this it is appropriate to consider the nature of the underlying incident and the injuries claimed and to craft a rule for discovering information specific to each. Temporal limitations may also be appropriate—for example, the court should consider whether photographs or messages posted years before an accident are likely to be germane to the litigation. Moreover, to the extent the account may contain sensitive or embarrassing materials of marginal relevance, the account holder can seek protection from the court (see CPLR 3103[a]). Here, for example, Supreme Court exempted from disclosure any photographs of plaintiff depicting nudity or romantic encounters.

---

[17] *Id.* at 664. *See also Gilbert v. Hackett*, 121 N.Y.S. 3d 638 (Mem) (App. Div. 2d Dep't May 6, 2020) (citing *Forman*).

. . . [T]he Appellate Division erred in modifying Supreme Court's order to further restrict disclosure of plaintiff's Facebook account, limiting discovery to only those photographs plaintiff intended to introduce at trial. With respect to the items Supreme Court ordered to be disclosed. . . defendant more than met his threshold burden of showing that plaintiff's Facebook account was reasonably likely to yield relevant evidence. At her deposition, plaintiff indicated that, during the period prior to the accident, she posted "a lot" of photographs showing her active lifestyle. Likewise, given plaintiff's acknowledged tendency to post photographs representative of her activities on Facebook, there was a basis to infer that photographs she posted after the accident might be reflective of her post-accident activities and/or limitations. The request for these photographs was reasonably calculated to yield evidence relevant to plaintiff's assertion that she could no longer engage in the activities she enjoyed before the accident and that she had become reclusive.

. . .

In sum, the Appellate Division erred in concluding that defendant had not met his threshold burden of showing that the materials from plaintiff's Facebook account that were ordered to be disclosed pursuant to Supreme Court's order were reasonably calculated to contain evidence "material and necessary" to the litigation.[18]

Thus, the old standard is "new" again in New York—and parties and attorneys should be aware. The courts of New York now make clear that the evaluation undertaken is on par with that utilized for realms outside of social media. For instance, the Appellate Division, First Department, held in 2018 that:

there was no showing that plaintiffs wilfully [*sic*] failed to comply with any discovery order, since they provided access to the infant plaintiff's social media accounts and cell phone records for a period of two months before the date on which she was allegedly attacked on defendant's premises to the present, which was a reasonable period of time. Defendant's demands for access to social media accounts for five years prior to the incident, and to cell phone records for two years prior to the incident, were overbroad and not reasonably tailored to obtain discovery relevant to the issues in the case.[19]

Furthermore, the Second Department of New York's Appellate Division has added:

[d]isclosure in civil actions is generally governed by CPLR 3101(a), which directs: "[t]here shall be full disclosure of all matter material and necessary in the prosecution or defense of an action, regardless of the burden of proof". . . However, "unlimited disclosure is not mandated,

---

[18] *Id.* at 664–668 (citations omitted).
[19] *Doe v. Bronx Prep. Charter Sch.*, 160 A.D.3d 591, 76 N.Y.S.3d 126, 127–128 (App. Div. 1st Dep't 2018) (citing *Forman*).

and the rules provide that the court may issue a protective order 'denying, limiting, conditioning or regulating the use of any disclosure device' to 'prevent unreasonable annoyance, expense, embarrassment, disadvantage, or other prejudice to any person or the courts'". . . . "The supervision of disclosure and the setting of reasonable terms and conditions therefor rests within the sound discretion of the trial court and, absent an improvident exercise of that discretion, its determination will not be disturbed."[20]

Another case deserves mention here. In a disorderly conduct prosecution, Twitter, Inc. sought to quash a subpoena issued to it seeking tweets and subscriber information related to defendant.[21] The Court, with assessment of the Stored Communications Act (SCA), denied the motion and ordered *in camera* production for review. Among other factors, tweets are not like private emails. If anything, the Court said perhaps they are like emails copied to others around the world.[22]

Consider, however, this further—nightmare—scenario. In *Lemon Juice v. Twitter, Inc.,*[23] an individual was arrested and falsely accused of posting the photograph of a child testifying in court against an alleged abuser and rapist, in violation of a court order. Worse, the person who actually created the fake site, posted a picture of Lemon Juice in the profile to make it look even more like Lemon Juice was the actual owner and operator of the handle. According to the court's opinion, it appears such actions may have been to frame Lemon Juice. The court held that Lemon Juice was entitled to learn the identity of the person who assumed his identity and posted it on Twitter. The criminal user of the handle named the handle ID close to that of the innocent civil plaintiff. The *Lemon Juice* Court held that the criminal handle user's action was so extreme and outrageous that it was beyond all bounds of decency.[24] Twitter was compelled to turn over information identifying the user of the specific handle, so they could be sued in a civil court by the wrongly accused innocent handle user. The privacy and anonymity of the criminal user yielded to the plaintiff's need for redress of the

---

[20] *Morrow v. Gallagher*, 81 N.Y.S.3d 491, 163 A.D.3d 804, 805 (App. Div. 2d Dep't 2018) (citations omitted) (citing *Forman*).

[21] *People v. Harris*, 36 Misc.3d 868, 949 N.Y.S.2d 590 (Crim. Ct. N.Y. County 2012) (Sciarrino, J.C.C.), *stay denied* Appellate Term, 1st Dep't (Sept. 7, 2012), *appeal dismissed* 39 Misc. 3d 142(A), 971 N.Y.S 2d 73 (Table) (App. Term. 1st Dep't 2013).

[22] *Id. See also Barres v. Cus Nashville, LLC*, 2010 WL 2265668 (M.D. Tenn. June 3, 2010) (U.S. Magistrate Judge offered to "friend" party for purpose of *in camera* review of Facebook account). However, the issue of criminal defendants' access to social media evidence in their defense through subpoenas to providers such as Twitter, Facebook and Instagram is far from settled, and providers still argue to quash under the SCA. The U.S. Supreme Court denied certiorari in a case in 2020, but the California Supreme Court granted review in June 2020 in the same case, with a very convoluted procedural history, where the trial court denied the platform providers' motion to quash, but the intermediate appellate court reversed. *See Facebook, Inc. v. Superior Court*, 259 Cal. Rptr. 3d 331 (Cal. Ct. Apps. 5th Dist. 2020); *cert. denied* — S.Ct.—, 2020 WL 2515495 (U.S. May 18, 2020); and *review granted* —P.3d—, 2020 WL 3096742 (Cal. June 10, 2020).

[23] 44 Misc.3d 1225(A), 997 N.Y.S.2d 669 (Table), 2014 WL 4287049 (Sup. Ct. Kings County Aug. 29, 2014).

[24] *Id.*

wrong against him. Even in the world before *Forman*, a reason for the disclosure was clearly established.[25]

Finally in this chapter section, let us evaluate the case of *Giacchetto v. Patchogue-Medford Union Free School District*.[26] In some ways, the *Giacchetto* Court seems to analogize *Forman* on the federal level—years before *Forman* was decided. "The fact that Defendant is seeking social networking information as opposed to traditional discovery materials does not change the Court's analysis. . . . The Court also notes that the 'fact that the information [Defendant] seeks is in an electronic file as opposed to a file cabinet does not give [it] the right to rummage through the entire file.'"[27]

Furthermore, the *Giacchetto* Court noted that:

> [s]ome courts have held that the private section of a Facebook account is only discoverable if the party seeking the information can make a threshold evidentiary showing that the plaintiff's public Facebook profile contains information that undermines the plaintiff's claims. . . . This approach can lead to results that are both too broad and too narrow. On the one hand, a plaintiff should not be required to turn over the private section of his or her Facebook profile (which may or may not contain relevant information) merely because the public section undermines the plaintiff's claims. On the other hand, a plaintiff should be required to review the private section and produce any relevant information, regardless of what is reflected in the public section. The Federal Rules of Civil Procedure do not require a party to prove the existence of relevant material before requesting it. Furthermore, this approach improperly shields from discovery the information of Facebook users who do not share any information publicly. For all of the foregoing reasons, the Court will conduct a traditional relevance analysis.[28]

That holding, of course, now looks very similar to the 2018 decision in *Forman* that eliminated the factual predicate requirements in New York State—applying more traditional analyses.

---

[25] *See also Sines v. Kessler*, 2018 WL 3730434 (N.D. Cal. Aug. 6, 2018) (Spero, Chief M.J.) (identity of "Jane Doe," accused of being white supremacist and possible organizer or encourager of the rally in Charlottesville, Virginia, was ordered disclosed to attorneys and limited persons in the case; however, under the Stored Communications Act, the content of electronic materials was not to be disclosed absent permission of a sender or recipient). *See also Facebook, Inc. v. Pepe*, —A. 3d —, 2020 WL 1870591, at * 6 & n. 34 (D.C. Ct. Apps. Jan. 14, 2020). *But see Strike 3 Holdings, LLC v. Doe*, 2019 WL 5446239 (D.N.J. Oct. 24, 2019) (in a copyright infringement case, court denied *ex parte* request for expedited discovery of "Doe" identities based on Rule 12(b)(6) pleading deficiencies of only identifying Internet Services Providers).

[26] 293 F.R.D. 112 (E.D.N.Y. 2013) (Tomlinson, M.J.).

[27] *Id.* at 114 (citing *EEOC v. Simply Storage Mgmt., LLC*, 270 F.R.D. 430, 434 (S.D. Ind. 2010) ("Discovery of [social networking postings] requires the application of basic discovery principles in a novel context")); *accord Mailhoit v. Home Depot U.S.A., Inc.*, 285 F.R.D. 566, 570 (C.D. Cal. 2012); *Howell v. Buckeye Ranch, Inc.*, 2012 WL 5265170, at *1 (S.D. Ohio Oct. 1, 2012)).

[28] 293 F.R.D. at 114 n.1 (citing, and seemingly distinguishing, *Tompkins v. Detroit Metro. Airport*; *Romano v. Steelcase, Inc.*; *Potts v. Dollar Tree Stores, Inc.*, 2013 WL 1176504, at *3 (M.D. Tenn. Mar. 20, 2013); *Keller v. Nat'l Farmers Union Property & Cas. Co.*, 2013 WL 27731, at *4 (D. Mont. Jan. 2, 2013)).

3.4 – The Law Pertaining to Discovery of eMail and Text Messages

Now when it comes to electronic mail (email), we have a number of decisions to choose from when discussing the matter. Let us look at the case of *Bower v. Mirvat El-Nady Bower.*[29] In this case, the court addressed a motion to compel production of emails stemming from a FRCP 45 subpoena served on Yahoo! and Google. The Court here also assessed the application of the Stored Communications Act, which governed the motion. The federal magistrate judge noted that pursuant to the statutory language, service providers such as Google and Yahoo! may not produce emails in response to civil subpoenas. Therefore, the plaintiff argued that the defendant, being a fugitive who did not appear, should have been deemed to consent to release of the emails by the service providers.[30] But, the Court also noted that there were no FRCP 34 requests for production, and therefore FRCP 37 sanctions were not before the court at that time.

The court could have ordered the defendant to produce the emails, since they were under her control, although held by the service providers, since the court has authority over litigants appearing before it. However, the Court ultimately disagreed with the plaintiff's positions and arguments—and denied the motion to compel. It was held that the defendant did not elect to participate in the judicial proceedings, and thus there was no basis for implied consent. The court's decision distinguished *Romano v. Steelcase* and *Flagg v. City of Detroit*, 252 F.R.D. 346 (E.D. Mich. 2008) (where public employees were warned there was no privacy in emails deemed the property of the city that might also be considered public records).[31]

Then there is the case of *Hausman v. Holland America Line-U.S.A.*[32] In *Hausman*, the court held that plaintiff, by clear and convincing evidence, had committed discovery misconduct that substantially interfered with the defendants' ability to fully prepare for trial. The judgment in favor of the plaintiff was vacated, and a new trial was ordered. Why?

In *Hausman*, the plaintiff had embarked on an eight-month, around-the-world cruise, when an automatic sliding door caused a head injury. At trial, the jury ruled in favor of the plaintiff. Thereafter, however, the plaintiff's former personal assistant approached the defendants and advised that the plaintiff had deleted or not disclosed emails relevant to the case, had tampered with witnesses, fabricated or exaggerated injuries, and testified falsely.[33]

The court held an evidentiary hearing, focusing on the former personal assistant's claims of email destruction or non-production. The personal assistant testified that upon receiving requests for categories of emails, the plaintiff spent days deleting emails and directed the assistant to do so, as well. The court found the plaintiff was not a credible witness at the

---

[29] 808 F. Supp. 2d 348 (D. Mass. 2011) (Dein, M.J.).
[30] *Id.*
[31] *Id. See also U.S. v. Finazzo*, No. 10–CR–457 (RRM)(RML), 2013 WL 619572 (E.D.N.Y. 2013) (no reasonable expectation of privacy, company policy).
[32] 2016 WL 51273 (W.D. Wash. Jan. 5, 2016) (Rothstein, D.J.).
[33] *Id.*

hearing, but that the former assistant was credible. The court did not believe the plaintiff forgot an email account, and if there was a routine deletion procedure followed (which the court doubted) that was to cease when the litigation started (or when the plaintiff knew or reasonably should have known to preserve evidence that was relevant, or evidence that if destroyed was foreseeable to prejudice the opposing party). In fact, from certain recovered messages, the court found evidence of concerted efforts to conceal an email account.[34]

Furthermore, the court held that: "[w]here discovery material is 'deliberately suppressed, its absence can be presumed to have inhibited the unearthing of further admissible evidence adverse to the withholder, that is, to have substantially interfered with the aggrieved party's trial preparation.'"[35] Although the plaintiff attempted to hide behind the retention of an e-discovery consultant, there remained uncertainty as to whether he had deleted emails before a court order, and before retention of the consultant—meaning that the consultant would not have obtained copies of all relevant emails in any event.[36]

Finally, the deleted emails could have been relevant to the lawsuit, in contravention of one of the plaintiff's arguments; and since the court found that the plaintiff deliberately deleted emails, "the Court must assume the deleted or non-disclosed emails were relevant."[37] Additionally, although defendants had threatened to seek sanctions five months before trial, they had not actually done so, and the court held they had not waived the issue. First, they were never made aware of the second email account. Second, the plaintiff only advised that emails were deleted through routine practice—he did not advise concerning his more nefarious practice of searching for and deleting emails containing the court-ordered search terms. The first activity might not have resulted in sanctions, but the latter certainly would have. The court held in kind: "[i]t is this trial judge's responsibility to ensure that a party is not the victim of a miscarriage of justice. . . . Based on the evidence presented at the evidentiary hearing, this trial judge concludes that a miscarriage of justice occurred in this case."[38]

Ultimately, the *Hausman* Court granted the defendants' motion to set aside the verdict and for a new trial. It was held that failure to disclose the requested materials can be misconduct under FRCP 60(b)(3).

Similarly, it should be noted that parties are entitled to seek, in discovery during litigation, text messages exchanged between other parties and witnesses or non-parties. The basis for such discovery is the same as that underlying the discovery of emails, social media, and paper for that matter.[39] Indeed, courts analyze discovery of text messages using the same mechanism

---

[34] *Id.*

[35] *Id.* at *10.

[36] *Id.*

[37] *Id.* at *11.

[38] *Id.* at *12.

[39] *See Walker v. Carter*, 2017 WL 3668585 (S.D.N.Y. July 12, 2017). Note that text messages have also been the basis for lawsuits, and a disputed issue for standing, in cases under the federal Telephone Consumer Protection Act (TCPA), and including *Melito v. Experian Mkting. Solutions*, 923 F. 3d 85 (2d Cir. 2019); and *Duguid v. Facebook, Inc.*, 926 F. 3d 1146 (9th Cir. 2019). *See* Shari Claire Lewis, *Standing in TCPA Cases: How Many Texts Are Enough*, N.Y. L.J. at 7 (Feb. 18, 2020).

as for paper or other forms of evidence—in the federal courts that is pursuant to FRCP 26 (discussed in Chapter 1).[40] Furthermore, text messages have been the subject of discovery and investigation, and have been utilized as evidence at trial, in both civil and criminal matters (as discussed further in Chapter 9).[41]

3.5 – Use of eMail to Communicate with Attorneys

Note separately, there is a related concern about the issue of attorney-client privilege and email communications with clients. Beware of the need for, among other things, an *expectation of privacy*. Work email systems *are not* generally private, particularly if the employer has a policy prohibiting private use of the work system, or advising that use of the system is monitored by the employer.[42] This could cause a loss of attorney-client privilege and/or work product protection in the communications exchanged on such a server—thus making them available for potential discovery and disclosure. Exceptions could be in the case of corporate litigation, where communications exist between attorneys and the client company's officers, directors, and employees—but the representation is of the entity, and the people speak for

---

[40] *See Ramos v. Hopele of Fort Lauderdale, LLC*, 2018 WL 1383188 (S.D. Fla. Mar. 19, 2018). *See also Hardy v. UPS Ground Freight, Inc.*, 2019 WL 3290346 (D. Mass. July 22, 2019) (citing to *Ramos*).

[41] *See Daugherty v. City & County of San Francisco*, 24 Cal.App.5th 928, 234 Cal.Rptr.3d 773 (Cal. Ct. App. 1st Dist. 2018); *Ellis v. State*, 517 S.W.3d 922 (Tex. Ct. App. 2017); *U.S. v. Cooke*, 853 F.3d 464, 474 (8th Cir. 2017).

[42] *See In re Asia Global Crossing, Ltd.*, 322 B.R. 247, 257–258 (Bankr. S.D.N.Y. 2005). *See also RE: DLO Enter., Inc. v. Innovative Chem. Prods. Group, LLC*, 2020 WL 2844497 (Del. Ct. Chanc. June 1, 2020) (unpublished) (Addressing the *Asia Global* factors when demand was made for e-mail communications with attorneys; "The first Asia Global factor focuses 'on the nature and specificity of the employer's policies regarding email use and monitoring' and weighs in favor of production 'when the employer has a clear policy banning or restricting personal use, where the employer informs employees that they have no right of personal privacy in work email communications, or where the employer advises employees that the employer monitors or reserves the right to monitor work email communications.'... '[A]n outright ban on personal use would likely end the privilege inquiry at the start,'... but a complete ban is not required.... The second factor focuses 'on the extent to which the employer adheres to or enforces its policies and the employee's knowledge of or reliance on deviations from the policy.'... If an employer has clearly and explicitly reserved the right to monitor work email, then the absence of past monitoring or a practice of intermittent or as-needed monitoring comports with the policy and does not undermine it. In that setting, 'evidence of actual monitoring would make an expectation of privacy even less reasonable.'... In a work email case, the third factor, which asks whether 'third parties have a right of access to the computer or e-mails,' is largely duplicative of the first and second factors.... '[B]y definition the employer has the technical ability to access the employee's work e-mail account.'... 'The third factor is most helpful when analyzing webmail or other electronic files that the employer has been able to intercept, recover, or otherwise obtain,' and considers what the employer had to do to obtain electronic files, 'such as whether the employer used forensic recovery techniques, deployed special monitoring software, or hacked the employee's accounts or files' due to password-protection, encryption, or deletion .... The final factor, regarding an employee's knowledge of the use and monitoring policies, favors the existence of a reasonable expectation of privacy if the employee lacked knowledge of the policies. 'If the employee had actual or constructive knowledge of the policy, then this factor favors production because any subjective expectation of privacy that the employee may have had is likely unreasonable.'... Importantly, '[d]ecisions have readily imputed knowledge of an employer's policy to officers and senior employees'") (footnotes and citations omitted).

the entity.[43] If, however, communications take place on a shared email account or an employer's email account between an employee and an attorney regarding a private legal matter of the employee, the hackles on the back of one's neck should rise. The same is true of communications between attorney and client when the client utilizes an email account shared with a spouse, friend, or family member.

Individuals should not just think about what may be more convenient. While it may be easier to utilize a work email, or a friend or family member's shared email account, attorneys should advise their individual clients to establish a private email of their own (Gmail, Hotmail, Outlook, Yahoo!, or the like).[44] Even an email account shared by *spouses* may cause a suffered loss of privilege—because the one spouse is made party to the conversation between the other spouse and an attorney. It will depend upon, in that circumstance, whether the two spouses are both seeking legal advice for the same matter and are not adverse to each other. But, it will also depend on whether the state in which the one spouse seeks the advice recognizes an absolute husband-wife (spousal) privilege.[45] Several states discuss the unsettled question whether the presence of a spouse destroys the privileges.[46] New York has held that presence of a spouse does destroy attorney-client privilege (and, thus, it is a fair read that presence of the attorney will destroy any spousal or marital communication privilege).[47] Ultimately, it is not worth the debate or the risk.

---

[43] *See, e.g.*, N.Y. Rules of Prof'l Conduct 1.13 (Organization as Client); A.B.A. Model Rules of Prof'l Conduct 1.13; Tex. Discip. Rules of Prof'l Conduct 1.12. Note also the issue of what constitutes a "communication seeking legal advice" – simply cc'ing an attorney on an e-mail does not usually create or grant privilege; and time-stamps on text message chains that at some point include additional persons could lead to privilege being granted to only some of the messages. *See* Devika Kewalramani & Peter Kimble, *Privilege, Discovery and Digital Technology: Remembering Other Warnings From 'Upjohn'*, N.Y. L.J. at 4 (Dec. 30, 2019) (discussing, *inter alia, Dolby Labs. Licensing v. Adobe*, 2019 WL 4082784 (N.D. Cal. Aug. 29, 2019); *Bruno v. Equifax Info. Servs.*) 2019 WL 633454 (E.D. Cal. Feb. 14, 2019)).

[44] *See* M. Hutter, *Using Employer's Server for Personal Communications: Privilege Protected?*, N.Y. L.J., at 3 (June 1, 2017).

[45] *See* Restatement of the Law, Third, The Law Governing Lawyers § 71 (comment b).

[46] *See Brownfield v. Hodous*, 82 Va. Cir. 315, 2011 WL 7493287 (Va. Cir. Ct. 2011) ("Given the unsettled state of this law and the relative scarcity of guiding case authority, the Court resolves the issue in favor of protecting the attorney-client privilege. Based on the facts and circumstances of this case, because Virginia does recognize the husband-wife privilege codified at Virginia Code § 8.01-398, the presence of Plaintiffs [*sic*] husband does not destroy the confidentiality required for the attorney-client privilege"); *Wesp v. Everson*, 33 P.3d 191 (Colo. 2001).

[47] *See In re Horowitz*, 16 Misc.3d 1106(A), 841 N.Y.S.2d 826 (Table) (Surr. Ct. Nassau County 2007) ("In some states, the presence of a spouse does not negate the confidentiality of an attorney-client communication, on the theory that the marital communications privilege is incorporated into the transaction. . . . This is not the law in New York. . . . In New York, confidential communications between spouses are protected from disclosure by CPLR 4502(b). Where a client discloses a prior attorney-client communication to a spouse, the disclosure is protected by CPLR 4502. . . . Nevertheless, communications between an attorney and client are generally not privileged in New York if the client's spouse is present at the time of the transaction. . . . The exception would be where the spouse is an agent of the client") (citing, *inter alia*, 9 Weinstein–Korn–Miller, N.Y. Civ. Prac. CPLR ¶ 4503.16; *Doe v. Poe*, 92 N.Y.2d 864 (1998)).

Additionally, the world has changed since the American Bar Association approved email communications in 1999. Some commentators caution that if certain material exchanged or discussed between attorney and client is highly sensitive, they may want to consider heightened encryption before emailing—or not email at all. An ABA Formal Opinion on this very issue requires similar considerations.[48]

3.6 – Issues with Discovery Demands and Objections

> *Pro se Litigant* – An individual representing their own legal interests without assistance from an attorney.

Briefly, but importantly, attorneys and *pro se* litigants[49] should take care how they craft discovery demands to opposing parties, and subpoenas to non-parties under state procedural rules and/or the Federal Rules. Also, be sure just the same to review and assess the demands received from opposing counsel. For non-attorneys reading this book, be alert for this issue requiring attention (especially if you are attempting to represent yourself in court); or be aware of the issues about which you should have a discussion with your legal counsel.

For instance, be mindful that: "'[o]bjections to discovery requests cannot be conclusory. Proper objections "show" or "specifically detail" why the disputed discovery request is improper.'. . . 'Boilerplate, generalized objections are inadequate and tantamount to making no objection at all.'"[50]

The *Dewidar* case concerned demands by AMTRAK for social media data of the plaintiff related to the case and the plaintiff's claims. The defendant sought "'[a]ll. . . social media data, including but not limited to data stored or exchanged on Facebook, Twitter, Instagram, Snapchat, WhatsApp from January 12, 2015, to the present. . . .' '[a]ll digital photographs, including digital photos and videos you took during your trip to San Diego, which gave rise to [the] lawsuit.'"[51] The plaintiff responded with: "Objection. Overbroad. [U]nduly burdensome. Privacy. Not relevant to the subject matter of this action nor to the discovery of admissible evidence. Defendants have the burden of establishing that the information sought will lead to the discovery of admissible evidence and the information is relevant."[52]

> *Without Prejudice* – In a legal proceeding, meaning without permanently closing the issue; a motion decided without prejudice may be brought to the court again if circumstances permit. Therefore, "with prejudice" means determined or decided with finality.

---

[48] *See* Am. Bar Ass'n Formal Op. 477R (May 2017); *In re. Reserve Fund Secur. & Deriv. Litig.*, 2011 WL 2039758 (S.D.N.Y. May 23, 2011).

[49] "Litigants or parties representing themselves in court without the assistance of an attorney are known as *pro se* litigants. '*Pro se*' is Latin for 'in one's own behalf.' The right to appear *pro se* in a civil case in federal court is defined by statute 28 U.S.C. § 1654." *Frequently Asked Questions About* Pro Se *Litigation*, U.S. District Court for the District of Massachusetts, available at: http://www.mad.uscourts.gov/general/pdf/prosefaqs.pdf.

[50] *Dewidar v. Nat'l R.R. Passenger Corp.*, 2018 WL 280023 (S.D. Ca. Jan. 3, 2018) (Brooks, M.J.) (citing cases).

[51] *Id.* at *5.

[52] *Id.*

The court found no basis for the plaintiff's objections, and further held that the requested material was directly related to the plaintiff's claims and sought information from a reasonable time period. The plaintiff was ordered to disclose all requested information.

In conclusion on this point, do not forget to attach a copy of the discovery demands that are challenged with any motion for a protective order, or any motion to compel, together with any responses served before relief is sought in court. Without that, the court may deny/dismiss the motion, usually without prejudice (meaning with the possibility of re-submitting), stating that the court cannot adequately assess the issue without seeing the demands in dispute.[53]

3.7 – Reminder Concerning the Importance of Preservation

Keep in the forefront of your mind that preservation is vitally important in this field—as discussed in Chapter 1. Even innocent, unintentional loss of electronic social media material can create problems.

Consider the case of *Gatto v. United Air Lines, Inc.*[54] In *Gatto*, the plaintiff deactivated his Facebook account after he was alerted it was accessed by an IP address unknown to him. However, United's counsel had already advised they utilized an authorization to have the account accessed. There was no way to recover the account, because Facebook had automatically deleted it fourteen days after deactivation. Defendants sought sanctions and argued that some of the photos in black and white that were salvaged before deactivation controverted the plaintiff's claim in the case.[55]

The court performed a spoliation analysis (under a prior iteration of the Federal Rules, before the current Rule 37(e) structure was in place). The court subsequently determined that the plaintiff had a duty to preserve the Facebook account, and then evaluated sanctions. Because the defendants were prejudiced by the lost information, the court found a spoliation inference (adverse inference) was appropriate to grant. However, no monetary sanctions or attorneys' fees were awarded, because the court held the plaintiff did not appear to have been motivated by fraudulent purpose or diversionary tactics, and there would not be unnecessary delay in the legal proceedings.[56]

3.8 – Law Enforcement and Social Media Searches

In 2014, the U.S. Supreme Court issued its decision in *Riley v. California*.[57] This decision—a 9-0 opinion—was important, holding that law enforcement personnel must have a warrant

---

[53] *See Gallipoli v. Delano*, Index No. 2580/10, 2012 WL 10900506 (Trial Court Order) (Sup. Ct. Dutchess County Oct. 1, 2012).

[54] 10-cv-1090-ES-SCM, 2013 WL 1285285 (D.N.J. Mar. 25, 2013) (Mannion, M.J.).

[55] *Id.*

[56] *Id.*

[57] 134 S.Ct. 2473, 573 U.S. 373 (2014).

under normal circumstances to search cell phones attendant to an arrest. In essence, the idea is that searching a mobile device is not the same as looking for a weapon when a suspect or accused is taken into custody. Neither a concern about an officer's safety nor a concern over the destruction/loss of evidence justified a failure to follow warrant procedures. The phone could be seized and secured in the meantime. Further, any concerns over "remote wiping" or data encryption were deemed to be largely inapplicable—either because only a rare case would involve police finding a phone with the lock not engaged, or only a few "anecdotal examples" discuss a third-party accessing and wiping the data from afar.[58] The decision held that cell phones, in actuality, are not just phones—they are mobile mini-computers (even though one phone in the case was a flip-phone and not a smartphone).[59]

The Supreme Court "doubled-down," so to speak, in this field with its 2018 decision in *Carpenter*, when the Court held cell site location information (CSLI) and searches utilizing such information fell under the Fourth Amendment, requiring probable cause and a search warrant.[60]

Consider the following excerpt from a 2017 decision of the U.S. Court of Appeals for the Third Circuit:

> In *Riley*, the Supreme Court held that officers' warrantless search of data stored on an individual's cell phone ran afoul of the Fourth Amendment, noting that the diversity and quantity of data stored on mobile phones today created a reasonable expectation of privacy therein. . . . However, *Riley* focused primarily on protecting the contents of cell phones, not metadata generated from cell phone usage. . . . This distinction is far from trivial; Fourth Amendment jurisprudence

---

[58] *Id.* at 2486-2487.

[59] *Id. See also People v. Weissman*, 46 Misc.3d 171, 997 N.Y.S.2d 602 (Crim. Ct. Kings County 2014) (suppressed evidence found in cellular phone records without warrant; no consent given, suspect was compelled to display photos) (citing *Riley*); *People v. Rodas*, 52 Misc.3d 1203(A), 38 N.Y.S.3d 832 (Table), at *5 (County Ct. Yates County 2016) (leaving phone at another's home shows the defendant did not consider it private, but the investigator followed correct procedure seeking warrant before searching phone and SIM card) (citing *Weissman*). *See also U.S. v. Green*, 954 F. 3d 1119 (8th Cir. 2020) (declining to extent *Riley* beyond searches attendant to an arrest to cover searches attendant to seizures by or pursuant to warrant). *But see Matter of Residence in Oakland, Ca.*, 354 F. Supp. 3d 1010 (N.D. Cal. 2019) ("technology is outpacing the law"; communicating passcodes is testimonial, and so are biometric features-faces, retinas, fingerprints-under the Fifth Amendment); *U.S. v. Wright*, 431 F. Supp. 3d 1175 (D. Nev. 2020); *but see Matter of Search Warrant Applic.*, 415 F. Supp. 3d 832 (N.D. Ill. 2019) (collecting cases cutting both ways; including, *inter alia, Minn. v. Diamond*, 905 N.W. 2d 870 (Minn. 2018) (fingerprint is physical characteristic not testimonial); *Commonwealth v. Baust*, 89 Va. Cir. 267 (Va. Cir. Ct. 2014) (holding similar)). *See also* Peter A. Crusco, *Warrant-Proof Encryption and Lawful Decryption*, N.Y.L.J. at 5 (Oct. 21, 2019).

[60] *Carpenter v. U.S.*, 138 S.Ct. 2206 (2018) (Roberts, C.J.). *See also U.S. v. Coles*, 2018 WL 3659934 (M.D. Pa. Aug. 2, 2018) (referencing *Carpenter* and holding, in part, there was probable cause for a warrant issued for CSLI by a local court state judge). *See also Standing Akimbo, LLC v. U.S. through I.R.S.*, 955 F.3d 1146, 1164-1165 (10th Cir. 2020) (declining to extend and distinguishing *Carpenter* based on voluntarily produced information under third-party doctrine (related to reports from the Enforcement Division's Marijuana Enforcement Tracking Reporting Compliance ('METRC') system), pursuant to precedents established in *Smith v. Maryland*, 442 U.S. 735, 740 (1979) and *U.S. v. Miller*, 425 U.S. 435, 442–44 (1976), where party had already voluntarily provided information to the government in order to legally conduct business).

has consistently protected only the contents of an individual's communications. . . . We recently emphasized this point in *United States v. Stanley*,. . . rejecting the argument that an individual had a reasonable expectation of privacy in his or her IP address routed through a third party's wireless router. Even though we acknowledged that obtaining an individual's IP address could roughly track his or her location, we reasoned that such records 'revealed only the path of the signal establishing this connection [and] revealed nothing about the content of the data carried by that signal.'. . . *Riley*'s holding is thus an application of the Fourth Amendment's protection of content. Goldstein does not argue that the CSLI [cell site location information] at issue here is content, nor would we find any such argument persuasive. . . . Accordingly, *Riley* provides little support for extending Fourth Amendment protections to historic CSLI.[61]

Thus, decisions like those in *Stimler* likely propelled the Supreme Court's iteration of an expanded jurisprudence. The following is a lengthy excerpt from Chief Justice John Roberts' majority opinion in *Carpenter*, so that one can see, evaluate, and discuss the majority's reasoning.

Carpenter v. United States (Supreme Court of the United States)

Roberts, Chief Justice

III

The question we confront today is how to apply the Fourth Amendment to a new phenomenon: the ability to chronicle a person's past movements through the record of his cell phone signals. Such tracking partakes of many of the qualities of the GPS monitoring we considered in *Jones*. Much like GPS tracking of a vehicle, cell phone location information is detailed, encyclopedic, and effortlessly compiled.

At the same time, the fact that the individual continuously reveals his location to his wireless carrier implicates the third-party principle of *Smith* and *Miller*. But while the third-party doctrine applies to telephone numbers and bank records, it is not clear whether its logic extends to the qualitatively different category of cell-site records. After all, when *Smith* was decided in 1979, few could have imagined a society in which a phone goes wherever its owner goes,

---

[61] *U.S. v. Stimler*, 864 F.3d 253, 264-65 (3d Cir. 2017) (citations/footnotes omitted); *granting rehearing and vacating opinion in part U.S. v. Goldstein*, 902 F.3d 411 (Mem) (3d Cir. 2018); *aff'd U.S. v. Goldstein*, 914 F.3d 200 (3d Cir. 2019) (Roth, J.) ("We granted Appellant Jay Goldstein's petition for rehearing to address the effect of the Supreme Court's recent decision in *Carpenter v. United States* on our prior panel decision, *United States v. Stimler*. In *Stimler*, we held that the District Court properly denied Goldstein's motion to suppress his cell site location information (CSLI) because Goldstein had no reasonable expectation of privacy in his CSLI, and, therefore, the government did not need probable cause to collect this data. *Carpenter* sets forth a new rule that defendants do in fact have a privacy interest in their CSLI, and the government must generally obtain a search warrant supported by probable cause to obtain this information. However, we still affirm the District Court's decision under the good faith exception to the exclusionary rule because the government had an objectively reasonable good faith belief that its conduct was legal when it acquired Goldstein's CSLI.") (Footnotes and citations omitted).

conveying to the wireless carrier not just dialed digits, but a detailed and comprehensive record of the person's movements.

We decline to extend *Smith* and *Miller* to cover these novel circumstances. Given the unique nature of cell phone location records, the fact that the information is held by a third party does not by itself overcome the user's claim to Fourth Amendment protection. Whether the Government employs its own surveillance technology as in *Jones* or leverages the technology of a wireless carrier, we hold that an individual maintains a legitimate expectation of privacy in the record of his physical movements as captured through CSLI. The location information obtained from Carpenter's wireless carriers was the product of a search.

A

A person does not surrender all Fourth Amendment protection by venturing into the public sphere. To the contrary, "what [one] seeks to preserve as private, even in an area accessible to the public, may be constitutionally protected.". . . A majority of this Court has already recognized that individuals have a reasonable expectation of privacy in the whole of their physical movements. . . . Prior to the digital age, law enforcement might have pursued a suspect for a brief stretch, but doing so "for any extended period of time was difficult and costly and therefore rarely undertaken.". . . For that reason, "society's expectation has been that law enforcement agents and others would not—and indeed, in the main, simply could not—secretly monitor and catalogue every single movement of an individual's car for a very long period.". . .

Allowing government access to cell-site records contravenes that expectation. Although such records are generated for commercial purposes, that distinction does not negate Carpenter's anticipation of privacy in his physical location. Mapping a cell phone's location over the course of 127 days provides an all-encompassing record of the holder's whereabouts. As with GPS information, the time-stamped data provides an intimate window into a person's life, revealing not only his particular movements, but through them his "familial, political, professional, religious, and sexual associations.". . . These location records "hold for many Americans the 'privacies of life.'". . . And like GPS monitoring, cell phone tracking is remarkably easy, cheap, and efficient compared to traditional investigative tools. With just the click of a button, the Government can access each carrier's deep repository of historical location information at practically no expense.

In fact, historical cell-site records present even greater privacy concerns than the GPS monitoring of a vehicle we considered in *Jones*. Unlike the bugged container in *Knotts* or the car in *Jones*, a cell phone—almost a "feature of human anatomy,". . .—tracks nearly exactly the movements of its owner. While individuals regularly leave their vehicles, they compulsively carry cell phones with them all the time. A cell phone faithfully follows its owner beyond public thoroughfares and into private residences, doctor's offices, political headquarters, and other potentially revealing locales. . . . Accordingly, when the Government tracks the location of a cell phone it achieves near perfect surveillance, as if it had attached an ankle monitor to the phone's user.

Moreover, the retrospective quality of the data here gives police access to a category of information otherwise unknowable. In the past, attempts to reconstruct a person's movements were limited by a dearth of records and the frailties of recollection. With access to CSLI, the Government can now travel back in time to retrace a person's whereabouts, subject only to the retention polices of the wireless carriers, which currently maintain records for up to five years. Critically, because location information is continually logged for all of the 400 million devices in the United States—not just those belonging to persons who might happen to come under investigation—this newfound tracking capacity runs against everyone. Unlike with the GPS device in *Jones*, police need not even know in advance whether they want to follow a particular individual, or when.

Whoever the suspect turns out to be, he has effectively been tailed every moment of every day for five years, and the police may—in the Government's view—call upon the results of that surveillance without regard to the constraints of the Fourth Amendment. Only the few without cell phones could escape this tireless and absolute surveillance.

The Government and Justice KENNEDY contend, however, that the collection of CSLI should be permitted because the data is less precise than GPS information. Not to worry, they maintain, because the location records did "not on their own suffice to place [Carpenter] at the crime scene"; they placed him within a wedge-shaped sector ranging from one-eighth to four square miles. . . . Yet the Court has already rejected the proposition that "inference insulates a search.". . . From the 127 days of location data it received, the Government could, in combination with other information, deduce a detailed log of Carpenter's movements, including when he was at the site of the robberies. And the Government thought the CSLI accurate enough to highlight it during the closing argument of his trial. . . .

At any rate, the rule the Court adopts "must take account of more sophisticated systems that are already in use or in development.". . . While the records in this case reflect the state of technology at the start of the decade, the accuracy of CSLI is rapidly approaching GPS-level precision. As the number of cell sites has proliferated, the geographic area covered by each cell sector has shrunk, particularly in urban areas. In addition, with new technology measuring the time and angle of signals hitting their towers, wireless carriers already have the capability to pinpoint a phone's location within 50 meters. . . .

Accordingly, when the Government accessed CSLI from the wireless carriers, it invaded Carpenter's reasonable expectation of privacy in the whole of his physical movements.

. . .

B

We therefore decline to extend *Smith* and *Miller* to the collection of CSLI. Given the unique nature of cell phone location information, the fact that the Government obtained the information from a third party does not overcome Carpenter's claim to Fourth Amendment protection. The Government's acquisition of the cell-site records was a search within the meaning of the Fourth Amendment.

* * *

Our decision today is a narrow one. We do not express a view on matters not before us: real-time CSLI or "tower dumps" (a download of information on all the devices that connected to a particular cell site during a particular interval). We do not disturb the application of *Smith* and *Miller* or call into question conventional surveillance techniques and tools, such as security cameras. Nor do we address other business records that might incidentally reveal location information. Further, our opinion does not consider other collection techniques involving foreign affairs or national security. As Justice Frankfurter noted when considering new innovations in airplanes and radios, the Court must tread carefully in such cases, to ensure that we do not "embarrass the future."

. . .

IV

. . .

As Justice Brandeis explained in his famous dissent, the Court is obligated—as "[s]ubtler and more far-reaching means of invading privacy have become available to the Government"—to ensure that the "progress of science" does not erode Fourth Amendment protections. . . . Here the progress of science has afforded law enforcement a powerful new tool to carry out its important responsibilities. At the same time, this tool risks Government encroachment of the sort the Framers, "after consulting the lessons of history," drafted the Fourth Amendment to prevent.

We decline to grant the state unrestricted access to a wireless carrier's database of physical location information. In light of the deeply revealing nature of CSLI, its depth, breadth, and comprehensive reach, and the inescapable and automatic nature of its collection, the fact that such information is gathered by a third party does not make it any less deserving of Fourth Amendment protection. The Government's acquisition of the cell-site records here was a search under that Amendment.

The judgment of the Court of Appeals is reversed, and the case is remanded for further proceedings consistent with this opinion.[62]

---

[62] *Carpenter v. U.S.*, 138 S.Ct. 2206, 2216–2223 (2018) (citations omitted) (citing, *inter alia*, *Riley*, 134 S.Ct. 2473; *Olmstead v. U.S.*, 277 U.S. 438, 473–474 (1928); *U.S. v. Di Re*, 332 U.S. 581, 595 (1948)).

Therefore, contrary to the Third Circuit's 2017 decision in *Stimler*, and although contained in a narrow holding, the Supreme Court found reason to expand Fourth Amendment jurisprudence to CSLI—including historic CSLI—with concomitant warrant and probable cause burdens placed on the government.

> **Border** – Any point of international crossing, entry or exit – including ports, airports, or the physical border line of a nation.

3.9 – Electronic Materials and the United States' Border

Be aware that during entry/re-entry to or exit from the nation at the borders of the United States, there is no probable cause necessary or warrant requirement for searches of electronic devices.[63] This presents a grave concern, especially for those traveling across the border with confidential, proprietary, or trade secret business information, or attorneys traveling with devices containing attorney-client privileged or work-product material. Guidance has been issued by an ethics opinion of the New York City Bar Association, and commentators have

---

[63] *See U.S. v. Touset*, 890 F.3d 1227 (11th Cir. 2018) (distinguishing *Riley*, and citing *U.S. v. Ramsey*, 431 U.S. 606 (1977)). *See also* U.S. Customs & Border Protection CBP Directive No. 3340-049A (Jan. 4, 2018) (Review Date: January 2021). *See also U.S. v. Haynes*, 2020 WL 109061 (S.D. Fla. Jan. 9, 2020) ("border searches require *no suspicion* at all") (emphasis by court). But, there is a Circuit Split, where the 9th Circuit has required reasonable suspicion to search a laptop, *U.S. v. Cotterman*, 709 F. 3d 952 (9th Cir. 2013). *See also U.S. v. Aigbekaen*, 943 F.3d 713, 720-721 (4th Cir. 2019) ("border search" exception is not without bounds, and "[i]ndeed, neither the Supreme Court nor this court has ever authorized a warrantless border search unrelated to the sovereign interests underpinning the exception, let alone nonroutine, intrusive searches like those at issue here. Rather, our decision in *Kolsuz* teaches that the Government may not 'invoke[ ] the border exception on behalf of its generalized interest in law enforcement and combatting crime.'... This restriction makes particularly good sense as applied to intrusive, non-routine forensic searches of modern digital devices, which store vast quantities of uniquely sensitive and intimate personal information ... yet cannot contain many forms of contraband, like drugs or firearms, the detection of which constitutes 'the strongest historic rationale for the border-search exception,'.... Accordingly, as we explained in *Kolsuz*,... to conduct such an intrusive and nonroutine search under the border search exception (that is, without a warrant), the Government must have individualized suspicion of an offense that bears some nexus to the border search exception's purposes of protecting national security, collecting duties, blocking the entry of unwanted persons, or disrupting efforts to export or import contraband.") (Citations and footnotes omitted; citing, *inter alia, U.S. v. Ramsey*). Additionally, in 2019, a federal district court in Massachusetts agreed with and cited to *Cotterman*, among others. *See Alasaad v. Nielsen*, 419 F.Supp.3d 142 (D. Mass. 2019), *appeals filed, Alasaad, et al. v. Duke, et al.*, No. 20-1077 (1st Cir. Jan. 28, 2020) & *Alasaad, et al. v. Wolf, et al.*, No. 20-1081 (1st Cir. Jan. 29, 2020). *See also People v. Perkins*, --- N.Y.S.3d ---, 2020 WL 3260954, at *1-*2 (Sup. Ct. App. Div. 2d Dep't June 17, 2020) (citing, *inter alia, U.S. v. Flores-Montano*, 541 U.S. 149 (2004); *U.S. v. Ramsey*, 431 U.S. 606 (1977); *U.S. v. Cano*, 934 F.3d 1002 (9th Cir. 2019); and *Aigbekaen*, 943 F.3d at 723); Debra Cassens Weiss, *Judge Rules Border Agents Need Reasonable Suspicion to Search Electronic Devices*, ABA Journal (online Nov. 13, 2019), *available at* https://www.abajournal.com/news/article/judge-rules-border-agents-need-reasonable-suspicion-to-search-electronic-devices.

published articles concerning this same matter.[64] Ultimately, businesspersons and/or attorneys may wish to travel without such information, but if that is not a possibility because the information/device is needed at the destination, then the New York City Bar opinion and other resources offer helpful guidance.

3.10 – Caution! User Agreements May Address Release of Information

One should be very careful before clicking on user agreements—agreements such as those asking if "I agree" before utilizing the services of a particular website or service provider. Clicking on "I agree" creates, in almost all circumstances, a binding contract (discussed further in Chapter 11). Among other things, there could be a waiver of Fourth Amendment constitutional rights (and the otherwise held expectation of privacy) and authorizations under the Stored Communications Act provisions, by agreeing to a particular online or technology company's Terms of Service—meaning that a user consents to a search of emails by law enforcement, or response to a subpoena or document demand in civil litigation.[65]

3.11 – Conclusion

As is clear, discovery of social media, text messages, and electronic mail is a growing field within the law. Discovery in many cases involves these sources of potential evidence. It is vitally important for attorneys to understand how to pursue such discovery. It is equally important, though, for clients and citizens not only to recognize the necessity of preserving information that may be relevant to a present or future litigation, but to also understand how

---

[64] *See* N.Y. City Bar Ass'n Formal Op. 2017-5 ("An Attorney's Ethical Duties Regarding U.S. Border Searches of Electronic Devices Containing Clients' Confidential Information") (reissued May 9, 2018) (an annotated update was added to the Opinion in May 2018, referencing CBP Directive No. 3340-049A, but stating that the guidance in the opinion holds nonetheless). But, note that according to the CBP Directive, at Section 5 - Procedures, while a "basic search" may be conducted with or without suspicion, an "advanced search" may only be conducted when "there is reasonable suspicion of activity in violation of the laws enforced or administered by CBP, or in which there is a national security concern, and with supervisory approval at the Grade 14 level or higher." *See* CBP Directive No. 3340-049A, at §§ 5.1.3 & 5.1.4. Furthermore, per the Agency's own directive, "The border search will include an examination of only the information that is resident upon the device and accessible through the device's operating system or through other software, tools, or applications. Officers may not intentionally use the device to access information that is solely stored remotely. To avoid retrieving or accessing information stored remotely and not otherwise present on the device, Officers will either request that the traveler disable connectivity to any network (e.g., by placing the device in airplane mode), or, where warranted by national security, law enforcement, officer safety, or other operational considerations, Officers will themselves disable network connectivity. Officers should also take care to ensure, throughout the course of a border search, that they do not take actions that would make any changes to the contents of the device." *Id.* at § 5.1.2.
[65] *See U.S. v. DiTomasso*, 56 F.Supp.3d 584 (S.D.N.Y. 2014).

material that is posted, texted, or emailed can impact on lives, searches and seizures, legal claims or defenses.[66]

Chapter End Questions

Note: Answers may be found on page 261.

1. Which of the following cases is considered to be the start, the seminal case, with regard to discovery of social media?
   a. *Quon v. City of Ontario*
   b. *Marbury v. Madison*
   c. *Losee v. Clute*
   d. *Romano v. Steelcase, Inc.*

2. What is the standard in New York for discovery of social media material, following the decision in <u>Forman v. Henkin</u>?
   a. Showing of a factual predicate
   b. Clear and present reasoning
   c. Material and necessary to prosecution or defense of an action
   d. Direct proof of a claim

3. Which statute/rule in New York governs disclosures in discovery, including those for electronic materials?
   a. CPLR 101
   b. CPLR 2001
   c. CPLR 3101
   d. CPLR 5003

4. When communicating with an attorney, an employee of a company with a private legal matter, or a husband and wife who are not both clients of the attorney, should do which of the following?
   a. Utilize their work e-mail systems, since there is an expectation of privacy in work e-mail systems
   b. Use a private e-mail account, to which only the husband and wife have access
   c. Utilze a private e-mail address only used and accessible by the attorney's client and no one else
   d. None of these selections

---

[66] For another summary overview of social media discovery, see also K. Dandy & G. Portera, *Guiding Principles on the Discoverability and Admissibility of Social Media*, 88 N.Y. ST. B.J. 26 (Jan. 2016). *See also* Frank Ready, *Forgetting Mobile Data Is 'Malpractice' But Remembering It Is No Fun*, N.Y. L.J. at 9 (Feb. 6, 2020).

5. Which U.S. Supreme Court decision held that cell site location information ("CSLI") and searches utilizing such information fell under the Fourth Amendment, requiring probable cause and a search warrant?

   a. *Romano v. Steelcase, Inc.*
   b. *Carpenter v. United States*
   c. *McCulloch v. Maryland*
   d. *Lazarus v. County of Sullivan*

6. Which of the following is true concerning electronic devices and international travel?

   a. U.S. Customs and Border Protection cannot access electronic devices for any reason
   b. U.S. Customs & Border Protection can only access and search electronic devices if they first have probable cause and obtain a warrant
   c. U.S. Customs and Border Protection can access and search electronic devices without a warrant, as there is no 4th Amendment protection at the international border
   d. None of these selections

7. True or False: When using social media, particularly postings that are public, there is an expectation that one has privacy in their communications, privacy that is protected by law.

   a. TRUE
   b. FALSE

8. True or False: Any social media accounts a party may have are automatically discoverable in full during a lawsuit.

   a. TRUE
   b. FALSE

9. True or False: When attendant to an arrest, police officers/law enforcement officers may search mobile phones and devices of a suspect without a warrant.

   a. TRUE
   b. FALSE

10. True or False: If utilizing a website that requires a click on an "I agree" button, providing terms of service, that potentially creates a binding contract.

    a. TRUE
    b. FALSE

CHAPTER 4

Ethical and Legal Considerations for Social Media Discovery: Attorneys

4.1 – Introduction—Attorney Ethical Obligations

Discovery is not the only area in which social media has an effect on the legal profession and legal landscape. With the explosion of social media, there has been an increase in ethical issues, as well. Some questions exist:

- Can an attorney view the MySpace or Facebook, or other social media pages of another party (not the lawyer's client) for information gathering?
- May an attorney either directly, or through another person, contact an unrepresented person through social networking?
- What should attorneys tell their clients about social media preservation? About any ESI preservation?

We will explore these questions in this chapter.

The mind of the reader may, however, raise two other questions while evaluating this material:

- Is an attorney permitted to research prospective jurors before, during, and/or after *voir dire* (jury selection)?
- What about judges? If they "friend" attorneys on social media, should they have to recuse themselves from cases? What if they have "friended" litigants?

We will answer these last two questions through the discussions contained in Chapters 5 and 6, respectively.

In the meantime, let us examine the first three questions, above, in more detail. One thing to remember, however, both in this chapter and all others of this book is that review of opinions issued by the ethics committees of bar associations is quite informative, and provides a wealth of knowledge and guidance. At the end of the day,

Ethics Opinion – Bar associations and other legal associations have committees or groups that issue written guidance or thoughts as to how specific actions might or might not violate ethics rules.

however, the ethics opinions are not binding authority or precedent in any court[1]—and an attorney may not rely on general guidance in an ethical opinion for complete absolution when an issue arises. Instead, the opinions help guide an attorney's conduct in a particular instance, in conforming what an attorney does with the ethical proscriptions in a jurisdiction—and for that purpose the ethics opinions are, in the author of this book's estimation, invaluable.

4.2 – Relevant Ethical Rules at Issue—Attorney Activities and Conduct

The ethical questions and issues raised in the introduction to this chapter concern a number of professional rules of conduct governing attorneys. In New York, for instance, consider N.Y. Rules of Professional Conduct 3.1, 3.3, 3.4, 3.5, 4.1, 4.2, 4.3, 5.3, and 8.4, among others. Similar rule frameworks exist in the other state jurisdictions of the United States, as well as in the Model Rules of the American Bar Association. The rules address, among other things, communication, truthfulness, misconduct, attorneys' responsibility for the conduct of non-lawyers, and communications with represented and unrepresented persons.

Many courts, rules, and commentators speak about the obligation placed on attorneys to provide competent representation, maintain requisite knowledge, and keep abreast of the benefits and risks of technology. Attorneys are, in other words, required to "stay up to date"[2]—and should further supervise those working in their offices—to ensure completion of obligations and, among other things, familiarity with client activities and systems, TAR (technology-assisted review), and keyword searches, and establish discovery plans considering preservation issues.

As is clear from bar association ethics opinions, the ethical rules do not only involve attorney use of social media, but all professional activities of an attorney—including any and

---

[1] *See Miano v. AC & R Advertising, Inc.*, 148 F.R.D. 68, 83 (S.D.N.Y. 1993) ("Although bar opinions are not binding on this Court, they are instructive in applying ethical rules to attorney conduct in litigation and provide guidance to attorneys themselves in conforming their conduct to ethical proscriptions"); *State ex rel. Nationwide Mut. Ins. Co. v. Karl*, 222 W.Va. 326, 331, 664 S.E.2d 667, 672 (W.Va. 2008) ("Nationwide has placed great emphasis upon a formal advisory ethics opinion, L.E.I. 99–01,. . . . At the outset we must note that opinions issued by the Lawyer Disciplinary Board are not binding upon this Court"); *Gafcon, Inc. v. Ponsor & Assoc.*, 98 Cal.App.4th 1388, 1414, 120 Cal.Rptr.2d 392, 411 (Cal. App. Ct. 4th Dis. 2002) ("We are not bound by an ethics opinion, and we need not adopt it in full for our holding in this case"); *Harleysville Ins. Co. v. Holding Funeral Home, Inc.*, 2017 WL 4368617, at *11 (W.D. Va. Oct. 2, 2017) ("Legal Ethics Opinions ('LEOs') issued by the Virginia State Bar Standing Committee on Legal Ethics, while not binding on this or any court, provide persuasive guidance regarding the propriety of attorneys' conduct").

[2] *See* N.Y. RULES OF PROF'L CONDUCT 1.1 & Cmt. 8.

all electronic activities (even advertising).[3] Certainly, attorneys should always be wary, and cautious, when utilizing new technology—and advise clients of same.[4] Failure to understand and properly secure new technologies could result in unintended disclosures, and potentially waiver of attorney-client privilege.

Attorneys should also always keep in mind, both when utilizing social media and when conducting themselves in any professional endeavor, to refer to the governing professional conduct rules of their jurisdiction. For instance, depending upon the content of their own social media pages, attorneys may have to include the disclaimer found in Rule of Professional Conduct 7.1 (concerning "Attorney Advertising," which is not discussed much further in this book).[5]

Attorneys, as officers of the court (even if one is not a litigator), have duties and responsibilities that are separate from being citizens. One court in Florida held:

> as a member of the Bar of this Court and The Florida Bar, [the attorney-litigant] cannot post on social media in connection with the practice of law in a manner that is prejudicial to the administration of justice. . . . Nor can he, in connection with the practice of law, knowingly, or through callous indifference, disparage, humiliate or discriminate against litigants, witnesses or other lawyers on any basis in his social media posts. . . . This is especially true when his posts are directed against defendants, deponents, or attorneys in two federal civil lawsuits pending before this Court. And, pursuant to the rules of this Court, [attorney's] social posts related to litigation in this Court must be within the bounds of cooperation, professionalism and civility. . . . The Court has carefully considered the social media posts made by [attorney], the arguments of all the parties in these two cases, and the applicable rules and law. This is a very serious matter with serious consequences.[6]

---

[3] *See* Am. Bar Ass'n Formal Op. 480 (Mar. 6, 2018) ("Confidentiality Obligations for Lawyer Blogging and Other Public Commentary") (addresses concerns regarding disclosure of client information under ABA Model Rule 1.6, impartiality and decorum of tribunals under Model Rule 3.5, and trial publicity under Model Rule 3.6); N.Y. City Bar Ass'n Op. 2015-3 (speaks to issue of attorneys who fall victim to scams; if attorneys fall prey to vicious scammers, or fail to secure client data, they could violate Rule 1.6; they must confirm an email is a legitimate prospective client; if funds are received, they have to make sure they are confirmed and honored by the bank, as with all funds in the trust account, per Rule 1.15—not just "check cleared"). Attorneys are certainly permitted to use social media, so long as the ethics rules are adhered to. *See, generally,* Janet Falk, *Social Media Activity For the Solo Attorney,* N.Y. L.J, at 3 (May 14, 2020).

[4] *See Harleysville Ins. Co. v. Holding Funeral Home, Inc.*, 2017 WL 1041600 (W.D. Va. Feb. 9, 2017), sustained in part and overruled in part by 2017 WL 4368617 (W.D. Va. Oct. 2, 2017), for a cautionary tale.

[5] For more concerning online activities and potential overlap with advertising, see N.Y. County Lawyers Ass'n Formal Op. 748 (2015) (addressing attorney use of LinkedIn and advising attorneys to use required disclaimer in N.Y. Rule 7.1; however, using "skills" or "endorsements" on the profile does not violate N.Y. Rule 7.4 but must be truthful and accurate); *see also* N.Y. City Bar Ass'n Formal Op. 2015-7 (2015) (also addressing attorney advertising, LinkedIn, and N.Y. RPC 7.1, 7.4, 7.5, *et al.*).

[6] *Leigh v. Avossa*, Civil No. 16-81612-CIV, Civil No. 16-81624-CIV, 2017 WL 2799617 (S.D. Fla. June 28, 2017) (citing Fla. Bar R. 4-8.4(d); Fla. Bar R. 4-8.4(d); S.D. Fla. L.R. 11.1(c)).

Additionally, as discussed previously in Chapter 2, an opinion of the American Bar Association, issued in 2017, begins in relevant part:

> A lawyer generally may transmit information relating to the representation of a client over the internet without violating the Model Rules of Professional Conduct where the lawyer has undertaken reasonable efforts to prevent inadvertent or unauthorized access. However, a lawyer may be required to take special security precautions to protect against the inadvertent or unauthorized disclosure of client information when required by an agreement with the client or by law, or when the nature of the information requires a higher degree of security.[7]

Therefore, it is very clear that attorneys have duties of competence and confidentiality.

If state and local authorities governing attorney conduct and obligations adopt the same standard as that in the ABA's opinion, there will be an increasing burden on attorneys to ensure proper security protocols—really an obligation that already exists when considered in light of current confidentiality and competency rules. Committee reports and publications issued by bar associations across the country, although advisory, again remain a good place to look for guidance material.

For instance, with regard to social media use by attorneys, the Guidelines published by the New York State Bar Association's Commercial & Federal Litigation Section provide fruitful direction.[8] Additionally, the California Bar Association issued an opinion in 2015 outlining skills that every attorney should have when handling e-discovery (especially considering that today most litigation potentially involves some form of ESI).[9]

4.3 – Attorney Contact/Research of Parties or Witnesses Using Social Media

Many state and national bar associations have issued opinions on different social media uses by attorneys.

For instance, when it comes to *adverse parties* in a legal action, one New York State Bar Association opinion held that an attorney may view the social media pages of another party, as long as the attorney does not "friend" the person, or have a third-party (paralegal, client, or other) do so.[10] This, of course, applies to represented parties. Use of *public* pages on the network is acceptable, and does not run afoul of Rules 4.1, 4.2, 4.3, 5.3(b)(1), or 8.4. Attorneys, and their staff, may view websites, as long as the viewing is public or passive—like reading a magazine. With represented parties attorneys must be much more careful concerning "friend"

---

[7] Am. Bar Ass'n Formal Op. 477R (2017) (revisiting issues addressed in Formal Op. 99-413 and Formal Op. 11-459).

[8] SOCIAL MEDIA ETHICS GUIDELINES OF THE COMMERCIAL & FEDERAL LITIGATION SECTION OF THE NEW YORK STATE BAR ASSOCIATION. For this chapter, see in particular *Guidelines*, 4.A, 4.B, 4.C, 4.D, 5.A, 5.B, 5.C, 5.D, & 5.E, and comments thereto (updated June 2019).

[9] Cal. Bar Ass'n Formal Op. 2015-193.

[10] N.Y.S. Bar Ass'n Op. 843 (2010).

requests or other such activities concerning the *private* pages. Attorneys and paralegals must follow the same ethical rules as for non-electronic contact with represented parties. For example, in New York an attorney may not contact a represented party without notice to that party's legal counsel.[11] Furthermore, as will be discussed in Chapter 5 concerning juries, some social media platforms utilize automatic notification procedures to advise a user when someone else views his or her profile. With regard to represented persons, New York ethics guidance holds that the automatic notification—particularly if an attorney is aware his or her viewing will result in the notification to the party, hence a communication—will violate the rules.[12]

When it comes to an unrepresented party, or witness, however, an attorney is free to contact the party, and to even send a "friend" request to the party on social media (and may have staff or another person do the same)—as long as the attorney identifies himself or herself and does not deceive the other party in any way.[13] In fact, in some jurisdictions, ethical authorities have held that until there is actual knowledge that a person is represented by legal counsel, direct contact is not prohibited.[14]

An attorney may also generally accept information about the non-public social media activity of another person, provided by a client—with some restriction. Here, we are concerned about those who take actions that attorneys could not undertake themselves. Thus, "[a] lawyer may review a represented person's non-public social media information provided to the lawyer by her client, as long as the lawyer did not cause or assist the client to: (i) inappropriately obtain non-public information from the represented person; (ii) invite the represented person to take action without the advice of his or her lawyer; or (iii) otherwise overreach with respect to the represented person."[15]

It is wise, though, to still be cautious even with unrepresented adverse witnesses. In the New York State Bar Association opinion referenced earlier (Opinion 843), the Committee cited to Philadelphia Bar Association opinion 2009-02 (March 2009), which held pursuant to a Pennsylvania ethics rule (similar to New York's Rule 8.4(c)) that the lawyer's intention to have a third party "friend" a witness to obtain access to non-public Facebook and MySpace pages and find impeachment material would violate the rule concerning attorney dishonesty, fraud, deceit, and misrepresentation. At this time, ethical guidance advises that the person contacted must understand who is seeking to "friend" him or her and whether that person has a role in the litigation in which the witness is taking part. If the attorney makes the contact, the attorney must use his or her real name and identity.

---

[11] N.Y. RULES OF PROF'L CONDUCT 4.2; NYSBA ComFed *Social Media Guidelines*, Guideline 4.C.

[12] *See* Comments to NYSBA ComFed *Social Media Guidelines*, Guideline 4.C & Comment (citing N.Y. City Bar Ass'n Formal Op. 2012-2 (May 30, 2012); N.Y. County Lawyers Ass'n Formal Op. 743 (May 18, 2011)).

[13] *See* N.Y. RULES OF PROF'L CONDUCT 4.3, 4.4, 5.3, 8.4; NYSBA ComFed *Social Media Guidelines*, Guideline 4.B, 4.D.

[14] *See* Comments to NYSBA ComFed *Social Media Guidelines*, Guideline 4.C & Comment; Oregon State Bar Comm. on Legal Ethics, Formal Op. 2013-189 (2013).

[15] *See* NYSBA ComFed *Social Media Guidelines* 5.D & Comment; *see also* N.Y. City Bar Ass'n Formal Op. 2002-3 (2002) (cited by ComFed Guideline 5.D Comment).

To emphasize the point here, an opinion of the New York City Bar Association held that when it comes to whether an attorney may, either directly or through another, contact an un-represented person through social networking, the answer is both Yes and No.[16] Neither an attorney nor his or her staff can employ trickery. There are non-deceptive means that may be utilized—such as an attorney using Facebook and "truthful" friending, or utilizing subpoenas to non-parties holding information.

If you are an attorney gathering information, or directing another to do so on your be-half—or if you are a member of the public contacted by an attorney (or someone working on behalf of an attorney) via social media—be aware there are important ethical limitations with which to comply. Of course, when contacting witnesses, *do not* create fake identities or fake social media accounts. That happened in Ohio, where a prosecutor created a fake Facebook page to contact two alibi witnesses in a murder case. The prosecutor, doubting the defendant's alibi, created a Facebook account pretending to be someone else known to the witnesses in the case. Using that Facebook account, the prosecutor discussed the alibi with one of the witnesses.[17] The prosecutor claimed he was going to turn over copies of the messages to the defense, but he did not; and after printing the messages, he deleted the Facebook account.[18] When the messages were later discovered in the file, they were turned over to the defense by the new prosecutor on the case, and the case was turned over to the Attorney General. The former prosecutor was fired, and in a February 2016 opinion the Ohio Supreme Court issued a stayed one-year suspension from the practice of law. One justice dissented, citing the former prosecutor's "glaring disdain for the ethical responsibilities this court imposes on all attorneys in this state."[19] That justice would have imposed an indefinite suspension.

Next, consider a case in New Jersey. There, attorneys faced a ruling by the New Jersey Supreme Court that the New Jersey Office of Attorney Ethics, an arm of the state judiciary, could file and pursue charges.[20] A paralegal in the defense firm "friended" the plaintiff after the plaintiff changed Facebook privacy settings. The plaintiff discovered this when the para-legal was disclosed as a witness. The "friending" took place in 2008, and the two attorneys for the defense argued they were unaware of privacy settings at the time of the activities.[21] Nevertheless, the Supreme Court of New Jersey, citing the fact that there were no reported

---

[16] N.Y. City Bar Ass'n Formal Op. 2010-2 (2010).

[17] *See* D. Weiss, *Ex-prosecutor Gets Stayed Suspension for Using Fake Facebook Account to Contact Alibi Witnesses*, ABA JOURNAL (online Mar. 2, 2016), available at: http://www.abajournal.com/news/article/ex_prosecutor_gets_stayed_suspension_for_using_fake_facebook_account_to_con/; *Disciplinary Counsel v. Brockler*, 145 Ohio St.3d 270, 48 N.E.3d 557 (Ohio 2016). *See also Disciplinary Counsel v. Spinazze*, —N.E. 3d—, 2020 WL 1264176 (Ohio Mar. 17, 2020) (citing *Brockler*) ("attorneys who serve as prosecutors 'are authorized to enforce the law and administer justice' and 'must meet or exceed the highest ethical standards imposed on our profession').

[18] *Id.*

[19] *Disciplinary Counsel*, 145 Ohio St.3d at 276, 48 N.E.3d at 562.

[20] *See Robertelli v. N.J. Off. of Atty. Ethics*, 224 N.J. 470, 134 A.3d 963 (N.J. 2016); S. Hodge, *For Lawyers, Facebook Friending Your Opponent Can Lead to Trouble*, FORBES (online May 6, 2016), available at: https://www.forbes.com/sites/legalnewsline/2016/05/06/for-lawyers-facebook-friendingyour-opponent-can-lead-to-trouble/#23a39ce0df5a.

[21] *Id.*

cases on point, held that the State Office could pursue the charges, with accusations that the lawyers communicated with a represented party without proper consent, and engaged in conduct involving dishonesty, fraud, deceit, or misrepresentation—among other ethics charges.[22]

Now, consider the following in our present world of the 24-hour news cycle and non-stop social media. Amazingly, a partner at a large law firm in Chicago had to appear before the Chief District Judge of the Northern District of Illinois, in November 2015, to face potential sanctions for tweeting photos during a trial.[23] The attorney claimed to have forgotten the rules against photos and videos in the courtroom, although they were posted on a board outside the courtroom.[24] A panel of judges later determined that the attorney would have to attend an education seminar on social media and its ethical implications for lawyers, pay a $5,000 fine to the Chicago Bar Association, and provide 50 hours of service to the Court's legal assistance desk for *pro se* litigants.[25]

This last case in particular illustrates that social media has become so ubiquitous and ever-present in some people's lives, including attorneys, that one must be very cautious about the contexts in which they post, and what they post. Some attorneys also claim to use social media sites for client development and advertising—which implicates its own set of ethical rules (for example, see the NYSBA ComFed *Social Media Guidelines* 2.A, 2.B, 2.C, 2.D and 2.E-not further addressed in this book).

4.4 – Attorney Advice to Clients Concerning Their Social Media

Remember the discussions elsewhere in this chapter and this book. Be very cautious before any material is deleted or removed from social media sites. Furthermore, as an attorney, be very cautious before advising clients to remove any material from social media sites.[26] Attorneys should remind clients to preserve all information in "native format." Again, native format is the form that the post, communication, or electronic document is in when created—the form maintaining and preserving the metadata. It is the format that is key to preserving not only the information visible on its face but also the metadata.

> *Native Format* – The original format an electronic item is in when first created.

> *Metadata* – Data about data; for example when a document was created, the author, the version, the format (Excel, Word, e-mail), etc.

---

[22] *Id. See also* Charles Toutant, *Not a 'Friend': Lawyer Faces Admonition Over Facebook 'Friending' to Gather Info*, N.Y.L.J. at 5 (May 19, 2020) (addressing the *Robertelli* matter).

[23] *See Submission to the Court & Exec. Comm., In re. Vincent Paul Schmeltz*, 2015 WL 12670761 (N.D. Ill. Dec. 4, 2015); Claire Bushey, *Tweeting Barnes & Thornburg Lawyer Rapped by Federal Court*, CRAIN'S CHICAGO BUSINESS (Dec. 10, 2015).

[24] *Id.*

[25] *Id.*

[26] *See* NYSBA ComFed *Social Media Guidelines* 5.A.

However, attorneys in New York may certainly advise clients concerning what they should/ should not post to social media, privacy settings, what existing postings they may or may not remove, and the implications of social media posts during the pendency of a particular case.[27] The same rules apply here as to legal advice in other contexts—Rules of Professional Conduct 3.1, 3.3, 3.4, 4.1, 4.2, 8.4.[28] As always, remain mindful of litigation hold requirements.[29] An attorney may technically advise a client to take down material from social media as long as there is no violation of a duty to preserve, no spoliation, and especially if the substance of the post is preserved in cyberspace or on the client's computer.

Attorneys may advise clients that items can be removed, in our adversarial system, as long as safeguards are followed and preservation is accomplished in some usable form, maintaining all necessary data and metadata. The removal cannot violate evidentiary preservation duties.[30] In light of the above, a best practice would still be to never recommend that a client take anything down, wholesale, from a private or public page once posted—particularly if litigation is commenced or reasonably anticipated, since the attorney is likely not going to be able to monitor the client's every action on the site.

If an attorney ever does advise a client to remove posts, the best practice is to put all advice into a detailed writing, addressed to the client, and keep a copy for the file. Maintain a clear record of what was advised and, just as importantly, what the attorney advised against. In such a circumstance, again, it is vital to ensure that the attorney advise the client specifically about the original data, including metadata, and the need to preserve it in a format that is accessible should issues arise in discovery.

In the end, though, it is submitted to the reader that the best course of action is to place a hold on all material, stop posting but keep the account active (even if that means logging-in periodically so the account is not shut down by the provider for inactivity),[31] and not post anything new unless it is completely unrelated to anything involving the case (which requires a case-by-case assessment).

Some commentators advise that pursuant to ethics rules and opinions in New York, Philadelphia, and elsewhere, not only is it ethical to advise a client to remove material from the social media site (again, as long as there is no obstruction of justice, and both metadata and original information is preserved), but it is also ethical to advise a client in the posting of material that may be *beneficial* to the client or that tends to show the client in a good light. Such posts cannot be false, of course, and if there are ever subsequent questions at a deposition,

---

[27] *See* N.Y. County Lawyers Ass'n Formal Op. 745 (July 2, 2013) (The opinion only addresses civil cases, and advises attorneys in criminal cases may have different ethical considerations).

[28] *See also* NYSBA ComFed *Social Media Guidelines* 5.A, 5.B, 5.C.

[29] *See QK Healthcare, Inc. v. Forest Labs., Inc.*, Index No. 117407/09, 2013 N.Y. Misc. LEXIS 2008 (Sup. Ct. N.Y. County May 8, 2013).

[30] *See* N.Y. RULE OF PROF'L CONDUCT 3.4 & Comment 1; NYSBA ComFed *Social Media Guidelines* 5.A, 5.B, 5.C.

[31] *See, e.g., Gatto v. United Airlines, Inc.*, 2013 WL 1285285 (D.N.J. Mar. 25, 2013). And be sure to understand that once some platforms delete/remove/modify material it is gone forever, so beware violation of RPC 3.4. *See* Christina M. Jordan, *Discovery of Social Media Evidence in Legal Proceedings*, ABA Litigation, vol. 45, no. 1, at 3 (Fall 2019).

or other proceeding, about any changes made to the account, those must be answered truthfully.[32] Nevertheless, extreme caution and care must be taken, and both attorney and client should educate themselves prior to action.

Most importantly, *do not* ever advise or undertake a blanket "clean-up" of social media pages. Something just like that happened in Virginia. An attorney in a wrongful death case had his paralegal advise the client husband to "clean up" photos from his Facebook account. Most photos were eventually recovered, and the jury was told about the photos that were scrubbed. While the plaintiff prevailed in the case, the judge ordered a $722,000 sanction to be paid for opposing side's legal fees—of which the plaintiff's attorney paid $544,000. In addition, the attorney resigned from his law firm and faced ethics charges for his conduct that violated ethics rules and for misconduct.[33]

Finally, attorneys should be sure to advise clients whether contacting others via social media is permitted during specific legal proceedings. For example, in the unreported New York case of *People v. Gonzalez*,[34] the defendant had an order of protection against her and was charged with criminal contempt in the second degree. The order included the restriction that she was not to communicate with or contact the protected party by electronic or other means.[35] Thereafter, the defendant "tagged" the protected party (and included comments) in two Facebook posts, and the protected party received notification of the posts from Facebook (recall our earlier discussion of automatic notifications—on that social media platform if someone is "tagged" on another's post, the subject usually receives a notification).

The *Gonzalez* Court determined that the tag on Facebook was a communication by electronic means, and therefore the court denied the defendant's motion to dismiss the charge against her (related to violation of the order) as facially insufficient.[36] The court also cited to a 2014 decision of the New York Court of Appeals, which held that a defendant's contact with a witness via Facebook messages was essentially like email. That case concerned a prosecution for witness tampering.[37]

4.5 – Disbarment Is a Potential Penalty for Egregious Violations

Attorneys take care! Obligations under the rules and caselaw are serious business—both for social media activity and any other conduct undertaken by an attorney. Repeated failures, or

---

[32] *See* J. Cohen & J. Bernard, *Changing a Criminal Client's Social Media Pages*, N.Y. L.J. at 3, 9 (Dec. 8, 2015); *see also* NYSBA ComFed *Social Media Guidelines*, Guideline 5.B.

[33] *See* M. Neil, *Lawyer Who Urged Client to Clean Up Facebook Page Has Paid $544K Legal Bill; Now Faces Ethics Case*, ABA JOURNAL (online July 11, 2013), available at: http://www.abajournal.com/news/article/lawyer_who_encouraged_client_to_clean_up_facebook_page_paid_544k_legal_bill; the issues resulted from actions in the litigation in *Allied Concrete Co. v. Lester*, 285 Va. 295, 736 S.E.2d 699 (Va. 2013).

[34] Index No. 15-6081M, NYLJ 1202746987949 (Sup. Ct. Westch. County Jan. 4, 2016).

[35] *Id.*

[36] *Id.*

[37] *Id.* (citing *People v. Horton*, 24 N.Y.3d 985 (2014)).

egregious behavior, can result in disciplinary sanctions up to and including disbarment. Take the case of a 35-year veteran attorney who was disbarred for serial, and serious, violations of the rules of procedure. In 22 cases over seven years, the attorney abused the rules concerning subpoenas, made frequent misrepresentations in motions to compel and for contempt, frequently noted depositions in the wrong venue, failed to make good-faith attempts to resolve discovery disputes and then prevaricated, misrepresented the contents of court orders, and misrepresented the required terms of expert compensation.[38] The disbarred attorney also made misrepresentations to the targets of subpoenas, as well as to the courts on motions to compel.[39] According to the decision, it appeared those were not innocent or unintentional mistakes.[40]

4.6 – Conclusion

As is clear from the material in this chapter, attorneys must familiarize themselves with the rules of professional conduct, and related ethical opinions, in their jurisdiction. While attorneys may utilize social media and electronic resources to research witnesses and opposing parties, their activities are neither unfettered nor unbounded. Within the rules of the respective jurisdictions, attorneys are limited as to whom they may contact, when they may do so, and what methods they may employ. It would behoove them to know the playing field before starting the game.[41]

Chapter End Questions

Note: Answers may be found on page 262.

1. Which of the following rules of professional conduct govern attorneys' use of social media and social media discovery?
   a. N.Y. RPC 3.1 & 3.3
   b. N.Y. RPC 4.1 & 4.2
   c. N.Y. RPC 4.3 & 5.3
   d. All of these Rules

---

[38] *Attorney Grievance Comm'n v. Mixter*, 441 Md. 416 (2015).

[39] *Id.*

[40] *See also* Charles S. Fax, *Serial Abuse of Discovery Rules Warrants Disbarment*, ABA LITIGATION NEWS (Vol. 41, No. 1, Fall 2015).

[41] For instance, attorneys who create social media posts (either as private persons or in their professional roles), may create ethical conflicts under the Rules. *See* S. Klevens & A. Clair, *Conflicts May Lurk In Social Media – Think Before Posting*, N.Y. L.J. at 5 (Aug 27, 2019) (citing, *inter alia,* Rule 1.7 & D.C. Ethics Opinion 370 (Nov. 2016)).

2. Which of the following may attorneys do with regard to online research concerning parties or witnesses who are represented by legal counsel?

   a. Send direct friend requests to see the private pages of the party or witness.

   b. Use another person, such as a paralegal, to "friend" the party or witness in order to see private pages.

   c. View public pages only, as if passively viewing a magazine.

   d. None of these may be done ethically.

3. Which of the following may an attorney advise a client to do concerning social media posts during or in anticipation of a lawsuit?

   a. Advise the client to take down, delete and destroy all information on the social media account.

   b. Advise the client to close/delete the account immediately, before any information is permanently saved.

   c. Advise the client to allow the social media account to go dormant/unused so that the service provider deletes the account and all material before it can be used in the lawsuit.

   d. None of these selections are ethical.

4. What is an important take-away point from the New York case of People v. Gonzalez?

   a. Social media is never a source of good information in a case.

   b. Social media evidence is almost never allowed in a case.

   c. Courts always take judicial notice of social media evidence.

   d. Depending on the terms of an order of protection, contact over social media could violate the order and result in punishment.

5. Which of the following is a potential sanction faced by attorneys who commit egregious violations of the ethical rules?

   a. Suspension from practice for a period of time.

   b. Disbarment.

   c. Imposition of sanctions such as fines or community service.

   d. All of above are potential sanctions, depending on the level and frequency of violations.

6. Which of the following ethical opinions provides guidance for attorneys in the area of social media and discovery?

   a. Cal. Bar Ass'n Formal Op. 2015-193

   b. N.Y.S. Bar Ass'n Op. 843.

   c. Phila. Bar Ass'n Op. 2009-02

   d. All of these opinions.

7. True or False: Attorneys, as officers of the court, have obligations and restrictions on their online activities that are different from those of an ordinary Citizen.

    a. TRUE

    b. FALSE

8. True or False: Attorneys may not "friend" or send "friend requests" to any witness in a legal action in which they represent a party.

    a. TRUE

    b. FALSE

9. True or False: Attorneys may not "friend" or send "friend requests" to the other party, even if that party is not represented by an attorney (is pro se).

    a. TRUE

    b. FALSE

10. True or False: An attorney may not advise clients to post positive information about themselves on social media while a litigation is pending.

    a. TRUE

    b. FALSE

CHAPTER 5

Ethical and Legal Considerations for Social Media: Jurors and Juries

5.1 – Introduction

Attorneys and litigants are not the only ones to whom rules and laws concerning social media are applicable when it comes to legal proceedings. Anyone who receives a jury summons, anyone who teaches about civic obligations and jury service, any attorneys who work as litigators selecting juries in state or federal trials, and any paralegals[1] assisting those attorneys should read this chapter and take its information to heart.

Some jurors across the country have taken to the Internet and social media before and during deliberations, and sometimes even during *voir dire*, whether looking to Wikipedia for definitions, conducting other outside research, reviewing news reports, or communicating with others. There have been requests for mistrials—which are not often granted. Social media is changing our legal system, and presenting new and different challenges to which the system must adapt and address.

> *Juror* – A member of the public chosen to sit in judgment on a legal matter; no legal degree or education required. Jurors are the judges of the facts, while trial judges are the judges of the law.

> *Voir dire* – The process of questioning potential jurors and selecting the jury panel that will hear and decide the facts and outcome of a case when either a civil or criminal jury trial is held.

---

[1] Paralegals have professional ethical responsibilities, in that they cannot do anything on behalf of an attorney that the attorneys are prohibited from doing. Further, attorneys should take care to supervise the activities of all non-attorneys acting on their behalf, to ensure they are not running afoul of the ethics rules. *See e.g.,* N.Y. Rules of Prof'l Conduct 5.3; ABA Model Rules of Prof'l Conduct 5.3; Indiana Rules of Prof'l Conduct 5.3; Wyoming R. State Bar, Att'y Cond. & Prac. 5.3.

> *Jury Instructions* – Information given by judges to jurors concerning their obligations, and the law(s) applicable to the case before them.

5.2 – The Background and Rules/Instructions for Jurors

Starting in the first decade of the new millennium (the 2000s), federal courts began employing new model or proposed jury instructions for civil and criminal cases. The instructions direct jurors not to research matters before them on the Internet or social media.[2] Jurors are also instructed not to use BlackBerry, iPhone (or other cell phone), email, Twitter, Facebook, LinkedIn, YouTube, or other technology to communicate with others during trial or deliberations.[3] The instructions are to be given at the start of trial, during trial, and before deliberations, and the federal district judges and federal magistrate judges conducting the trials are, of course, free to adopt, adapt, or modify the language of specific model or proposed guideline instructions as needed and appropriate, and in consultation with counsel for the parties at charge conferences, after receiving the proposed charges.

In May 2009, and again in February 2016, the State of New York similarly modified its preliminary jury admonitions in criminal matters to include specific mention of the prohibition on outside research of legal or factual issues by jurors, and the prohibition on juror communications with anyone via any means, including social media.[4] These kinds of preliminary instructions to a jury in a criminal case are actually required by New York State Criminal Procedure Law,[5] while the remainder of the jury instructions—as in the federal system—are advisory and guidelines. Several other states have adopted similar guideline jury instructions, admonishing jurors about their use of the Internet or social media to either conduct research or discuss a case prior to verdict—and in some cases during a recess in jury selection before a jury is even empaneled.[6]

---

[2] *See e.g.*, Manual of Model Criminal Jury Instructions for the District Courts of the Ninth Circuit, Model Instruction 1.8—Conduct of the Jury; Model Instruction 2.1—Cautionary Instruction—First Recess (2010 ed.); Pattern Jury Instructions (Criminal Cases)—*Fifth Circuit*, at 1.01—Preliminary Instructions, *Conduct of the Jury*; Proposed Model Jury Instructions: The Use of Electronic Technology to Conduct Research on or Communicate About a Case (2012 ed.), available at: uscourts.gov/sites/default/files/jury-instructions.pdf; s*ee also* Pattern Criminal Jury Instructions for the District Courts of the First Circuit—*District of Maine Internet Site Edition*, at 1.07—Conduct of the Jury, paragraph Sixth ("do not do any research on the internet about anything in the case or consult blogs or dictionaries or other reference materials, and do not make any investigation about the case on your own") (updated June 24, 2019), available at: http://www.med.uscourts.gov/pdf/crpjilinks.pdf; *Oracle Am., Inc. v. Google Inc.*, 172 F.Supp.3d 1100, 1102 (N.D. Cal. 2016) (citing Judicial Conference Committee on Court Administration and Case Management, *Proposed Model Jury Instructions: The Use of Electronic Technology to Conduct Research on or Communicate about a Case* (June 2012)).
[3] *Id.*
[4] *See* N.Y. Preliminary Jury Instructions—Criminal, Required Jury Admonitions (May 2009 & Feb. 2016), *available at*: nycourts.gov/judges/cji/1-General/CJI2d.Jury_Admonitions.pdf.
[5] *See* N.Y. C.P.L. 270.40.
[6] *See, inter alia*, North Dakota Civil Jury Instructions, Introductory Instructions, C-1.57 Duties of Jurors and Conduct, *available at*: http://ndpji.casemakerlibra.com/home/browserulesnew.aspx?login=ndpjiautologin@ndpji.com&st=NDJI&pd=f94b144bfc22f02764859890477b0c802eed39054268e3e318c6706fb4f243bb&bookid=970

5.3 – Some Jurors Ignore the Rules/Instructions, Creating Legal Issues

Disregard for the instructions given to a jury reflects poorly on the administration of justice and may serve to call a verdict into question in the "court of public opinion," even if the courts of law affirm the jury's verdict on legal grounds. Furthermore, there are times when severe juror misconduct can result in a mistrial and need for a new trial. Certainly, if nothing else, the conduct at least requires additional time and resources to address motions by counsel for parties seeking mistrials or other curative actions by judges overseeing the cases.

However, it is important to keep in mind that just because a juror tweets during a trial, it does not necessarily mean that a mistrial is automatically warranted. In the federal district court for the Southern District of New York, during a criminal trial in which the defendants were accused of conspiracy to commit immigration fraud, two jurors tweeted the following— *during* the trial: "Add in just one song & dance number and this federal case would rival anything I've seen on broadway #juryduty rocks," "The bloody courtroom is freezing," "one of the defense [attorneys] had the balls in his summation to say that his client was just trying to help poor refugees from rural communities who didn't have the education to get asylum on their own. At $11k a pop," and "these people prey on the fear and ignorance of applicants. it's [sic] horrible."[7]

The defense argued that the tweets provided evidence of bias by the jurors, disregard for the judge's instructions, and a violation of the defendants' Sixth Amendment right to a trial by an impartial jury. Following the return of a guilty verdict, the defense sought a new trial. The trial judge, though, denied the defenses' request for relief.[8] The jurors in question in this matter were Jurors 2 and 10. As discussed by the Court, Juror 10 was dismissed from service because of her tweets, which rose to the level of improper and inappropriate conduct.[9] Juror 2, however, was not found to have violated rules and was therefore not dismissed.[10]

The judge, further, did not grant a mistrial, stating that regardless of the particular issue, the court must assess the "likely prejudicial impact" of the potential juror misconduct. The judge thereafter held that jurors who tweet during a trial may in certain circumstances threaten a defendant's Sixth Amendment right to trial by an impartial jury; however, according to the court Juror 2 did not fail to honestly answer *voir dire* questions during the selection process in

---

6&p=8838; OKLAHOMA UNIFORM JURY INSTRUCTIONS—CIVIL, Instruction No. 1.0—Use of Electronic Devices and Research Prohibited, *available at*: http://www.oscn.net/applications/oscn/deliverdocument.asp?id=479101& hits=483+474+472+73+64+62+; COLORADO JURY INSTRUCTIONS CRIMINAL, B:02—Admonition Prior to Recess During Jury Selection; B:06—Admonition About Conduct During Trial (2019), available at: https://www.courts .state.co.us/Courts/Supreme_Court/Committees/Committee.cfm?Committee_ID=9. For more on jury instructions, including whether courts should admonish jurors that their social media (public) may be monitored by attorneys and consultants, see Mark Berman, *Social Media, Jurors and Jury Instructions*, N.Y.L.J. at 5 (Nov. 5, 2019).
[7] *U.S. v. Liu*, 69 F.Supp.3d 374 (S.D.N.Y. 2014).
[8] *Id.*
[9] *Id.*
[10] *Id.*

the way they were posed. Additionally, Juror 2 did not violate the court's instructions because Juror 2 did not tweet about *facts and circumstances* in the case, and Juror 2 was not failing to pay attention because she was concerned about her novel-writing. In fact, the court believed if anything, Juror 2 demonstrated command of the facts and rapt attention.[11] Although the court did note that vigilance is warranted in this new social media age, it was found that Juror 2 in the *Liu* case was not biased or inattentive, and the court denied the defendants' Rule 33 motion for a new trial.

In another case, it was discovered, after the juror was seated on the panel, that he had been less than forthright in answering questions about his feelings concerning matters in the case.[12] The discovery occurred because social media posts that the juror had previously made came to light. A motion was made to remove the juror, and the trial court subsequently denied the motion. On appeal, the majority of the appellate court affirmed.[13] However, the dissent wrote strongly about how the social media posts demonstrated the juror was not forthright in answering *voir dire* during jury selection and therefore should have been removed from the trial panel. The dissenting judge would have quashed the judgment and remanded for a new trial.[14]

Thus, we see that juror social media activity—whether before or during a trial—can create both anticipated and unforeseen consequences and problems for the legal system.

In response, an article in 2016 discussed legislation that had been proposed in California, which would have authorized judges to fine jurors up to $1,500.00 for social media or Internet use that violated instructions, leading to mistrials or overturned convictions.[15] Those sanctions aside, jurors are usually reprimanded and/or dismissed from service if the violation is serious enough. The article also reported that a 2011 state law in California makes improper electronic and wireless communications or research by jurors punishable by contempt of court—carrying with it more severe potential punishments for the juror.[16]

---

[11] *Id. See also United States v. Aiyer*, — F.Supp.3d —, 2020 WL 223619 at *4 (S.D.N.Y. Jan. 15, 2020) ("The fourth category of allegations related to Juror No. 4's social media use also does not raise any concerns that necessitate a post-verdict inquiry. In certain circumstances, a juror who connects with other jurors over social media or makes public comments about the trial over social media may threaten a defendant's Sixth Amendment right to an impartial jury .... However, the use of social media alone is not, without more, prejudicial to the defendant because '[a] mistrial or other remedial measure is required only if juror misconduct and actual prejudice are found.'... Nothing that Juror No. 4 discussed during his mid-trial podcasts could be construed as indicating any bias towards the defendant and it did not amount to prejudice.") (citing, *inter alia, Liu*, 69 F.Supp.3d at 385-386).

[12] *See R.J. Reynolds Tobacco Co. v. Allen*, 228 So.3d 684 (Fla. Dist. Ct. Apps. 1st Dist. 2017) (rehearing then granted), *review denied* 2018 WL 798394 (Fla. Feb. 9, 2018).

[13] *Id.*

[14] *Id. See also U.S. v. Stone*,—F. 3d—, 2020 WL 1892360 (D.D.C. Apr. 16, 2020) (Defense had multiple opportunities to question, examine, strike potential juror; absence of due diligence and juror open to questions; will not upset verdict).

[15] *California Considers Fining Jurors Who Google or Tweet about Cases*, THE GUARDIAN (Apr. 24, 2016), available at: https://www.theguardian.com/us-news/2016/apr/24/california-jurors-social-media-fines-twitter-google-jury-duty.

[16] *Id. See also Steiner v. Superior Court*, 220 Cal.App.4th 1479, 1492–1493 (Cal. Ct. App. 2d Dis. 2013); CAL. CODE CIV. P. § 611; CAL. PENAL CODE § 166(a)(6). Although it is interesting to note that the appellate court in the

There have been several cases, though, in which courts have imposed the severe remedy of throwing out convictions and granting new trials because of juror misconduct online. Take a case in Arkansas, where the conviction of the defendant for capital murder and aggravated robbery—and a concomitant sentence of death—was reversed and remanded, with instructions for a new trial to be held.[17] The trial court denied the defense's requests for relief. The Arkansas Supreme Court, on appeal, noted that one of the things that was quite disturbing was the fact that Juror 2—the juror in question—not only tweeted during the trial but had continued to tweet *after* the court raised questions about the juror's tweeting.[18] The Supreme Court stated that although defendants are "entitled to a fair trial, not a perfect trial," in the case at issue the juror had admitted to disregarding the trial court's instructions and prohibitions. The Court continued:

> Appellant was denied a fair trial in this case where Juror 2 disregarded the circuit court's instructions and tweeted about the case and Juror 1 slept through part of the trial.

> Finally, we take this opportunity to recognize the wide array of possible juror misconduct that might result when jurors have unrestricted access to their mobile phones during a trial. Most mobile phones now allow instant access to a myriad of information. Not only can jurors access Facebook, Twitter, or other social media sites, but they can also access news sites that might have information about a case. There is also the possibility that a juror could conduct research about many aspects of a case. Thus, we refer to the Supreme Court Committee on Criminal Practice and the Supreme Court Committee on Civil Practice for consideration of the question of whether jurors' access to mobile phones should be limited during a trial.[19]

Therefore, because a juror refused to adhere to the court's instructions and could not resist the apparent ever-present temptation to engage in online activities, a defendant's conviction was thrown out, and the prosecution, defense, and court system—together with a new panel of jurors—had to expend additional time, energy, and resources to retry the case.

Consider also an unpublished decision in the California case *People v. Fernandez*.[20] The case was a criminal matter in which a juror disregarded a standing instruction of the court and used the Internet to research an issue related to the case on which the juror was serving. The defendant, an investment firm CEO, was on trial accused of defrauding an investor, but the defendant blamed the accountant working for the firm.[21] That was an issue in the case. One of the jurors—the foreperson, actually—looked up the accountant on the Internet and found

---

Steiner case determined it was a violation of the constitutional free speech rights of an attorney when the trial court directed said attorney to remove descriptions of two prior cases in which the attorney had been successful so that jurors who disregarded the trial court's prohibitions against outside research would not see them. The trial court had other tools to address juror misconduct if a juror were disobedient.

[17] *Dimas-Martinez v. State*, 2011 Ark. 515, 385 S.W.3d 238 (Ark. 2011).

[18] *Id.* at 247.

[19] *Id.* at 247–249.

[20] 2016 WL 146562 (Cal. Ct. App. 4th Dis. Jan. 13, 2016) (unpublished).

[21] *Id.* at *1.

information concerning a prior matter in which it appeared the defendant CEO and accountant had defrauded an investor. The information did not come out at the defendant CEO's trial before the jury on which this juror sat, but undoubtedly impacted the thinking of the juror in question.

The trial judge, after interviewing all jurors, held there was no actual prejudice and sustained the conviction of the defendant.[22] The appellate court, however, held that the juror's actions created a presumption of prejudice and reversed the conviction, returning the case to the trial court for a new trial. "In cases of juror misconduct by receiving extraneous material, prejudice is presumed. If the material, 'judged objectively, is inherently and substantially likely to have influenced the juror,' the prejudice cannot be rebutted. It is a per se reversal .... So we must reverse the judgment of conviction and remand the matter for a new trial."[23]

Once more, because of a juror's disregard for the instructions of the court and surrender to the siren call of the ever-present Internet, the prosecution, defense, and court—together with a new panel of jurors—had to expend additional time, energy, and resources to retry the case.[24]

Finally, with regard to this discussion we have a case out of New York State, from October 2019, wherein that state's highest court (the Court of Appeals) issued a unanimous decision reversing a murder conviction and ordering a new trial—with "full observance and enforcement of the cardinal right of a defendant to a fair trial" even in the face of what was argued to be overwhelming proof of guilt.[25] *People v. Neulander* can be included in the growing list of cases where juror misconduct involving text messages, social media, online research, or other electronic/online activities is leading to reversals of convictions. Because of its great importance to the topic of this chapter, the text of the *Neulander* decision is provided herein, in its entirety, for greater consideration.

People v. Neulander (New York State Court of Appeals)

Wilson, Associate Judge

The issue before us is whether the undisputed juror misconduct in this case warrants a reversal of the judgment convicting Dr. Neulander of murder and evidence tampering. The Appellate Division concluded that the trial court abused its discretion by denying his CPL 330.30 motion to set aside the verdict against him based on that juror misconduct. We agree that, on this record, he is entitled to a new trial. "Nothing is more basic to the criminal process than the right of an

---

[22] *Id.*

[23] *Id.* (citation omitted).

[24] How to deal with "rogue" or "stealth" jurors has become a hot topic; and the stakes for the Justice System have not been higher. *See* Gail Prudenti, *Dealing with Rogue and Stealth Jurors,* N.Y.L.J. at 6 (Feb. 3, 2020). *See also* M. Berman, *Social Media, Jurors, supra.*

[25] *People v. Neulander,* 34 N.Y.3d 110, 115, 111 N.Y.S.3d 259, 305 (2019).

accused to a trial by an impartial jury" (*People v. Branch*, 46 N.Y.2d 645, 652, 415 N.Y.S.2d 985, 389 N.E.2d 467 [1979]). We therefore affirm.

In April 2015, a jury convicted Dr. Neulander of murdering his wife and tampering with physical evidence. Throughout the trial, one of the jurors, Juror 12, sent and received hundreds of text messages about the case. Certain text messages sent and received by Juror 12 were troublesome and inconsistent with the trial court's repeated instructions not to discuss the case with any person and to report any attempts by anyone to discuss the case with a juror. Juror 12 also accessed local media websites that were covering the trial extensively. In order to hide her misconduct, Juror 12 lied under oath to the court, deceived the People and the court by providing a false affidavit and tendering doctored text message exchanges in support of that affidavit, selectively deleted other text messages she deemed "problematic," and deleted her now-irretrievable internet browsing history. The cumulative effect of Juror 12's extreme deception and dishonesty compels us to conclude that her "improper conduct ... may have affected a substantial right of defendant" (CPL 330.30[2]).

Defendant moved, pursuant to CPL 330.30, to set aside the verdict based on juror misconduct. Upon conclusion of a fact-finding hearing, the trial court determined that Juror 12 consciously engaged in misconduct and prevarication, but nonetheless believed her misconduct did not render the trial unfair. The Appellate Division, with two Justices dissenting, reversed the judgment and granted a new trial, with the majority observing that "every defendant has a right to be tried by jurors who follow the court's instructions, do not lie in sworn affidavits about their misconduct during the trial, and do not make substantial efforts to conceal and erase their misconduct when the court conducts an inquiry with respect thereto" (*People v. Neulander*, 162 A.D.3d 1763, 1768, 80 N.Y.S.3d 791 [4th Dept. 2018]). We agree that the extensiveness and egregiousness of the disregard, deception, and dissembling occurring here leave no alternative but to reverse the judgment of conviction and remit for a new trial and compel us to affirm publicly the importance of juror honesty.

Under CPL 330.30, a verdict should be set aside if "improper conduct by a juror ... may have affected a substantial right of the defendant and ... was not known to the defendant prior to the rendition of the verdict." Of course, "not every misstep by a juror rises to the inherently prejudicial level at which reversal is required automatically" (*People v. Brown*, 48 N.Y.2d 388, 394, 423 N.Y.S.2d 461, 399 N.E.2d 51 [1979]); "[e]ach case must be examined on its unique facts" (*People v. Clark*, 81 N.Y.2d 913, 914, 597 N.Y.S.2d 646, 613 N.E.2d 552 [1993]). Under the extraordinary circumstances of this case, we conclude that Juror 12's behavior "may have affected a substantial right of defendant." This is not a case of stray texts or inadvertent misstatements. The record plainly supports the findings of both lower courts that Juror 12's conduct disregarded the court's plentiful instructions as to outside communications and when such conduct was brought to light, the juror was deliberately and repetitively untruthful. During the third day of jury deliberations, when the court first examined Juror 12 as to whether she had violated the court's instructions concerning outside communications, she insisted that she had not, which later proved to be false. When, after the jury had rendered its verdict, an alternate juror advised the court, by affidavit, that Juror 12 had engaged in improper conduct, Juror 12 secretly and selectively deleted numerous text messages which she believed to be "problematic," and presented to the People the remaining portions of the text message exchanges as if they were complete recitations of the communications. Juror 12 thereby induced the People to submit to the court her exculpatory affidavit, attaching the fabricated versions of her conversations with no indication that she had doctored them.

When, at the CPL 330.30 hearing, she was examined about her selective deletions, which were uncovered through a forensic examination of her cell phone, Juror 12 gave several answers evidencing a continued lack of candor. Although she deleted her entire Internet browsing history and could not explain why, when confronted with evidence that she had accessed a website providing daily trial coverage, she testified that she read nothing about the trial and probably accessed the site to read an article on cheerleading. The People and defense counsel stipulated, however, that no such article appeared on that site during the relevant time. She several times accessed a second news website providing trial coverage, as to which she offered no explanation.

In this case, a sworn juror, when examined by the court about the breadth of her outside communications was repeatedly and deliberately untruthful about the scope of that misconduct and affirmatively sought to conceal evidence of her misconduct. That extraordinary and dishonest behavior by a juror purposefully selected to be a fair and objective arbiter of the facts in the case causes irredeemable injury to the judicial system and the public's confidence in it. "Jurors, of course, do not live in capsules" (*Brown*, 48 N.Y.2d at 393, 423 N.Y.S.2d 461, 399 N.E.2d 51) and cannot be isolated during their service from the outside world, including their friends and families. However, they must be expected, at the very minimum, to obey the admonishments of the trial court, report attempts by others trying to influence their oath to be objective, and to be forthcoming during court inquiries into their conduct as a juror. Juror 12's blatant disregard for the court's instructions coupled with her purposeful dishonesty and deception when her actions and good faith as a juror came into question vitiate the premise that Juror 12 was fair and impartial and lead us to conclude that a new trial is required. On this record, the cumulative effect of Juror 12's misconduct, deceit, and destruction of evidence—conduct which obfuscated the full extent of her misconduct—compels us to agree with the Appellate Division that Juror 12's "improper conduct ... may have affected a substantial right of defendant" (CPL 330.30[2]), and, therefore, the trial court abused its discretion in declining to set aside the verdict.

The People contend that even if Juror 12 engaged in misconduct, "that misconduct is significantly outweighed by the substantial proof of guilt presented at trial." However, "[t]he right to a fair trial is self-standing and proof of guilt, however overwhelming, can never be permitted to negate this right" (*People v. Crimmins*, 36 N.Y.2d 230, 238, 367 N.Y.S.2d 213, 326 N.E.2d 787 [1975]). The "public at large is entitled to the assurance that there shall be full observance and enforcement of the cardinal right of a defendant to a fair trial" (*id.*). Affirming a conviction where a juror engaged in dishonesty of this magnitude would not discharge our "overriding responsibility" to ensure the public's confidence in the fairness of trials (*id.*).

Accordingly, the order of the Appellate Division should be affirmed.

Chief Judge DiFiore and Judges Rivera, Stein, Fahey, Garcia, and Feinman concur.

Order affirmed.[26]

5.4 – Jurors Who Are Social Media "Friends" with Litigants/Parties or Others

In Chapter 6 of this book, we address a question that arises when judges are social media friends with attorneys, litigants, and litigants' families. The same issue can be a concern for jurors.

In a 2016 case out of Indiana, a criminal defendant was convicted of rape, and the defense moved for a mistrial due to juror misconduct.[27] During *voir dire*, one juror had denied knowing the victim or her family. However, it turned out that a relative of the victim was a Facebook "friend" of that very juror.[28] The juror argued that as a Realtor, she had 1,000 friends, most for networking, never recognized the name, did not recognize the victim when she testified, and did not know the family.[29]

The trial court denied the defense's motion for mistrial and determined that the juror had been truthful during *voir dire*, thus providing no basis for mistrial. On appeal, the court of appeals affirmed, "noting the novel issue involving a juror's 'expansive list of Facebook friends.'" Thereafter, the Supreme Court of Indiana affirmed and adopted the opinion of the Court of Appeals: the number of Facebook (or other social media) friendships one has may be so numerous as to not necessarily form the basis for what one may consider a "friendship" in the physical, non-virtual world. If the social media "friendship" is no more than merely appearing on a list of hundreds or thousands of contacts in a social media account, without more direct contact, it will likely not present a problem in cases such as those similar to *Slaybaugh*.

The *Slaybaugh* case stands as an important lesson for developing as much information as possible in *voir dire* if you are a litigator, and to ask questions that are both detailed and expansive—providing opportunity to obtain specific information of use, yet also casting a wide-enough net so as not to miss information being sought. In the *Slaybaugh* case, though, that may not have been possible; since the juror had so many friends, he or she testified they were not aware of the particular ones who were related to the victim in the case on which the

---

[26] *Id.*

[27] *Slaybaugh v. State*, 47 N.E.3d 607 (Ind. 2016).

[28] *Id.*

[29] *Id.*

juror was serving. No amount of questions, however constructed, could have garnered further information in *voir dire*.

Then we can examine the case of *Sluss v. Commonwealth of Kentucky*.[30] The Supreme Court of Kentucky addressed the issue of a juror who was Facebook "friends" with the victim's mother. The case was, according to the court, a matter of first impression in their jurisdiction. The accusation of the defendant on appeal (among 15 other issues raised) was that two jurors had lied during *voir dire* concerning their use of social media—specifically Facebook.

In *Sluss*, the facts of the crime—a motor vehicle accident that killed an 11-year-old—were of such a public nature that many potential jurors were stricken. The court established an extensive *voir dire* process, during which as many as 50 potential jurors were excused.[31] The questions in *voir dire* (both of the panel and individual potential jurors) did concern whether potential jurors had knowledge of the case and/or knew the victim or the victim's family. Of the two jurors in question, one went on to serve as the foreperson of the jury. She answered that she had seen something on social media about the case; she was not asked specifically if she was friends with the victim or victim's family and did not volunteer any information.[32] The other juror in question responded that she was not on Facebook.[33]

Following trial, the trial court was presented with evidence that both of the jurors were Facebook "friends" with the victim's mother. The problem, then, appeared to be that the jurors were not honest during *voir dire* and thus created an issue of bias.[34] As discussed by the Supreme Court:

> The problem in this case, however, is the jurors also appear to have made misstatements during voir dire. During individual voir dire, [the one juror] responded that she was not on Facebook when asked by the trial court, strongly suggesting that she was not even a member of the website. The Appellant thus had no reason to explore the extent of the Facebook interaction between [the] juror [ ] and [the victim's mother] because he could reasonably believe, given her statement made under oath, that [the] juror [ ] was not a member of the website. Further, both juror[s] did not respond during general voir dire to the question about knowing any of the victims or their families. The Appellant thus had no reason to explore the extent of any possible relationship between them because he could reasonably believe, given that they were under oath, that they did not know the mother. In essence, Appellant was foreclosed from conducting proper voir dire of the jurors.[35]

However, the Court stated: "[t]his is further complicated by the fact that a person can become 'friends' with people to whom the person has no actual connection, such as celebrities

---

[30] 381 S.W.3d 215 (Ky. 2012) (distinguished by *Gore v. Commonwealth*, 2015 WL 4984397 (Ky. Aug. 20, 2015) (wherein no evidence presented to support petition to challenge juror *voir dire*)).

[31] 381 S.W.3d at 221–222.

[32] *Id.*

[33] *Id.*

[34] *Id.* at 223–224.

[35] *Id.*

and politicians. . . . Thus, a Facebook member may be 'friends' with someone in a strictly artificial sense."[36] The Court cited to the instance of Lady Gaga and her 10 million Facebook "friends."[37]

> *Strike a Potential Juror* – When a party/attorney eliminates a potential juror from sitting on a case; either a cause challenge (i.e. bias) or peremptory challenge (no reason necessarily needed or given). Challenges are limited in number when peremptory.

The Supreme Court remanded the case to the Circuit Court to determine the extent to which two jurors answered *voir dire* questions truthfully concerning Facebook and the extent of exposure to the Facebook account of the victim's mother[38]—and therefore going to the heart of whether the defendant was tried by a fair and impartial jury.

On remand, the Circuit Court in *Sluss* held two hearings and determined that the two jurors should not have been struck for cause. Therefore, the court determined that the defendant was not entitled to a new trial.[39] On the appeal from that decision, the Supreme Court addressed the Facebook issue, as well as the other 15 issues raised on the first appeal, and reversed the conviction and sentence—but on grounds other than the Facebook issue—so we have nothing further to elucidate our discussion in that regard.[40]

5.5 – Preserving Challenge to a Juror

Here, we pause for a brief side note related to our discussion in this chapter—the procedure if there is a challenge to a juror's service.

If an attorney believes there is reason to strike a juror, then

> In order to preserve the argument that a trial court committed reversible error by failing to strike a juror a litigant must do the following. First, the litigant must move to strike the problematic juror for cause and be denied the strike by the trial court. Then, the litigant must use a peremptory strike to remove the juror from the venire and show in writing on the strike sheet that the peremptory strike was used for that juror, and exhaust all other peremptory strikes. Next, the litigant is required to clearly write **on her strike sheet** the juror she would have used a peremptory strike on had she not been forced to use the strike on the juror that she believes should have been struck for cause. By requiring this strict compliance with *Gabbard*, we now overrule *Sluss* [*v. Commonwealth*, 450 S.W.3d 279 (Ky. 2014)] prospectively, only insofar as it holds that stating would-be peremptory strikes orally on the record constitutes substantial

---

[36] 381 S.W.3d at 224.

[37] *Id.* at 222 & n.8.

[38] *Id.*

[39] 450 S.W.3d 279 (Ky. 2014) (overruled on other grounds by *Floyd v. Neal*, 590 S.W. 3d 245 (Ky. 2019)).

[40] *Id.* For more on tips for effective *voir dire* in the age of social media, see Anne E. McClellan, *#JuryDuty: How to Use Social Media to Research and Monitor Jurors*, ABA Journal (online Apr. 20, 2020), *available at* https://www.americanbar.org/groups/litigation/committees/trial-practice/articles/2020/spring2020-jury-duty-social-media-monitor-research-jurors/.

compliance with *Gabbard* and is therefore sufficient to preserve the error .... Requiring both sides to make their peremptory strikes concrete by writing them down prior to the parties discussing their strikes with the court safeguards the fairness of this process.

The next requirement to preserve a for cause strike error has never been addressed directly by this Court. Specifically, that the number of jurors a litigant identifies on her strike sheet must be the same number of jurors the litigant originally moved to strike for cause. Failure to abide by this rule will render the error unpreserved.

...

The final box a litigant must check in order to preserve a for cause strike error is to make her would-be peremptory strikes known before the jury is empaneled. And, as already discussed, at least one of the jurors identified by the litigant must ultimately sit on the jury. We also reiterate our previous holding that all of the preceding rules apply in both civil and criminal jury trials.[41]

That is the procedure in Kentucky.

In New York,

[a]ll issues of fact or law arising on the challenge must be tried and determined by the court. If the challenge is allowed, the court must exclude the person challenged from service. An erroneous ruling by the court allowing a challenge for cause by the people does not constitute reversible error unless the people have exhausted their peremptory challenges at the time or exhaust them before the selection of the jury is complete. An erroneous ruling by the court denying a challenge for cause by the defendant does not constitute reversible error unless the defendant has exhausted his peremptory challenges at the time or, if he has not, he peremptorily challenges such prospective juror and his peremptory challenges are exhausted before the selection of the jury is complete.[42]

In civil cases, again, raise the issues that arise to the trial court and make objections as needed.

Always remember the record, and preserve all issues for appeal by raising them in the trial court. If an issue is not placed on the record, it *has not* been preserved for any appellate review. That means an attorney or the assisting paralegal should ensure that the court reporter's record at the end of the trial day reflects any and all issues arising that day for which a party wishes to preserve an issue for appellate review. *Important:* if the court reporter is not present, and typing, during jury *voir dire*, during a conference in chambers, or during a side-bar conference in the courtroom, the matter discussed *is not* on any record and *is not* preserved for appeal. The attorney must then take the affirmative step of raising the issue with the trial judge, with the court reporter present, so that the matter is preserved.

---

[41] *Floyd v. Neal*, 590 S.W.3d 245, 250-252 (Ky. 2019) (emphasis by Court).

[42] N.Y. CPL § 270.20(2); *People v. Tieman*, 132 A.D.3d 703 (App. Div. 2d Dep't 2015).

5.6 – What Contact May Attorneys Have with Potential Jurors? What Information May Attorneys Gather in Voir Dire (Jury Selection) and Beyond?

The answer to the first question—what contact may attorneys have with potential jurors—is relatively easy: *none*. However, attorneys are permitted to research background information on potential jurors utilizing the Internet and social media, as will be addressed further in a moment. So, the question of "what constitutes contact with a potential juror" is slightly more involved, because as long as an attorney complies with ethical requirements, attorneys and paralegals may gather any public information that is available about a potential juror.

The guidance for attorneys is similar to that for attorneys utilizing social media to research witnesses or parties—*except* an attorney *may not have contact in any way* or cause another person (such as a paralegal) to contact or send a message that would be received by a juror or potential juror (including a "friend" request).[43] The New York County Lawyers Association, in Opinion 743, advises that attorneys (and anyone supervised by or working with an attorney) shall have no contact, no "friending," and no tweets with potential jurors. A juror must not become aware, or be made aware, of the monitoring an attorney or a paralegal is conducting as part of *voir dire*. Furthermore, a lawyer may not engage in deceit or misrepresentations, or cause others to do so on the attorney's behalf, in order to garner information from or about potential jurors. Finally, if an attorney becomes aware of misconduct or deliberations in violation of a court's instructions, the attorney is under an ethical obligation to advise the court under Rule 3.5 *before* the attorney engages in further activity.[44]

The American Bar Association (ABA), in its Formal Opinion 466, stated in part:

Unless limited by law or court order, a lawyer may review a juror's or potential juror's Internet presence, which may include postings by the juror or potential juror in advance of and during a trial, but a lawyer may not communicate directly or through another with a juror or potential juror.

A lawyer may not, either personally or through another, send an access request to a juror's electronic social media. An access request is a communication to a juror asking the juror for information that the juror has not made public and that would be the type of ex parte communication prohibited by Model Rule 3.5(b).

The fact that a juror or a potential juror may become aware that a lawyer is reviewing his Internet presence when a network setting notifies the juror of such does not constitute a communication from the lawyer in violation of Rule 3.5(b).[45]

---

[43] *See* N.Y. County Lawyers Ass'n Opinion 743 (2011). *See also* N.Y. City Bar Ass'n Opinion 2012–2 (guidance concerning the limits on an attorney using social media to research potential and sitting jurors).

[44] *Id. See also* NYSBA ComFed *Social Media Guidelines* 6.E (Juror Misconduct).

[45] American Bar Ass'n Formal Opinion 466 (4/24/2014); ABA MODEL RULES OF PROF'L CONDUCT 3.5, 4.4; ABA Ethics Tip – February 2017 (June 7, 2019) (collecting opinions & guidance).

Interestingly, the ABA Opinion appears to permit communication that is "automated," although it might permit a juror to know that they were viewed because it does not bar "automated" or passive responses to a juror—those not otherwise written or affirmatively initiated by an attorney or someone on the attorney's behalf and those not an explicit "friend" request.

ABA Opinion 466 imposes restrictions on attorneys in that a lawyer may not personally or through another send an access request (or "friend" request) to a juror. The Opinion stated that "would be akin to driving down the juror's street, stopping the car, getting out, and asking the juror for permission to look inside the juror's house because the lawyer cannot see enough when just driving past."[46] That is barred.

Some commentators have criticized one particular aspect of ABA Opinion 466, however, and one must be aware of the divide. In 2014, then co-chairs of the Social Media Committee of the New York State Bar Association Commercial & Federal Litigation Section (Mark A. Berman, Esq. and Ignatius Grande, Esq.), together with retired Federal Magistrate Judge Ronald J. Hedges, wrote an article stating that they openly disagreed with ABA Opinion 466.[47] In their view—which the author of this book shares—Opinion 466 is not sufficient to guide an attorney in searching the social media activities of a juror. The authors instead cite to NYCLA's Opinion 743, ABA Model Rule 1.1, and NYSBA *ComFed Social Media Ethics Guidelines* 5.B.[48] When searching the online activity of potential jurors, attorneys must ensure that the network they are utilizing does not send alerts to those potential jurors that the attorney has viewed their profile. Some online service providers and social media providers have settings whereby users are alerted whenever someone accesses or passively reviews their profile pages or postings. Although ABA Opinion 466 opines that such a passive "contact" or alert, without more, is not a violation, most others agree that an attorney or paralegal must make an affirmative change to settings—either disabling the notification feature on their end (in their own account settings) or setting the feature to advise that "anonymous" has viewed the potential juror's site.[49]

Generally, as a best practice, an attorney should not search a juror's social media in any fashion that provides a response or contact to the juror (automated or otherwise) that the attorney was looking at their account. An attorney must keep abreast of changes in the law and its practice, including technological developments and updates.[50]

---

[46] *Id.*

[47] M. Berman, I. Grande & R. Hedges, *Why ABA Opinion on Jurors and Social Media Falls Short*, 86 N.Y. St. Bar J. 52 (Sept. 2014) (originally printed in N.Y. L.J. (May 5, 2014)).

[48] *Id.*

[49] *See, generally, id.*; *see also Oracle Am., Inc. v. Google Inc.*, 172 F.Supp.3d 1100, 1105–1106 (N.D. Cal. 2016) (discussing different social media sites, settings features, and whether they notify account holders when others view the profile); Social Media Ethics Guidelines of the Commercial & Federal Litigation Section of the New York State Bar Association, Guideline 6.B (2019) (and comments).

[50] *See* ABA Model Rule 1.1; N.Y. Rule of Prof'l Conduct 1.1.

A few commentators believe that failing to perform some kind of background check on potential jurors through social media is close to malpractice. However, that opinion is certainly not uniform and will largely depend on time and resources. In addition, some courts and jurisdictions have placed an obligation on attorneys to perform at least some online research regarding juror backgrounds before a jury is empaneled and sworn, or else the issue of juror nondisclosure during *voir dire* is waived.[51] Other courts, however, have put restrictions on online research seeking the background of potential jurors—either prohibiting it or placing rules on the undertaking.[52] Therefore, as always, one should be familiar with the rules and procedures in their jurisdiction.

In addition, checking on jurors' backgrounds on social media raises the concomitant ethical issues discussed above. Again, some argue court admonitions to potential jurors should make them aware that attorneys and parties may be looking into the juror's background, social activities, political views, and even prior articles or blogs, simply because nothing is private on the Internet. The aim of this chapter and other resources is to do just that. No specific contact by attorneys, or those working for them, however, should inform the juror.

While it should not dissuade potential jurors from exercising their rights of free speech, it is something that potential jurors may find to be of interest, and may wish to be aware of during the course of their jury service.

5.7 – Conclusion

Therefore, information posted on public sites, or sites that the public may access, are fairly used in litigation information-gathering and to research the backgrounds of potential jurors in *voir dire*. Jurors and the general public should be aware of this. However, take care if you are an attorney gathering information or directing another to do so on your behalf. There are important ethical limitations with which to comply. It is important that attorneys and their

---

[51] *See Spence v. BNSF Railway Co.*, 547 S.W.3d 769 (Mo. 2018) (citing *Johnson v. McCullough*, 306 S.W.3d 551 (Mo. 2010) (obligation exists for "reasonable investigation" at least with regard to prior litigation history of potential jurors, and search of Case.net). *See also* MISSOURI SUPREME COURT RULE 69.025—Juror Nondisclosure.

[52] *See U.S. v. Norwood*, 2014 WL 1796644 (E.D. Mich. May 6, 2014) (citing, *inter alia*, *U.S. v. Kilpatrick*, 2012 WL 3237147 (E.D. Mich. Aug. 7, 2012); *U.S. v. Stone*, 2012 WL 113479 (E.D. Mich. Jan. 13, 2012)) (finding that an anonymous jury panel was required given that defendants were accused of undertaking violent and dangerous crimes as part of a criminal enterprise in the community; the alleged need for social media monitoring of the jury panel was outweighed by the competing concerns, and the jury venire was asked to complete lengthy questionnaires so that bias could be determined during *voir dire* to assure defendants a fair trial); *Oracle Am., Inc.*, 172 F.Supp.3d 1100 (holding that the parties either needed to voluntarily agree to a ban on online research regarding juror backgrounds or they had to follow specific procedures set by the Court concerning disclosure of their activities to each other and the jury venire; the Court's expressed concerns included the danger of personal appeals to jurors based on interests discovered in Internet searches, the privacy concerns of jurors, and the concern that if attorneys searched the jurors and the jurors found out, the jurors might then disregard the Court's instructions and perform Internet research on the attorneys, parties, and Court). *See also U.S. v. Chin*, 913 F. 3d 251 (1st Cir. 2019) (" 'Juror names and addresses must be made public' in 'the absence … of particularized findings reasonably justifying non–disclosure' ").

paralegals strictly comply with all ethical guidance and technological requirements, both during *voir dire* and during trial (through deliberations and verdict), both so as not to risk violation, a mistrial and/or ethical penalties, and to provide potential bases/foundation for appeal.[53]

Chapter End Questions

Note: Answers may be found on page 262.

1. When the Court gives jurors information on the law that they are to use in a case, that is called:

    a. Instructions
    b. Implications
    c. Innoculations
    d. Impressions

2. With regard to instructions given to jurors in federal cases concerning outside social media and Internet research, it is recommended that the district judge or magistrate judge provide the instructions at what stage of the trial?

    a. At the start of trial.
    b. During trial and before breaks.
    c. At the end, before deliberations.
    d. At all of these times.

3. Which of the following jurisdictions have specific model instructions judges may utilize to instruct jurors on social media and Internet research?

    a. Fifth and Ninth Circuit U.S. Courts of Appeals
    b. States of North Dakota and Oklahoma
    c. States of Colorado and New York
    d. All of these jurisdictions.

4. Which of the following states considered proposed legislation to punish jurors for social media activity that resulted in a mistrial?

    a. New York
    b. Kansas
    c. California
    d. New Jersey

---

[53] *See, e.g.,* Social Media Ethics Guidelines of the Commercial & Federal Litigation Section of the New York State Bar Association, Guidelines, 6.A, 6.B, 6.C, 6.D, & 6.E, and comments thereto (updated June 2019). *See also, inter alia,* N.Y. Rules of Prof'l Conduct 3.5, 4.1, 5.3, 8.4; Colo. Bar Ass'n Ethics Comm., Formal Op. 127 (2015); D.C. Bar Legal Ethics Comm., Formal Op. 371 (2016); N.Y. City Bar Ass'n Comm. on Prof'l Ethics, Formal Op. 2012–2 (2012).

5. What is an attorney required to do under the ethical rules if the attorney learns, through viewing public social media pages, that a juror is posting about a case or performing outside research?

   a. Nothing, since the attorney is not supposed to have contact with the juror.
   b. The attorney may now ethically send a "friend" request to the juror so that the attorney can continue to monitor and investigate all activity of the juror.
   c. The attorney is required to advise the court and opposing counsel immediately of the alleged impropriety before taking any other action.
   d. None of these options are viable or acceptable.

6. True or False: Jurors are permitted to research information about a case on the Internet or social media.

   a. TRUE
   b. FALSE

7. True or False: If jurors ignore the rules, and engage in social media use during trial or deliberations, it results in an automatic mistrial.

   a. TRUE
   b. FALSE

8. True or False: Attorneys may "friend" jurors, who are not represented by counsel in the case, in order to learn information about the jurors during jury selection.

   a. TRUE
   b. FALSE

9. True or False: Once a juror is selected for a trial, attorneys may then send them "friend" requests to monitor their social media activities during trial.

   a. TRUE
   b. FALSE

10. True or False: In no jurisdiction is an attorney required to perform a search of a potential juror's background on any social media or electronic site during jury selection.

    a. TRUE
    b. FALSE

Ethical and Legal Considerations for Social Media: Judges*

6.1 – Introduction

Judges of the New York State Court System are either elected or appointed depending on the office; other states also employ procedures of election or appointment; judges of the federal courts of the United States, whether Article I or Article III,

> *Article I & Article III Judges* –
> Federal judges whose positions were created pursuant to those Articles of the United States Constitution.

are all appointed. They all take an oath/affirmation,[1] don the black robe, and ascend the bench to preside. Now, for the big questions. Not how to decide a particular case. No, for the purposes of this chapter the big questions are: *What do judges do about their social media accounts and posts? And, do they result in a judge having to recuse in more cases than the average judge not on social media?* The answer is "it depends."

What is the content of the particular social media post? What is its relation to a case pending before the judge, if any? Who are the "friends" of the judge on social media, and what is the nature of the communications with those "friends," if any?

---

*The author of this text published a prior version of this discussion. *See* M. Fox, *Objection, Your Honor (To Your Social Media Activity?)*, 90 N.Y. ST. BAR J. 42 (Mar./Apr. 2018); reprinted in THE MAGISTRATE, Vol. 58, No. 3 (Summer 2018) (*Journal of The New York State Magistrates Association*).

[1] *See* 28 U.S.C. § 453 (oath of office taken by all United States Justices and Judges); *see also* N.Y.S. Const. Art. XIII, § 1 (oath of office taken by the New York State Executive, Legislators and Justices/Judges). It is interesting to note that once the oath is taken and filed along with the commission, it is a public record, and a judge need not produce a certified copy of same to satisfy a litigant that the oath was properly taken, and to discharge judicial duties (*In re Anthony*, 481 B.R. 602, 613–614 (D. Neb. 2012)). It is further interesting to note that had New Yorkers answered 2017 ballot Proposition 1 in the affirmative, and called for a Constitutional Convention, the Delegates to said Convention would have taken the State oath of office. *See* Notes of Decisions to N.Y.S. Const. Art. XIII, § 1; Op.Atty. Gen. 202 (1938).

These, and other inquiries, form the basis for evaluating any challenge to the social media posts of a judicial officer.[2] In addition, look to the codes of ethics: "A judge should maintain and enforce high standards of conduct and should personally observe those standards, so that the integrity and independence of the judiciary may be preserved. . . . A judge should respect and comply with the law and should act at all times in a manner that promotes public confidence in the integrity and impartiality of the judiciary."[3]

6.2 – Can a Judge Have a Social Media Account?

First of all, a judge is not forbidden from having social media accounts. New York ethics opinions have specifically held that there is nothing inherently improper about a judge utilizing social media.[4] However, these opinions have also advised that judges should take care concerning appearances of impropriety, should stay abreast of changes in technology that may impact the judge's duties under the rules, and should consider whether online connections and friendships in combination with any other factors create a circumstance for recusal.[5] Judges should evaluate if they can be fair and impartial—such as in situations where the contact is happenstance or coincidental, similar to being members of the same professional, civic, or social organizations as an attorney appearing before them. Indeed, the Maryland Judicial Ethics Committee once stated it best: "[a]ttorneys are neither obligated nor expected to retire to hermitage upon becoming a judge."[6]

The ethical authorities are also clear, though, that while judges are entitled to participate in social media, concomitant warnings exist regarding the care that should be taken when posting, with judges reminded to consider that posts rarely stay within the circle of connections in the original post, and that judges must maintain the integrity and impartiality of the

---

[2] Keep in mind that, with regard to attorneys, there are numerous rules and guidelines to be aware of before embarking on social media. *See, e.g.*, SOCIAL MEDIA ETHICS GUIDELINES OF THE COMMERCIAL & FEDERAL LITIGATION SECTION OF THE NEW YORK STATE BAR ASSOCIATION (updated 2019), available at: nysba.org/app/uploads/2020/02/NYSBA-Social-Media-Ethics-Guidelines-Final-6-20-19.pdf (last visited June 8, 2020). For two cautionary tales, *see* D. Weiss, *Penn State Frat Prosecutor Faces Ethics Hearing Over Fake Facebook Page, Texts to Judges*, ABA JOURNAL (online, Aug. 22, 2017), available at: http://www.abajournal.com/news/article/penn_state_frat_prosecutor_faces_ethics_hearing_over_fake_facebook_page_all (last visited Nov. 29, 2017); D. Weiss, *CBS Fires Lawyer Over Facebook Posts Calling Vegas Shooting Victims Likely "Republican Gun Toters,"* ABA JOURNAL (online, Oct. 2, 2017), available at: http://www.abajournal.com/news/article/cbs_fires_lawyer_over_facebook_comments_calling_vegas_victims_likely_republ/ (last visited Nov. 29, 2017). Separately, there are concerns about making clear to jurors that, while they are serving and deliberating, they are not to utilize social media. *See U.S. v. Ganias*, 755 F.3d 125, 132–133 (2d Cir. 2014).

[3] Code of Conduct for United States Judges, Canons 1 & 2(A). *See also* 22 N.Y. Comp. Codes R. & Regs. §§ 100.1, 100.2(A), 100.3. *See, generally*, A. Kaufman, *Judicial Ethics: The Less-Often Asked Questions*, 64 WASH. L. REV. 851, 854 (1989) ("Indeed, the basic rule of the Code of Conduct, the one to which all other rules are mere commentary, reflects this concern: judges should avoid not only impropriety but also the appearance of impropriety in all things relating to their office").

[4] *See* New York Opinions 08–176 (2009) and 11–125 (2011). *See also* ABA Formal Op. 462 (2013).

[5] *Id.*

[6] *See* Maryland Judicial Ethics Committee Op. Request No. 2012–07 (2012).

judiciary and avoid the appearance of impropriety, as outlined in Rule 1.2 of the Model Code.[7] For instance, in Kentucky prosecutors sought to have a judge removed from criminal cases because of the judge's Facebook posts.[8] The judge had posted comments to Facebook after the prosecutor's office took an appeal from the judge having dismissed an entire jury panel because it contained forty white jurors and one black juror. The judge posted statements critical of an appeal that he characterized as seeking to have "all-white juries"; he posted that his having granted a motion by a defendant striking the almost entirely Caucasian panel was not racist; and the judge further posted "calling people on racist language doesn't make me a racist either."[9] This ignited a firestorm of controversy, concerning whether the judge needed to recuse himself because of racist views. The judge completely denied any racial bias, but the social media postings contributed to a wider readership of the judge's views and comments.

Judges should also be mindful of how social media could result in *ex parte* communications. In some jurisdictions, a judge should not seek or accept social media "friendships" with attorneys who are likely to appear before the judges; and if the judge is social media "friends" with attorneys, the judge must disclose the relationship and potentially "unfriend" the attorneys or recuse from the case.[10] However, other jurisdictions are not as severe. For instance, in New York in May 2013, a judge asked the Advisory Committee on Judicial Ethics whether the judge had to recuse from a criminal trial at request of defendant's attorney or defendant because the judge was Facebook friends with parents or guardians of minors allegedly affected by the defendant's activities.[11] As discussed later in this chapter, the Opinion held that Facebook friend status *alone* was not sufficient for recusal, but the judge needed to consider if the relationship with the parents or guardians was really "mere 'acquaintance,'" not requiring recusal.[12]

New York Opinions 08–176 and 11–125 have held that there is nothing inherently improper about a judge utilizing social media, although the Opinions advised that judges should take care about appearances of impropriety, and warned that judges should consider whether online connections and friendships in combination with any other factors create a circumstance for recusal. A judge should evaluate if he or she can be fair and impartial—especially in situations where the contact is happenstance or coincidental, similar to being members of the same professional, civic, or social organizations.[13]

Generally speaking, if a judge is "friends" with a party/litigant, it is likely a reasonable assessment that the judge would have to recuse himself or herself from the case.[14]

---

[7] ABA Formal Ethics Opinion 462 (Feb. 21, 2013).

[8] J. Gershman, *Prosecutors Want Judge Off Criminal Cases Because of Facebook Posts*, THE WALL STREET JOURNAL (Nov. 18, 2015).

[9] *Id.*

[10] *See* Massachusetts CJE Opinion No. 2016–01 (2/16/16).

[11] New York Judicial Ethics Opinion 13–39 (5/28/13).

[12] *Id.*

[13] For more, see Singh, *Friend Request Denied: Judicial Ethics & Social Media*, 7 CASE W. RESERVE J.L. TECH. & INTERNET 153 (2016).

[14] *See* NYSBA ComFed *Social Media Guideline* 7: "A lawyer shall not communicate with a judicial officer over social media if the lawyer intends to influence the judicial officer in the performance of his or her official duties."

6.3 – Potential Impact of a Judge's Active Use of Social Media

What results when a judge affirmatively posts to social media and those posts raise eyebrows? Any number of outcomes, dependent upon the severity of the incident and the jurisdiction where it occurs.

Take the case of an Ohio Supreme Court Justice, who at the time of his posts was also a candidate for Governor in Ohio, making headlines for two separate social media posts. In one, he spoke out against the NFL players who were kneeling during the National Anthem, criticizing the "draft dodging millionaire athletes disrespect[ing] the veterans."[15] In a second post some months later, in the wake of the sexual harassment scandal reaching Senator Al Franken's office, the judge posted the following, in relation to his campaign for governor: "[to] save my opponents some research time. . . . In the last fifty years I was sexually intimate with approximately 50 very attractive females. . . . Now can we get back to discussing legalizing marijuana and opening the state hospital network to combat the opioid crisis."[16]

Given backlash, including from the state's chief justice, the social media post was deleted—but the judge later posted new comments.[17] The Ohio Chief Justice was quoted as stating: "No words can convey my shock,. . . . This gross disrespect for women shakes the public's confidence in the integrity of the judiciary."[18] The judge in question, following much outcry, issued a heartfelt apology on social media and radio.[19]

Then there is the case of a Gwinnett County, Georgia, magistrate judge. The judge posted on social media, in the wake of the Charlottesville protests, a severe criticism of the protesters seeking removal of Southern Confederate monuments, calling them "snowflakes," and stating, with regard to monuments in Richmond, Virginia, they "all . . . have people on horses whose asses face North. PERFECT!"[20]

For a cautionary tale, *see Commonwealth v. Grove*, No. 1934 MDA 2016, 2017 WL 2859006 (Super. Ct. Pa. July 5, 2017) (alleged *ex parte* communications with counsel, and discovery demand for them in special proceeding following conviction).

[15] J. Delk, *Ohio Judge Slams Cleveland Browns Players for Protesting During National Anthem*, THE HILL (online, Aug. 25, 2017), available at: http://thehill.com/homenews/347974-ohio-supreme-court-justice-criticizes-cleveland-browns-players-national-anthem (last visited Sept. 29, 2018) (the article notes that the draft ended in 1973, and therefore none of the NFL players are actually draft-dodgers).

[16] L. Bever & M. Eltagouri, *Ohio Governor Candidate Apologizes for Boasting of Sexual History with "50 Very Attractive Females,"* THE WASHINGTON POST (online Nov. 18, 2017), available at: https://www.washingtonpost.com/news/politics/wp/2017/11/17/ohio-governor-candidate-boasts-of-sexual-history-with-approximately-50-very-attractive-females/?utm_term=.dea65ee74828 (last visited Sept. 29, 2018). The social media post included details describing several women such that they might be identifiable, and those descriptions are not repeated here.

[17] *Id.*

[18] *Id.*

[19] *See O'Neill Apologizes for Controversial Social Media Post*, WCBE.org (Nov. 20, 2017), available at: http://www.wcbe.org/post/oneill-apologizes-controversial-social-media-post.

[20] D. Weiss, *Magistrate Judge Retires After Facebook Comments About "Snowflakes" and "Nut Cases,"* ABA JOURNAL (online Aug. 18, 2017), available at: http://www.abajournal.com/news/article/

The judge made a follow-up post comparing the protesters wanting to take down the monuments to ISIS destroying history. The Chief Magistrate Judge initially suspended the judge and issued a statement that the Judicial Canons and court policies place a requirement on judicial officers to act in manners that ensure the public views judges as fair and impartial. When judges violate those rules, according to the chief magistrate judge, it is "a matter of utmost concern."[21] The judge in question later offered his resignation/retirement.[22]

Several other incidents—sitting judges, judicial candidates, and nominees posting on all manner of social and political topics—have made headlines and raised questions.[23] In one matter before the Florida Third District Court of Appeal in 2017, the court addressed a law firm's petition for a trial judge to recuse from a case because she was Facebook friends with an attorney for a non-party who had entered an appearance. The firm contended that the judge would be influenced by the social media friendship.[24] Now, we understand from the earlier discussion that in New York, and other jurisdictions, the mere occurrence of a "friendship" on social media is not sufficient to force recusal of a judge—any more than a "friendship" or "acquaintance relationship" that develops through repeated contacts at bar association and social functions would force recusal. However, according to one news article, of the 11 states that have issued rules/guidance for judges on social media, Florida's are the most restrictive.[25] The Third District Court of Appeal, though, in a decision issued in August 2017, broke with a prior Florida appellate court ruling and ethics opinion, finding recusal was not required.

The Court first noted that, stemming from pre-social media days, "as a general matter,. . . . 'allegations of mere "friendship" with an attorney or an interested party have been deemed insufficient to disqualify a judge.'"[26] That very much looks to be in line with New York and

magistrate_judge_retires_after_his_facebook_comments_about_nut_cases_tearin/ (last visited Sept. 28, 2018).

[21] C. O'Brien, *Gwinnett Magistrate Judge Suspended Over Facebook Post About Charlottesville, Va. Protesters*, GWINNETT DAILY POST (online Aug. 15, 2017), available at: http://www.gwinnettdailypost.com/local/gwinnett-magistrate-judge-suspended-over-facebook-post-about-charlottesville-va/article_672ba1c8-07be-5edd-9e67-6a66be9bfc1d.html (last visited Sept. 28, 2018).

[22] *See* D. Weiss, *Magistrate Judge Retires after Facebook Comments About "Snowflakes" and "Nut Cases," supra.*

[23] For brief descriptions of other instances, *see* C. Ampel, *Watch Your Mouth, Your Honor: Lessons for Judges on Social Media*, N.Y. L.J. (Aug. 28, 2017), available at: https://www.law.com/newyorklawjournal/almID/12027966 79106/?slreturn=20171029093202.

[24] D. Weiss, *Appeals Court Considers Removal of Judge Who Is Facebook Friends with Lawyer*, ABA JOURNAL (online, July 31, 2017), available at: http://www.abajournal.com/news/article/appeals_court_considers_removal_of_judge_who_is_facebook_friends_with_lawye.

[25] *See* D. Weiss, *Judge Shouldn't Be Booted from Case Because of Facebook Friendship with Lawyer, Appeals Court Rules*, ABA JOURNAL (online, Aug. 25, 2017), available at: http://www.abajournal.com/news/article/judge_shouldnt_bc_booted_from_case_because_of_facebook_friendship_with_lawy.

[26] *Law Offices of Herssein & Herssein, P.A. v. U. Serv. Auto. Assoc.*, 229 So.3d 408 (Fla. App. 3d Dist. 2017) (citing *Smith v. Santa Rosa Island Auth.*, 729 So.2d 944, 946 (Fla. App. 1st Dist. 1998); *MacKenzie v. Super Kids Bargain Store, Inc.*, 565 So.2d 1332, 1338 (Fla. 1990) ("There are countless factors which may cause some members of the community to think that a judge would be biased in favor of a litigant or counsel for a litigant, e.g., friendship, member of the same church or religious congregation, neighbors, former classmates or fraternity brothers. However, such allegations have been found legally insufficient when asserted in a motion for disqualification")).

others. The Florida court then recounted a prior decision of the Fourth District Court of Appeal in the *Domville* case (which cited Florida Judicial Ethics Advisory Committee Opinion 2009–20), which decision held that a judge who was Facebook friends with a prosecutor on a case had to recuse.[27] The *Herssein* Court also discussed the Fifth District Court of Appeal decision in *Chace v. Loisel*, in which that court held a trial judge was required to recuse from a matrimonial action when the judge sent a friend request to the litigant-wife during the pendency of the case (which the wife did reject), although the *Chace* Court at the same time cast doubt on the Fourth District's *Domville* holding.[28] Of course, the result in the *Chace* decision appears very appropriate, given the judge's direct and improper communications to and interest in a specific party appearing before the judge, and such a recusal result is not inconsistent with what would likely result in other jurisdictions.

Florida's Third District Court of Appeal, in *Herssein*, ultimately held that "'[a] Facebook friendship does not necessarily signify the existence of a close relationship.' . . . 'some people have thousands of Facebook "friends."'"[29] The *Herssein* Court also reasoned that "Facebook members often cannot recall every person they have accepted as 'friends' or who have accepted them as 'friends.'"[30] Finally, the Court stated: "many Facebook 'friends' are selected based upon Facebook's data-mining technology rather than personal interactions."[31] Therefore, the *Herssein* Court concluded as follows:

> To be sure, some of a member's Facebook "friends" are undoubtedly friends in the classic sense of person for whom the member feels particular affection and loyalty. The point is, however, many are not. . . . In fairness to the Fourth District's decision in Domville and the Judicial Ethics Advisory Committee's 2009 opinion, electronic social media is evolving at an exponential rate. Acceptance as a Facebook "friend" may well once have given the impression of close friendship and affiliation. Currently, however, the degree of intimacy among Facebook "friends" varies greatly. The designation of a person as a "friend" on Facebook does not differentiate between a close friend and a distant acquaintance. Because a "friend" on a social networking website is not necessarily a friend in the traditional sense of the word, we hold that the mere fact that a judge is a Facebook "friend" with a lawyer for a potential party or witness, without more, does not provide a basis for a well-grounded fear that the judge cannot be impartial or that the judge is under the influence of the Facebook "friend." On this point we respectfully acknowledge we are in conflict with the opinion of our sister court in Domville. Petition denied.[32]

---

[27] *Herssein*, 229 So.3d at 410 (citing *Domville v. State*, 103 So.3d 184 (Fla. App. 4th Dist. 2012); Fla. JEAC Op. 2009–20 (Nov. 17, 2009)).

[28] *Herssein*, 229 So.3d at 410 (citing *Chace v. Loisel*, 170 So.3d 802, 803–04 (Fla. App. 5th Dist. 2014)).

[29] *Herssein*, 229 So.3d at 411 (citing *Sluss v. Commonwealth*, 381 S.W.3d 215, 222 (Ky. 2012); *State v. Madden*, No. M2012-02473-CCA-R3-CD, 2014 WL 931031, at *1–*2 (Tenn. Crim. App. Mar. 11, 2014)).

[30] *Herssein*, 229 So.3d at 411 (citing *Furey v. Temple Univ.*, 884 F.Supp.2d 223, 241 (E.D. Pa. 2012); *Slaybaugh v. State*, 47 N.E.3d 607, 608 (Ind. 2016)).

[31] *Herssein*, 229 So.3d at 411-412.

[32] *Id.*

The law firm appealed the matter to the Florida Supreme Court,[33] and the state's highest court issued a decision affirming the Court of Appeal, and disapproving the *Domville v. State* decision.[34]

In 2020 in Ohio, the Supreme Court of that state issued a decision fully evaluating a situation where a trial judge hearing a criminal matter continued to preside while being Facebook "friends" with the victim. Because of the importance of the decision, it is set forth here in almost its entirety (footnotes are omitted) for consideration:

In re Kerenyi (Supreme Court of Ohio)

O'Connor, Chief Justice

{¶ 1} Defendant Kendall Richards has filed an affidavit pursuant to R.C. 2701.03 seeking to disqualify Judge Mark Kerenyi from the above-referenced case. According to Judge Kerenyi, after Mr. Richards entered a guilty plea—but prior to sentencing—the underlying case was transferred to Judge Kerenyi's docket because the original judge had retired and that judge's successor had a conflict.

Allegations in the affidavit

{¶ 2} Mr. Richards claims that Judge Kerenyi is biased against him and that the judge's removal is necessary to avoid any appearance of partiality. According to Mr. Richards, the judge and the alleged victim have significant connections, including that (1) the judge and the victim are Facebook "friends," (2) the victim is a former county commissioner who has personal and political connections to Judge Kerenyi and influence over the common pleas court, (3) the victim contributed to Judge Kerenyi's campaign for judicial office, and (4) the victim's son is a director of a county agency and in that capacity worked closely with Judge Kerenyi for over a decade. Mr. Richards also alleges that Judge Kerenyi and Mr. Richards's attorney have engaged in "numerous ex parte communications" and that the judge has subjected Mr. Richards to "coercive tactics, arbitrary change of pleas, and due process deprivations."

{¶ 3} Judge Kerenyi filed a response to the affidavit in which he denies any bias against Mr. Richards and denies having any special relationship with the victim. The judge states that he has "unfriended" the victim on Facebook and that regardless, the judge has over 1,400 Facebook "friends," so that status does not convey any significant connection with the court. The judge acknowledges that almost 40 years ago, the victim served as a county commissioner—although Judge Kerenyi states that he was previously unaware of that fact. The judge further acknowledges that he and the victim are members of the same political party and have occasionally

---

[33] *See Law Offices of Herssein and Herssein, P.A. v. United Servs. Auto. Ass'n,* 2017 WL 6281070 (Fla. Dec. 11, 2017); *Law Firm Asks: Does Facebook Friendship Disqualify Judge?,* CBS NEWS (Oct. 18, 2017), available at: http://dfw.cbslocal.com/2017/10/18/does-facebook-friendship-disqualify-judge/.

[34] 271 So.3d 889 (Fla. 2018); *rehearing denied* 2018 WL 7136575 (Fla. Dec. 12, 2018).

attended the same party-sponsored events. But the judge denies that the victim contributed to the judge's campaign for judicial office and denies that the victim has any influence over the court. The judge also acknowledges that the victim's son is the director of the county children-services agency and that the judge, while he served as a magistrate in the juvenile and probate court, occasionally presided over matters involving that agency. The judge states, however, that he is not unduly influenced by the position of the victim's son. Finally, Judge Kerenyi expressly denies that he engaged in any ex parte communications with Mr. Richards's attorney and states that an assistant prosecutor has been present for all hearings and meetings in the underlying case. The judge also denies that he subjected Mr. Richards to any coercive tactics regarding any attempt to change his plea.

Disqualification standard

{¶ 4} In disqualification requests, "[t]he term 'bias or prejudice' 'implies a hostile feeling or spirit of ill-will or undue friendship or favoritism toward one of the litigants or his attorney, with the formation of a fixed anticipatory judgment on the part of the judge, as contradistinguished from an open state of mind which will be governed by the law and the facts.'" *In re Disqualification of O'Neill*, 100 Ohio St.3d 1232, 2002-Ohio-7479, 798 N.E.2d 17, ¶ 14, quoting *State ex rel. Pratt v. Weygandt*, 164 Ohio St. 463, 469, 132 N.E.2d 191 (1956). "The proper test for determining whether a judge's participation in a case presents an appearance of impropriety is * * * an objective one. A judge should step aside or be removed if a reasonable and objective observer would harbor serious doubts about the judge's impartiality." *In re Disqualification of Lewis*, 117 Ohio St.3d 1227, 2004-Ohio-7359, 884 N.E.2d 1082, ¶ 8. "The reasonable observer is presumed to be fully informed of all the relevant facts in the record—not isolated facts divorced from their larger context." *In re Disqualification of Gall*, 135 Ohio St.3d 1283, 2013-Ohio-1319, 986 N.E.2d 1005, ¶ 6. Finally, in considering a disqualification request, "[a] judge is presumed to follow the law and not to be biased, and the appearance of bias or prejudice must be compelling to overcome these presumptions." *In re Disqualification of George*, 100 Ohio St.3d 1241, 2003-Ohio-5489, 798 N.E.2d 23, ¶ 5.

Merits of the affidavit of disqualification

{¶ 5} For the reasons explained below, Mr. Richards has not established that Judge Kerenyi is biased or that an appearance of partiality exists.

{¶ 6} First, "the more intimate the relationship between a judge and a person who is involved in a pending proceeding, the more acute is the concern that the judge may be tempted to depart from the expected judicial detachment or to reasonably appear to have done so." *In re Disqualification of Shuff*, 117 Ohio St.3d 1230, 2004-Ohio-7355, 884 N.E.2d 1084, ¶ 6. Based on this record, Mr. Richards has not offered convincing evidence of a significant professional, political, or personal relationship between Judge Kerenyi and the victim that would suggest that the judge could be tempted to depart from his expected judicial neutrality.

{¶ 7} Standing alone, a judge's Facebook "friendship" with a lawyer, litigant, or other person appearing before the judge does not automatically require the judge's disqualification. *See* Board of Commissioners on Grievances and Discipline Op. No. 10-007, syllabus (Dec. 3, 2010) ("A judge may be a 'friend' on a social networking site with a lawyer who appears as counsel in a case before the judge. * * * There is no bright-line rule: not all social relationships, online or otherwise, require a judge's disqualification"). It is long settled that "the mere existence of a friendship between a judge and an attorney or between a judge and a party will not disqualify the judge from cases involving that attorney or party." *In re Disqualification of Bressler*, 81 Ohio St.3d 1215, 688 N.E.2d 517 (1997). Because not every relationship characterized as a friendship provides a basis for disqualification, "there is no reason that Facebook 'friendships'—which regularly involve strangers—should be singled out and subjected to a per se rule of disqualification." *Law Offices of Herssein & Herssein, P.A., v. United Servs. Auto. Assn.*, 271 So.3d 889, 899 (Fla.2018). Therefore, the same principles that apply to a judge's in-person social relationships apply to the judge's online "friendships," and determining whether a judge should preside over a case involving a Facebook "friend" requires assessing the nature and scope of that particular relationship, combined with all other relevant factors.

{¶ 8} Here, Judge Kerenyi states that he has over 1,400 Facebook "friends" and that he has "unfriended" the victim "as well as any of his family members on Facebook." If Judge Kerenyi knew that he and the victim were Facebook "friends" when the underlying case was transferred to the judge's docket, the judge probably should have also disclosed that online connection at that time. Regardless, the mere fact that the victim was formerly one of the judge's many Facebook "friends" does not create the appearance that the victim is in a special position to influence the court or cast doubt on Judge Kerenyi's ability to act impartially in the underlying case. And Judge Kerenyi's unfriending the victim should alleviate any concerns that the judge may be in a position to view the victim's Facebook posts or activity on the judge's newsfeed.

{¶ 9} Likewise, none of the other allegations in Mr. Richards's affidavit establish that the judge has a close relationship with the victim. That the judge and the victim are active members of the same political party, without more, does not suggest the appearance of partiality. *See In re Disqualification of Gallagher*, 155 Ohio St.3d 1251, 2018-Ohio-5428, 120 N.E.3d 853 (a litigant's involvement in local party politics was insufficient to warrant disqualification of judges from the same political party). That the victim served as a county commissioner almost 40 years ago does not show that the victim continues to have influence over the common pleas court. *Compare In re Disqualification of Corrigan*, 110 Ohio St.3d 1217, 2005-Ohio-7153, 850 N.E.2d 720 (county trial judges disqualified from a case involving a county commissioner who wielded considerable influence over the court's funding and who played a leadership role in local politics). And Mr. Richards failed to substantiate his allegation that the victim contributed to the judge's campaign—although such information is publicly available—and failed to sufficiently explain how Judge Kerenyi's former professional relationship with the victim's son creates an appearance of partiality. In a disqualification request, the burden "falls on the affiant to submit specific allegations and facts to demonstrate that disqualification is warranted." *In re Disqualification of DiGiacomo*, 157 Ohio St.3d 1225, 2019-Ohio-4432, 134 N.E.3d 1231, ¶ 4. Mr. Richards has failed to meet that burden here.

{¶ 10} Second, "[a]n alleged ex parte communication constitutes grounds for disqualification when there is 'proof that the communication * * * addressed substantive matters in the pending case.' " (Ellipsis sic.) *In re Disqualification of Forsthoefel*, 135 Ohio St.3d 1316, 2013-Ohio-2292, 989 N.E.2d 62, ¶ 7, quoting *In re Disqualification of Calabrese*, 100 Ohio St.3d 1224, 2002-Ohio-7475, 798 N.E.2d 10, ¶ 2. "The allegations must be substantiated and consist of something more than hearsay or speculation." Id. Mr. Richards has not substantiated his allegation that Judge Kerenyi engaged in any improper ex parte communications with Mr. Richards's attorney. Indeed, the evidence that Mr. Richards submitted with his affidavit—a letter from his attorney—indicates that the assistant prosecutor was present for all meetings regarding Mr. Richards's sentence.

{¶ 11} Third, Mr. Richards's vague allegation that Judge Kerenyi engaged in "coercive tactics" relating to an "arbitrary change of plea" does not require Judge Kerenyi's removal, especially considering that Mr. Richards entered his plea before a different judge. "[V]ague, unsubstantiated allegations of the affidavit are insufficient on their face for a finding of bias or prejudice." *In re Disqualification of Walker*, 36 Ohio St.3d 606, 522 N.E.2d 460 (1988).

{¶ 12} Given the lack of evidence indicating that Judge Kerenyi has a close personal, political, or professional relationship with the victim, a reasonable and objective observer would have no basis to question Judge Kerenyi's ability to preside over the remainder of this case.

{¶ 13} The affidavit of disqualification is denied. The case may proceed before Judge Kerenyi.[35]

Separately, in Memphis, Tennessee, a state criminal court judge was publicly reprimanded as the penalty for sharing, among other things, an article on Facebook that referred to some

---

[35] *In re Disqualification of Kerenyi (State v. Richards)*, — N.E.3d —, No. 19-AP-147, 2020 WL 1465955 (Mem) (Ohio Jan. 16, 2020) (footnotes omitted). The *Kerenyi* Court noted, however, that judges must be just as careful with online friendships as with real-life friendships; and, further, that use of social media and having online friendships can be misunderstood or require disqualification because of the creation of an appearance of impropriety that cannot be overcome. *Kerenyi*, 2020 WL 1465955, at *2 n.1 (citing *State v. Thomas*, 376 P.3d 184 (N.M. 2016); and *In re Paternity of B.J.M.*, 386 Wis.2d 267, 925 N.W.2d 580 (Wis. Ct. of Apps. 2019), *review granted* 388 Wis.2d 657, 933 N.W.2d 489 (Table) (Wis. 2019) (trial judge was disqualified by the Court of Appeals on review where the judge friended a litigant on Facebook, undisclosed, while motions were pending, and the judge could see the party's social media postings related to the litigation)). *See also* Mark Flores, *Facebook "Friendship" Leads to Judge's Disqualification*, ABA Litigation, vol. 45, no. 1, at 8 (Fall 2019) (citing and discussing *In re. Paternity of B.J.M.*).

immigrants as "foreign mud."[36] The Tennessee Board of Judicial Conduct issued a formal letter of reprimand[37] to the judge on November 15, 2019, although the Board was clear to say there was no proof the judge in question, himself, made any racist, anti-immigrant, or anti-Semitic statements. However, the letter noted the subject matter of the posts and articles shared/re-shared by the judge—subject matter that was as relevant in 2016, 2017, 2018, and 2019 as it will be in 2020 and beyond—included material addressing, *inter alia*, concern for the credibility of some federal agencies, a position about athletes kneeling during the National Anthem, opposition to Democratic platform ideas and showing favor toward Donald Trump, and positions as to police shooting deaths, Black Lives Matter, and media biases.[38] In light of the judge's violations of Canons 1 and 2 of the Tennessee Code of Judicial Conduct, but because the judge had no history of violations, a public reprimand was issued.[39] According to the *ABA Journal*, the judge acknowledged he should not have shared the problematic posts, and "agreed to complete an educational program addressing ethical issues and the use of social media."[40]

In New York, three opinions/proceedings warrant inclusion in this chapter. First, as referenced earlier in this chapter, in 2013 a New York State judge asked the Advisory Committee on Judicial Ethics whether the judge had to recuse from a criminal trial at the request of the defendant's attorney or defendant because the judge was Facebook friends with the parents or guardians of minors allegedly affected by defendant's activities.[41] The Opinion held Facebook friend status alone was not sufficient for recusal, and the judge's impartiality was not reasonably in question. There was thus no appearance of impropriety.[42] However, the Opinion provided warning:

> The Committee believes that the mere status of being a "Facebook friend," without more, is an insufficient basis to require recusal. Nor does the Committee believe that a judge's impartiality may reasonably be questioned (see 22 NYCRR 100.3[E][1]) or that there is an appearance of impropriety (see 22 NYCRR 100.2[A]) based solely on having previously "friended" certain individuals who are now involved in some manner in a pending action. . . . As the Committee noted in Opinion 11–125, interpersonal relationships are varied, fact-dependent, and unique to the individuals involved. Therefore, the Committee can provide only general guidelines to assist judges who ultimately must determine the nature of their own specific relationships with

---

[36] Debra Cassens Weiss, *Judge Who Shared "Foreign Mud" Article on Facebook is Reprimanded for Partisan Posts*, ABA Journal (online Nov. 20, 2019), *available at* https://www.abajournal.com/news/article/judge-who-shared-foreign-mud-article-on-facebook-is-reprimanded-for-partisan-posts.

[37] Letter of Reprimand *available at* http://www.tncourts.gov/sites/default/files/docs/lammey_reprimand_letter_only_2019_11_18.pdf.

[38] *Id.*

[39] *Id.*

[40] D. Weiss, *Judge Who Shared "Foreign Mud" Article, supra.*

[41] *See* N.Y. Judicial Ethics Opinion 13–39 (2013), available at: https://www.nycourts.gov/ip/judicialethics/opinions/13-39.htm.

[42] *Id.*

particular individuals and their ethical obligations resulting from those relationships. With respect to social media relationships, the Committee could not "discern anything inherently inappropriate about a judge joining and making use of a social network" (Opinion 08–176). However, the judge "should be mindful of the appearance created when he/she establishes a connection with an attorney or anyone else appearing in the judge's court through a social network... [and] must, therefore, consider whether any such online connections, alone or in combination with other facts, rise to the level of a... relationship requiring disclosure and/or recusal" (id.). If, after reading Opinions 11–125 and 08–176, you remain confident that your relationship with these parents or guardians is that of a mere "acquaintance" within the meaning of Opinion 11–125, recusal is not required. However, the Committee recommends that you make a record, such as a memorandum to the file, of the basis for your conclusion . . .[43]

Next, in December 2016, the New York State Commission on Judicial Conduct issued its Determination in a rather extreme case—*In the Matter of the Proceeding Pursuant to Section 44, Subdivision 4, of the Judiciary Law in Relation to Lisa J. Whitmarsh, A Justice of the Morristown Town Court, St. Lawrence County.*[44] In that case, the town justice had posted comments to Facebook concerning an ongoing prosecution before a different town court. The posts were made between March 13 and March 28, 2016. The Determination set forth, in part, the following, which is provided at length for discussion and analysis on this important subject:

In the Matter of the Proceeding Pursuant to Section 44, Subdivision 4, of the Judiciary Law in Relation to Lisa J. Whitmarsh, A Justice of the Morristown Town Court, St. Lawrence County (New York Commission on Judicial Conduct)

Respondent had approximately 352 Facebook ""friends." [*sic*] Respondent's Facebook account privacy settings were set to "Public," meaning that any internet user, with or without a Facebook account, could view content posted on her Facebook page . . . On March 13, 2016, respondent posted a comment to her publicly viewable Facebook account, . . . criticizing the investigation and prosecution of [defendant]. Respondent commented, *inter alia*, that she felt "disgust for a select few," that [defendant] had been charged with a felony rather than a misdemeanor because of a ""personal vendetta," that the investigation was the product of "CORRUPTION" caused by "personal friends calling in personal favors," and that [defendant] had "[a]bsolutely" no criminal intent . . . Respondent's post also referred to her judicial position, stating, "When the town board attempted to remove a Judge position - I stood up for my Co-Judge. When there is a charge, I feel is an abuse of the Penal Law - I WILL stand up for [DEFENDANT]" [sic] [emphasis in original] . . . Other Facebook users posted comments on respondent's Facebook page, commending respondent's statements in her post of March 13, 2016, and/or criticizing the

---

[43] *Id.*
[44] 2016 WL 7743777 (N.Y. Com. Jud. Cond. Dec. 28, 2016).

prosecution of [defendant]. The first Facebook user to comment was Morristown Town Court Clerk [ ], who posted the following on March 13, 2016, at 7:58 AM: "Thank you Judge [ ]! You hit the nail on the head." Respondent did not delete the court clerk's comment, which was viewable by the public . . . On March 23, 2016, a local news outlet posted an article on its website reporting on respondent's Facebook comments . . . and re-printed respondent's Facebook post of March 13, 2016, in its entirety . . . On March 28, 2016, respondent removed all postings concerning the [ ] matter from her Facebook page after receiving a letter from [the] District Attorney [ ] questioning the propriety of her comments and requesting her recusal from all matters involving the District Attorney's office.[45]

The Commission determined that the respondent town justice had "violated Sections 100.1, 100.2(A), 100.2(C) and 100.3(B)(8) of the Rules Governing Judicial Conduct . . . and should be disciplined for cause, pursuant to Article 6, Section 22, subdivision a, of the New York State Constitution and Section 44, subdivision 1, of the Judiciary Law."[46] The Commission further determined that posts to Facebook are public, cannot be considered private in any sense, and, thus, "[a]ccordingly, a judge who uses Facebook or any other online social network 'should ... recognize the public nature of anything he/she places on a social network page and tailor any postings accordingly.'"[47] Within the Determination, the Commission forcefully and correctly pointed out that:

[w]hile the ease of electronic communication may encourage informality, it can also, as we are frequently reminded, foster an illusory sense of privacy and enable too-hasty communications that, once posted, are surprisingly permanent. For judges, who are held to "standards of conduct more stringent than those acceptable for others" . . . and must expect a heightened degree of public scrutiny, internet-based social networks can be a minefield of "ethical traps for the unwary."[48]

Ultimately, the town judge was given an admonition as a sanction.

Finally, one should consider a 2017 case from the Town of Floyd (Oneida County), New York. There, the town justice, who had been serving since 1999, agreed to resign on December 31, 2017, as part of a stipulation with the New York State Commission on Judicial Conduct. According to reports, it appears that the judge (who also served as a justice in two village courts) gave the appearance of bias favoring law enforcement, and criticized public

---

[45] *Id.*

[46] *Id.*

[47] *Id.* (citing N.Y. Opinion 08–176).

[48] *Id.* (citing, *inter alia*, J. Browning, *Why Can't We Be Friends? Judges' Use of Social Media*, 68 U. MIAMI L. REV. 487, 511 (2014)).

officials and gun regulations, on Facebook.[49] More detail was not available, although the remarks, it was reported, were made in November 2017.[50]

As compared to the other cases discussed in this chapter, the outcome of the case in Oneida County was most severe—much more than that in the Morristown Town Court matter above. In this Town of Floyd matter the judge not only stepped down from the bench but the judge agreed not to run for or hold a judicial position again.[51] The Commission's Administrator also provided guidance, which sums up this topic perfectly, and should be taken to heart by all elected and appointed judges.[52] Basically, judges, regardless of whether or not they are posting to social media, cannot appear to favor or oppose particular political or social groups, and must beware concerning how their conduct reflects on their office.[53]

6.4 – Conclusion

Although judges can use social media, and certainly may have "friends" both on and off social media, judicial officers (as well as candidates for judicial office, judicial nominees, and attorneys) should be very cautious when it comes to social media friendships and postings.[54] Judges and attorneys are not only private citizens; they are also public officers and officers of the court.[55] They are under solemn oath to safeguard and uphold the integrity and impartiality of the law and the justice system—of maintaining the solid foundation upon which the Third

---

[49] J. Velasquez, *Town Justice Resigns after Probe of His Facebook Remarks*, NEW YORK LAW JOURNAL (Dec. 18, 2017), available at: https://www.law.com/newyorklawjournal/sites/newyorklawjournal/2017/12/18/town-justice-resigns-after-probe-of-his-facebook-remarks/ (last visited Dec. 19, 2017).

[50] *Id.*

[51] *Id.*

[52] *Id.*

[53] *Id.*

[54] Judges should further be cautious before using social media or the Internet to research cases, attorneys, parties, or potential jurors. For a full discussion, *see* ABA Formal Opinion 478 (Dec. 8, 2017) (citing, *inter alia*, Rule 2.9(C) of the Model Code of Judicial Conduct, which prohibits online research and information gathering about a juror or party).

[55] In addition to social media, judges should be cautious when communicating by any method, whether electronic or analog—always remembering their offices, and their professional and ethical obligations. For instance, in 2020 a Louisiana state court judge resigned from the bench after a firestorm of criticism—including from the state's governor—following her having used the "N-word" several times in angry text messages that concerned an alleged former romantic relationship. *See* David J. Mitchell, *Judge Resigns after Racist Texts, Affair with Top Deputy; Says She Quit to 'Stop the Madness'*, The Advocate (online Feb. 27, 2020), *available at* https://www.theadvocate.com/baton_rouge/news/communities/ascension/article_03fe9402-597c-11ea-846d-57c40c14944e.html.

Branch of government is built.[56] That sacred duty requires more than a passing thought before hitting "Tweet" or "Share."[57,58]

Chapter End Questions

Note: Answers may be found on page 263.

1. Which of the following ethical opinions provides guidance for judges in the area of social media?
   a. NY Op. 11-125
   b. Maryland Op. 2012-07
   c. Massachusetts Op. 2016-01
   d. All of these selections

2. Which is of the following is a potential discipline a judge might receive for misuse of social media communications?
   a. Letter of Censure
   b. Removal from the Bench
   c. Both of these selections
   d. Neither of these selections

3. Which of the following is a judge permitted to do on social media?
   a. Comment on cases positively or negatively when they are conducted before a colleague on the same court.
   b. Take positions on political matters, and disparage the opposite positions.
   c. Romantically pursue litigants or attorneys appearing before the judge.
   d. None of these selections are ethical or appropriate for judges to undertake on or off of social media.

---

[56] *See* M. Fox, *The Legal Profession—Attorneys & Courts—Bulwark Against Injustice*, PERSPECTIVE, at 7 (Spring 2018) (*Journal of the Young Lawyers Section of the New York State Bar Association*).

[57] As stated by the Gwinnett County Chief Magistrate Judge in accepting the resignation of one of the judges discussed above, even though the U.S. Constitution protects the expression of opinion by all citizens, "Judges are held to a more stringent standard by the Judicial Canons." T. Estep, *Gwinnett Judge Resigns after Controversial Confederate Monument Posts*, THE ATLANTA JOURNAL-CONSTITUTION (online, Aug. 16, 2017), available at: http://www.ajc.com/news/local-govt--politics/gwinnett-judge-resigns-after-controversial-confederate-monument-posts/eO1o7wMa0zUNoa7IGz4YZK/.

[58] However, just as others in society continue to use social media, so, too, do judges. As a result, new issues continue to occur. In July 2018, a Nebraska federal district judge caused upset by posting a statement on Twitter comparing a group of law clerks to the Spanish Inquisition in the wake of the group's involvement in the overhaul of federal sexual harassment policies. *See* M. Mitchell, *Judge Who Stirred Controversy with Tweet Unlikely to Face Discipline*, Experts Say, N.Y. L.J. at 2 (July 25, 2018).

4. As of the time this book was written, 11 states had issued formal guidance for judges' use of social media. Which of those states had the most restrictive rules?

   a. Florida
   b. New York
   c. Massachusetts
   d. Oklahoma

5. True or False: It is never appropriate for judges to have social media accounts.

   a. TRUE
   b. FALSE

6. True or False: A judge being "friends" with an attorney on social media - with an attorney appearing before the judge - is grounds for recusal (for the judge to take himself/herself off the case).

   a. TRUE
   b. FALSE

7. True or False: A judge having communications on social media with a litigant appearing before the judge requires recusal (the judge removing himself/herself from the case).

   a. TRUE
   b. FALSE

8. True or False: Social media provides a good way for attorneys to directly communicate points to a judge in a pending case.

   a. TRUE
   b. FALSE

9. True or False: The First Amendment protections of the U.S. Constitution do not apply to judges.

   a. TRUE
   b. FALSE

10. True or False: Judges are permitted to use the Internet and social media to look up any information they need or want about a party or juror, outside of that provided during trial.

    a. TRUE
    b. FALSE

Examples of Societal Concerns and Social Media

7.1 – Teenagers and Driving While Texting

We live in a world of nearly non-stop technological communication. For example, as far back as October 2010, CNN reported that Nielsen had conducted a study. It does not just appear that teenagers text every waking minute; apparently they do. Back then it was estimated at an average of 3,300+ texts per month.[1] Teenagers are not alone in non-stop texting. In June 2011, more than *196 billion* text messages were sent or received in the United States.[2] That was in just one month. And, that was an increase of 50% from June 2009.[3] Imagine what those same numbers would be today, with multiple additional social media platforms.

The ABA further reported, back in September 2014, that according to Nielsen's annual *State of the Media: The Social Media Report 2012*, Americans spent *121 billion minutes* on social media sites in *July 2012*—388 minutes (6½ hours) *per person* (if *every person* in the United States were on social media) just that month of July![4]

These numbers may be "fun" to read in the abstract, trying to imagine the sheer magnitude. But, behind these numbers is a very serious problem—teens and their non-stop use of mobile devices, teens and their social media activity. As of 2013, it was reported that texting and driving was the leading cause of death for teen

---

[1] B. Parr, *Average Teen Sends 3,339 Texts Per Month* (Mashable), available at: http://www.cnn.com/2010/TECH/mobile/10/15/teen.texting.mashable/index.html.

[2] According to http://www.distraction.gov/content/get-the-facts/facts-and-statistics.html.

[3] *Id.* (reported by the official U.S. Government distracted driving website, citing to CTIA The Wireless Association).

[4] A. Scafuri, *Tips for Young Lawyers: Privacy Issues in Discovery of Social Media* (ABA Section of Litigation), available at: https://www.americanbar.org/groups/litigation/committees/commercial-business/articles/2014/tips-for-young-lawyers-privacy-issues-discovery-social-media.html.

drivers in the United States.[5] Re-read that sentence. Were it not for mobile devices and social media, potentially thousands of teenagers would not be dead or injured because of texting and driving.[6] In January 2018, it was reported that distracted drivers—including those texting and driving—caused 10% percent of fatal accidents and 15% of accidents resulting in injury.[7] Furthermore, 9% of the drivers who were between the ages of 15 and 19 involved in fatal accidents were driving while distracted.[8] The raw numbers? According to the *TeenSafe*, in 2015, almost 3,500 people were killed and 391,000 injured because of distracted drivers on our roads—of those, 11 teenagers die each day.[9] Thousands of families inexorably altered forever.

Even worse than these numbers are the following two—although 94% of teenagers know and admit that driving while distracted is dangerous, 35% admitted they have driven while texting on their mobile devices.[10]

As mentioned in Chapter 8, state laws are clear that *drivers* are responsible for controlling their vehicles and seeing what there is to be seen while driving on roadways.[11] Teenagers must be educated that it is not just dangerous to text and drive; it is *illegal*. Forty-eight states, and Washington, D.C., Puerto Rico, Guam, and the U.S. Virgin Islands have banned texting and driving.[12] If a driver injures or kills someone while driving, the least of his or her worries will be responding to a friend's text message. Further, if the teen is 18 or older, he or she is an adult and will face adult criminal charges. For instance, there is the reported case in Massachusetts of a driver who at the time of his conviction was 18. Of what charge was he convicted? Homicide. He had texted and driven, crossed the double yellow line on the road, and killed an oncoming driver while seriously injuring the passenger in the oncoming vehicle. The sentence imposed was two years *in jail*, with at least one year actually served for this homicide—at 18-years-old.[13] No message is worth that.[14]

---

[5] D. Ricks, *Study: Texting While Driving Now Leading Cause of Death for Teen Drivers* (Newsday, May 8, 2013), available at: https://www.newsday.com/news/nation/study-texting-while-driving-now-leading-cause-of-death-for-teen-drivers-1.5226036.

[6] *See, generally,* M. Taggart & B. Robinson, *The Very Real Dangers of Texting While Driving*, HUFFINGTON POST (Apr. 17, 2017), available at: https://www.huffingtonpost.com/entry/the-very-real-dangers-of-texting-while-driving us_58eeddd4e4b0156697224c60.

[7] *Teens Texting and Driving: Facts and Statistics* (TeenSafe Jan. 8, 2018), available at: https://www.teensafe.com/blog/teens-texting-and-driving-facts-and-statistics/.

[8] *Id.*

[9] *Id.*

[10] *Id.*

[11] *See* N.Y. VEHICLE & TRAFFIC LAW §§ 1225–c, 1225–d.

[12] *Cellular Phone Use and Texting While Driving Laws,* NAT'L CONF. OF STATE LEGISLATURES (May 29, 2019), *available at* ncsl.org/research/transportation/cellular-phone-use-and-texting-while-driving-laws.aspx.

[13] M. Taggart & B. Robinson, *supra*, *The Very Real Dangers of Texting While Driving*.

[14] *See, generally,* AT&T, Verizon, Sprint and T-Mobile "It Can Wait" campaigns, available at: https://www.itcan-wait.com/, https://www.verizon.com/about/news/vzw/2013/05/it-can-wait, and https://www.cbsnews.com/news/big-4-cell-phone-carriers-to-team-in-ads-against-texting-while-driving/.

While we may be aware of the risks of texting and driving at any age, there are two other more insidious problems involving teens and social media—bullying and sexting.[15]

> **Sexting** – Sexually explicit pictures or messages sent on mobile phones or devices. Not generally illegal unless involving minors, harassment, etc.

7.2 – Teenagers and "Sexting"

"Sexting" has become a uniquely popular activity in the United States. One U.S. District Court, citing a 2008 study of The National Campaign to Prevent Teen and Unplanned Pregnancy, wrote that it was "suggested that 22% of teen girls and 18% of teen boys ('teens' are defined as ages 13 to 19) and 36% of young adult women and 31% of young adult men ('young adults' are defined as ages 20–26) are sending or posting nude or semi-nude images of themselves."[16] But what those teens, in particular, do not understand is that by sending nude or semi-nude photos of themselves, or their underage friends, they are actually engaging in transmission of *child pornography*—with all attendant legal consequences.[17] Indeed, the *Nash* case in Alabama concerned the prosecution of a 22-year-old man for his possession of nude photographs sent by his 16-year-old girlfriend (interestingly, in Alabama the relationship itself was legal—the age of consent is 16).[18] Despite having a legal romantic relationship under state law, the boyfriend's possession of the photographs sent by text message was in violation of federal law. The boyfriend was sentenced to 60 months in prison.[19]

Additionally, the judge in the *Nash* case cited to other serious instances:

> [f]or example, an eighteen-year-old Florida man was convicted of distributing child pornography after sending nude pictures of his sixteen-year-old ex-girlfriend to her friends and family in a fit of anger over their breakup (the girl had taken the photos of herself). An eighteen-year-old Cincinnati woman committed suicide after nude pictures she had sent her boyfriend were circulated around her high school.[20]

---

[15] Sexting is defined, alternatively, as: "the sending of sexually explicit messages or images by cell phone" *U.S. v. Nash*, 1 F.Supp.3d 1240, 1241 n.1 (N.D. Ala. 2014) (citing *Merriam-Webster, m-w.com*); "[T]he practice of sending or posting sexually suggestive text messages and images, including nude or semi-nude photographs via cell phones or over the Internet." *U.S. v. Ashworth*, 2015 WL 5168340, at *1 n.1 (U.S. Navy-Marine Corps Ct. Crim. Apps. Sept. 3, 2015) (citing R. Patel, *Taking It Easy on Teen Pornographers: States Respond to Minors' Sexting*, 13 J. High Tech. L. 574, 575 (2013) (quoting *Miller v. Skumanick*, 605 F.Supp.2d 634, 637 (M.D. Pa. 2009))).

[16] *Nash*, 1 F.Supp.3d at 1245. *See also U.S. v. Henson*, 705 Fed.Appx. 348, 353 (6th Cir. 2017) (for additional disturbing numbers and information regarding underage teenage engagement in "sexting").

[17] *See U.S. v. Wiley*, 2017 WL 3430250, at *12 (U.S. Navy-Marine Corps. Ct. Crim. Apps. Aug. 10, 2017) ("our sister court recently concluded that 'a child can commit the offense of producing child pornography[,]' noting that child pornography is contraband, and finding '[t]he plain language of the offense has no exception that would allow children to produce and distribute child pornography, even when the images are of themselves.'") (citing and quoting *U.S. v. Thomas*, 2016 CCA LEXIS 551, at *10 (A. Ct. Crim. Apps. Sept. 9, 2016)).

[18] *Nash*, 1 F.Supp.3d at 1245.

[19] *Id.*

[20] *Id.* at 1245–1246 (citing E. Eraker, *Stemming Sexting: Sensible Legal Approaches to Teenagers' Exchange of Self-Produced Pornography*, 25 Berkeley Tech. L.J. 555, 559–60 (2010)).

In Colorado, there was the case of *T.B.* A criminal matter, T.B. was accused of raping two teenage girls. In the course of the investigation police found photographs of the juvenile girls on the phone of T.B., the juvenile male defendant. The juvenile male, T.B., was then also prosecuted for possession of child pornography—but the exploitation of a child charges were separated from the other charges. A jury acquitted the defendant of the severed charges. At a bench trial, the defendant was convicted (found delinquent) of two counts of sexual exploitation of a child.[21] He was sentenced to "two concurrent two-year terms of sex offender probation, and [required] to register as a sex offender."[22]

On appeal, the defendant argued, *inter alia*, that what we will colloquially call the anti-sexting statutes in Colorado (the statutes prohibiting sexually exploitive materials of minors) were not intended to apply to minors/juveniles possessing photos of other minors/juveniles.[23] The majority of the intermediate appellate court disagreed, stating, in part, "the sexual exploitation of a child act criminalizes teen sexting when it meets the enumerated elements of the statute. These elements are clear and unambiguous. Although the consequences for a convicted teenager may be substantial, as pointed out in the dissent, when the evidence satisfies the elements of the statute, we must apply the statute as written."[24] The Colorado Supreme Court agreed with the majority, and in 2019 affirmed the decision, with the U.S. Supreme Court denying certiorari/review in 2020.

Finally, take the case of E.G. in the State of Washington. E.G., a minor/juvenile, was found (adjudicated) to have committed the offense of second degree dealing in depictions of a minor engaged in sexually explicit conduct.[25] The defendant attempted to argue that the statutes under which he was prosecuted were overbroad and violations of First Amendment speech protection. The Court disagreed. "While it is inherently 'dangerous' to regulate any form of expression, certain categories of expression are exempt from First Amendment protections. . . . One such category is child pornography. In light of the State's interest in safeguarding the physical and psychological well-being of minors, the United States Supreme Court determined that *all* child pornography is exempt from First Amendment protection."[26] The Court also rejected the defendant's argument that there should be an exception for self-produced child pornography because the goal of protecting minors from exploitation by others is otherwise not served. The Court found that would be a far too easy exculpation for pornographers who could simply find willing minors to self-produce images and thereafter distribute them.[27]

---

[21] *People In Interest of T.B.*, 452 P.3d 36, (Colo. Ct. Apps., Div. VI 2016), *aff'd* 445 P.3d 1049 (Colo. 2019), *cert. denied* 140 S.Ct. 876 (2020) (review by the Colorado Supreme Court included "[w]hether the court of appeals misconstrued section 18–6–403(3)(b.5) in holding the evidence was sufficient to support the [juvenile defendant's] adjudication for sexual exploitation of a child").

[22] *Id.*

[23] *Id.* at *6-*7.

[24] *Id.* at *7.

[25] *State v. E.G.*, 194 Wash.App. 457, 377 P.3d 272 (Wash. Ct. Apps. Div. 3 2016), *aff'd*, *State v. Gray*, 189 Wash. 2d 334 (2017).

[26] *Id.* at 463, 275–276 (emphasis in original) (citing *New York v. Ferber*, 458 U.S. 747, 754–755, 764–765 (1982); *State v. Luther*, 157 Wash.2d 63, 70–71, 134 P.3d 205 (2006)).

[27] *State v. E.G.*, 194 Wash.App. at 463–464, 377 P.3d at 286.

Of even more interest, however, was the defendant's argument that the statutes were void for vagueness because, since perhaps 20% of teenagers engage in sexting and could be subject to prosecution, the law is neither clear nor understood by approximately one-fifth of the population.[28] The Court, however, noted two problems with the argument. First, of course, is the maxim that ignorance of the law is no defense. "[I]gnorance of a law is not the same as ignorance of the meaning of a law. The fact that minors may not appreciate they are breaking a law is not proof that they do not understand it."[29] Additionally,

> [t]he fact that a large number of people do something is not the test of whether police or prosecutors arbitrarily enforce a law, nor does not tell us whether it is vague. . . . [T]he test of vagueness is whether an ordinary person would understand the meaning of the statute. . . . It is the burden of the challenger to establish that the statute is vague. . . . E.G. does not attempt to meet that burden in this appeal. He can point to no ambiguity in the text nor otherwise show that the statute is not understandable.[30]

E.G.'s adjudication was affirmed by the appellate court and State Supreme Court. It is also worth noting that E.G., according to a footnote in the decision, suffered from Asperger's Syndrome—but that also did not prevent his prosecution and adjudication—although the trial court mitigated the sentence to time served and registration as a sex offender.[31]

7.3 – Teenagers and Cyberbullying

With regard to bullying, this matter has become an insidious and disgraceful result of teenage use of social media in all aspects of their lives. Teenagers utilize all kinds of social media platforms and even have their own websites. It is true that bullying has existed for as long as we have had schools. However, no longer can a teenager leave school and not face a bully again until the next school day. Now, bullies and their friends can reach, intimidate, humiliate, and ostracize a bullied student at 1 p.m. on a Tuesday or 10 p.m. on a Sunday. There is no escaping it for the victims. Legal consequences have resulted.

> **Cyberbullying** – An act of bullying that takes place online or through electronic means.

Consider the case of a gay teenager at Rutgers University, whose roommate recorded, and used social media to broadcast to his friends, the teen's romantic encounter with another person of the same sex. Not long after the teen became aware of what the roommate had done, he committed suicide by jumping from the George Washington Bridge, which connects New York City and New Jersey over the Hudson River.[32]

---

[28] *Id.* at 464–465, 276–277.

[29] *Id.* at 466, 277.

[30] *Id.* (citations omitted).

[31] *Id.* at 461, 274 n.2.

[32] *See* Tyler Clementi Foundation, available at: https://tylerclementi.org/tylers-story/; P. McGeehan, *Conviction Thrown Out for Ex-Rutgers Student in Tyler Clementi Case*, N.Y. Times (Sept. 9, 2016), available at: https://www.nytimes.com/2016/09/10/nyregion/conviction-thrown-out-for-rutgers-student-in-tyler-clementi-case.html (new trial ordered on 10 charges).

We can also examine the case of *D.C. v. R.R.*,[33] out of California. In the case, D.C. brought an "action against the other students and their parents, alleging a statutory claim under California's hate crimes laws . . . and common law claims for defamation and intentional infliction of emotional distress."[34] Plaintiff D.C. had his own website, because he was interested in a future career in the entertainment industry. It was alleged that defendants—fellow students—had posted messages on the plaintiff's website commenting on the plaintiff's sexual orientation and threatening harm.[35] The plaintiff, at the time, was 15-years-old.[36]

To illustrate the vulgarity and seriousness of cyberbullying, to demonstrate the vileness that teenagers apparently believe is "okay" and protected on social media, some of the actual postings, as relayed by the Court, will now be quoted here. According to the Court's decision, some posts read: "'Faggot, I'm going to kill you.' Another read, '[You need] a quick and painless death.' One student wrote, 'Fuck you in your fucking fuck hole.' . . . One post announced, 'You are now officially wanted dead or alive.'"[37] One of the defendants actually posted the following—again on the website of a 15-year-old:

> . . . I want to rip out your fucking heart and feed it to you. . . . I've . . . wanted to kill you. If I ever see you I'm . . . going to pound your head in with an ice pick. Fuck you, you dick-riding penis lover. I hope you burn in hell.[38]

After D.C. and his parents brought suit against defendant minor students and their parents, the defendant minor who posted the above comment sought to dismiss the claims against him under the anti-SLAPP statute (strategic lawsuit against public participation). The appellate court, affirming the trial court below, after a detailed analysis set forth in the opinion, determined the anti-SLAPP statute did not protect the defendant student's posting. "[D]efendants did not make the requisite showing that plaintiffs' complaint is subject to the anti-SLAPP statute. In particular, defendants did not demonstrate that the posted message is protected speech. Further, defendants contend the message was intended as 'jocular humor.' Assuming the message was a 'joke'—played by one teenager on another—it does not concern a 'public issue' under the statute."[39]

What teenagers, or minors, in our society do not seem to understand is that these kinds of posts are not acceptable, they are not "okay," and they are not "jocular humor"—not that anyone would likely really be under an illusion that such posts were meant as humor. Instead, this

---

[33] 182 Cal.App.4th 1190, 106 Cal.Rptr.3d 399 (Cal. App. 2010).

[34] *Id.* at 1199, 405.

[35] *Id.* at 1199, 404–405.

[36] *Id.*

[37] *Id.* at 1200, 406.

[38] *Id.* at 1199, 405.

[39] *Id.* at 1199, 405 (citing CAL. CODE CIV. P. § 425.16(b)(1), (e)(3)). The dissent disagreed with this holding, and following analysis, would actually have found this speech to be protected and addressing a matter of public interest. For further discussion and assessment of state anti-SLAPP provisions, see Amy Mattson, *Federal Court Renews Hostility Toward Anti-SLAPP Laws*, ABA Litigation, vol. 45, no.2, at 4-5 (Winter 2020).

conduct is illegal and has consequences beyond the legal system, affecting the lives and well-being of others, of the victims, and their families.

Fortunately here, the plaintiff did not commit suicide according to the reported case. However, the plaintiff must live the rest of his life recalling the hurtful words of others, posted on the Internet webpage for all to see. Additionally, plaintiff D.C. and his family, because of the advice they received from the police, withdrew D.C. from the school he was attending and re-located to Northern California and a new school. They relocated their entire lives because of the conduct of classmates. "As a consequence of defendants' conduct, plaintiffs suffered personal and emotional injury, loss of income, the payment of medical expenses, the cost of moving, expenses for traveling back and forth from their new residential location to Los Angeles in order to support D.C.'s professional career commitments, and the related cost of housing while staying in Los Angeles."[40]

Defendants, for their part, faced civil prosecution and liability, all because they thought it was acceptable to cyberbully a fellow juvenile student.

Then, there is the case of J.C., a minor student at Beverly Hills Unified School District.[41] J.C. and some friends went to a restaurant after school one day, and J.C. recorded several of them speaking badly about a classmate—C.C. On the video, J.C. could be heard encouraging her classmates to speak poorly of C.C. "One of Plaintiff's friends, R.S., calls C.C. a 'slut,' says that C.C. is 'spoiled,' . . . and uses profanity during the recording. . . . R.S. also says that C.C. is 'the ugliest piece of shit I've ever seen in my whole life.'"[42] These videos, and the comments heard on them, are not uncommon in our present world.

J.C. posted the video to YouTube, which made it available to the world for viewing. J.C. also contacted five to ten students about the video, as well as C.C.[43] The day after, school administrators were advised about the video, and they took disciplinary action against J.C.[44] Fortunately, C.C., the victim, did not commit suicide according to the reported decision, although she did suffer humiliation and hurt feelings.[45] J.C. was suspended from school, and her lawsuit (the subject of the court decision discussed here) failed. The Court held that the school district was able to discipline the student for off-campus conduct and that there was no violation of J.C.'s First Amendment rights.[46] The Court, however, did not take up the issue of cyberbullying and "[w]hether a student separately may be liable in tort for defamatory, derogatory, or threatening statements made about a classmate and published over the Internet."[47]

---

[40] *Id.* at 1201, 406 (although the school newspaper at the plaintiff's former school ran follow-up articles and actually disclosed plaintiff's new residence and school).

[41] *J.C. ex rel. R.C. v. Beverly Hills Unified Sch. Dist.*, 711 F.Supp.2d 1094 (C.D. Cal. 2010).

[42] *Id.* at 1098.

[43] *Id.*

[44] *Id.*

[45] *Id.*

[46] *Id.* Other courts have similarly held that students may be disciplined, or prosecuted, for activities off school grounds/off-campus, which affect school function. *See O.P-G v. State*, 290 So.3d 950 (Fla. Ct. App. 3d Dist. 2019).

[47] *Id.* at 1122 n.15 (citing to *D.C. v. R.R.*, 182 Cal.App.4th 1190, 106 Cal.Rptr.3d 399 (Cal. App. 2010)).

Finally, there is the very distressing case of Michelle Carter and Conrad Roy, from Massachusetts. The case stretches all the way back to the year 2014, when Ms. Carter was alleged to have encouraged Mr. Roy to commit suicide.[48] At the time of Mr. Roy's death he was eighteen, but Ms. Carter was seventeen and was charged with involuntary manslaughter as a youthful offender.[49] Both Ms. Carter and Mr. Roy had been romantically involved at various times between 2011 and 2014.[50] Mr. Roy was found by authorities to have committed suicide in his vehicle, through inhalation of carbon monoxide created from a gasoline-powered water pump inside the vehicle—it was not the first time he had attempted suicide.[51] Police reviewed Mr. Roy's electronic communications during the investigation into his death, and located many messages exchanged between Mr. Roy and Ms. Carter. In fact, the Court summed up the interactions under investigation, and the subject of Ms. Carter's indictment, as follows:

> The grand jury heard testimony and were presented with transcripts concerning the content of those text messages in the minutes, days, weeks, and months leading up to the defendant's suicide. The messages revealed that the defendant was aware of the victim's history of mental illness, and of his previous suicide attempt, and that much of the communication between the defendant and the victim focused on suicide. Specifically, the defendant encouraged the victim to kill himself, instructed him as to when and how he should kill himself, assuaged his concerns over killing himself, and chastised him when he delayed doing so. The theme of those text messages can be summed up in the phrase used by the defendant four times between July 11 and July 12, 2014 (the day on which the victim committed suicide): "You just [have] to do it."[52]

The Court also cited specific text conversations between Ms. Carter (the Defendant) and Mr. Roy (the Victim), several of which went like this:

> On July 11, 2014, at 5:13 P.M., the defendant sent the victim the following text message: "... Well in my opinion, I think u should do the generator because I don't know much about the pump and with a generator u can't fail"

> On July 12, 2014, between 4:25 A.M. and 4:34 A.M., they exchanged the following text messages:
> DEFENDANT: "So I guess you aren't gonna do it then, all that for nothing"
> DEFENDANT: "I'm just confused like you were so ready and determined"
> VICTIM: "I am gonna eventually"
> VICTIM: "I really don't know what I'm waiting for ... but I have everything lined up"
> DEFENDANT: "No, you're not, Conrad. Last night was it. You keep pushing it off and you say you'll do it but u never do. Its always gonna be that way if u don't take action"

---

[48] *See Commonwealth v. Carter*, 474 Mass. 624, 52 N.E.3d 1054 (2016).
[49] *Id.*
[50] *Id.* at 626, 1057.
[51] *Id.* at 625, 1056.
[52] *Id.* at 626-629, 1057-1058 (footnotes omitted).

DEFENDANT: "You're just making it harder on yourself by pushing it off, you just have to do it"

DEFENDANT: "Do u wanna do it now?"

VICTIM: "Is it too late?"

VICTIM: "Idkk it's already light outside"

VICTIM: "I'm gonna go back to sleep, love you I'll text you tomorrow"

DEFENDANT: "No? Its probably the best time now because everyone's sleeping. Just go somewhere in your truck. And no one's really out right now because it's an awkward time"

DEFENDANT: "If u don't do it now you're never gonna do it"

DEFENDANT: "And u can say you'll do it tomorrow but you probably won't"

…

At various times between July 4, 2014, and July 12, 2014, the defendant and the victim exchanged several text messages:

DEFENDANT: "You're gonna have to prove me wrong because I just don't think you really want this. You just keeps pushing it off to another night and say you'll do it but you never do"

…

DEFENDANT: "SEE THAT'S WHAT I MEAN. YOU KEEP PUSHING IT OFF! You just said you were gonna do it tonight and now you're saying eventually...."

…

DEFENDANT: "But I bet you're gonna be like 'oh, it didn't work because I didn't tape the tube right or something like that' ... I bet you're gonna say an excuse like that"

…

DEFENDANT: "Do you have the generator?"

VICTIM: "not yet lol"

DEFENDANT: "WELL WHEN ARE YOU GETTING IT"

…

DEFENDANT: "You better not be bull shiting me and saying you're gonna do this and then purposely get caught"

…

DEFENDANT: "You just need to do it Conrad or I'm gonna get you help"

DEFENDANT: "You can't keep doing this everyday"

VICTIM: "Okay I'm gonna do it today"

DEFENDANT: "Do you promise"

VICTIM: "I promise babe"

VICTIM: "I have to now"

DEFENDANT: "Like right now?"

VICTIM: "where do I go? :("

DEFENDANT: "And u can't break a promise. And *just go in a quiet parking lot or something*" [emphasis added by Court].[53]

---

[53] *Id.* at 626-628, 1057-1058, nn. 4-6 (spelling and grammar errors in original).

At a later time, according to evidence in court, Ms. Carter admitted her culpability to a friend regarding Mr. Roy's death, saying in a text message: "'Sam, [the victim's] death is my fault like honestly I could have stopped him I was on the phone with him and he got out of the [truck] because it was working and he got scared and I fucking told him to get back in Sam because I knew he would do it all over again the next day and I couldnt have him live the way he was living anymore I couldnt do it I wouldnt let him.'"[54]

Defendant Carter was indicted for involuntary manslaughter under the state's youthful offender statute because she was seventeen at the time of the crime, and the case made its first trip to the Massachusetts Supreme Judicial Court (that state's highest court) after defendant moved to dismiss the charge. The indictment stood, and defendant was thereafter convicted following a bench trial.[55] Defendant appealed the conviction back to the Supreme Judicial Court, which affirmed the conviction, rejected her arguments, among other things, that the conviction based on her words alone violated her First Amendment rights, and held: "[t]he evidence against the defendant proved that, by her wanton or reckless conduct, she caused the victim's death by suicide. Her conviction of involuntary manslaughter as a youthful offender is not legally or constitutionally infirm. The judgment is therefore affirmed."[56]

The United States Supreme Court subsequently denied Defendant Carter's petition for writ of certiorari.[57]

You might ask why a case such as this is discussed in this section of the book, when the courts did not specifically discuss "bullying" in the decisions. The case is included herein precisely because looking at the language of the text messages between the two youths, it seems that the defendant utilized the text messages and electronic communications to harass and push Mr. Roy into taking actions about which he appeared to have doubts or fears. Such looks like it falls under the umbrella of bullying an individual with cybertechnology, and is a sad case that stands as a warning to all.

7.4 – Revenge Pornography, and the Dark Underbelly of Social Media

"Revenge pornography" (also known as "revenge porn" or "nonconsensual pornography") is a problem/concern that has existed since the early 2000's, but which is finding its way into the news and public consciousness more and more each year. Revenge pornography is "a subset of nonconsensual pornography published for vengeful purposes. 'Nonconsensual pornography' may be defined generally as 'distribution of sexually graphic images of individuals

---

[54] *Id.* at 629, 1059 & n. 8 (spelling and grammar errors in original).
[55] *Commonwealth v. Carter*, 481 Mass. 352, 115 N.E.3d 559 (2019).
[56] *Id.* at 371, 574.
[57] *Carter v. Massachusetts*, 140 S. Ct. 910 (2020).

without their consent.' The term 'nonconsensual pornography' encompasses 'images originally obtained without consent (e.g., hidden recordings or recordings of sexual assaults) as well as images originally obtained with consent, usually within the context of a private or confidential relationship.'"[58] It does not address or concern itself solely with sexting, but with numerous other scenarios, as well, including those where individuals who take part in explicit photos or videos while engaging in private relationships have expectations that those images will remain private to the participants only.[59]

Some reports claim revenge pornography to be a prevalent concern in our society—"4% of U.S. internet users—roughly 10.4 million Americans—have been threatened with or experienced the posting of explicit images without their consent."[60] The psychological effects, of course, can be damaging and long-lasting—as can the professional effects; such was the case for a now-former Member of Congress.[61]

Revenge pornography (often done for vengeful, mean-spirited purposes) has been legislated in more than forty states (46 at last count), with New Jersey having been the first to do so in 2004;[62] and, again, it is important to understand that the illegal action can involve the dissemination of material obtained both without participants' knowledge, or with what was at one time consent for the images. "Generally, the crime involves images originally obtained without consent, such as by use of hidden cameras or victim coercion, and images originally obtained with consent, usually within the context of a private or confidential relationship. Once obtained, these images are subsequently distributed without consent."[63] As a matter of federal law, the Federal Trade Commission (FTC) has brought enforcement actions against individuals affiliated with companies believed to be involved with revenge (nonconsensual) pornography.[64]

---

[58] *State v. VanBuren*, 214 A.3d 791, 794-795 (Vt. 2019) (citing D. Citron & M. Franks, *Criminalizing Revenge Porn*, 49 Wake Forest L. Rev. 345, 346 (2014)).

[59] *See VanBuren*, 214 A.3d at 795 n.1.

[60] *Id.* at 795 (citation omitted).

[61] *State v. Casillas*, 938 N.W.2d 74, 84 & n.4 (Minn. Ct. Apps. 2019) (collecting scholarly articles on the subject). *See also* Lorelei Laird, *First Amendment Defense Claims Could Threaten "Revenge Pornography" Statutes*, ABA Journal (online Dec. 19, 2019), *available at* https://www.abajournal.com/web/article/first-amendment-defense-claims-could-threaten-revenge-pornography-statutes (discussing, among other things, the impact of revenge pornography causing the resignation of former Rep. Katie Hill, Member of Congress, after photos of her with another person were made public).

[62] *VanBuren*, 214 A.3d at 795 (citation omitted); *Miranda v. S. Country Cent. Sch. Dist.*, — F.Supp.3d —, 2020 WL 2563091 at *4 & n.3 (E.D.N.Y. May 21, 2020) (citing Cyber Civil Rights Initiative, *46 States + DC Have Revenge Porn Laws*, https://www.cybercivilrights.org/revenge-porn-laws/ (collecting state statutes)); Joseph J. Pangaro, Comment, *Hell Hath No Fury: Why First Amendment Scrutiny Has Led to Ineffective Revenge Porn Laws, and How to Change the Analytical Argument to Overcome This Issue*, 88 Temple L. Rev. 185 (2015).

[63] *People v. Austin*, — N.E.3d —, 2019 WL 5287962, at *3 (Ill. Oct. 18, 2019) (citing D. Citron & M. Franks, *Criminalizing Revenge Porn*, 49 Wake Forest L. Rev. 345, 346 (2014); Adrienne N. Kitchen, *The Need to Criminalize Revenge Porn: How a Law Protecting Victims Can Avoid Running Afoul of the First Amendment*, 90 Chi.-Kent L. Rev. 247, 247-48 (2015)).

[64] *Fed. Trade Comm'n v. EMP Media, Inc.*, — F.R.D. —, 2020 WL 1889014 (D. Nev. Apr. 9, 2020).

Over the years, though, a number of these state statutes have been challenged as violative of the First Amendment, with varying degrees of success. For example, in Vermont the State Supreme Court upheld that jurisdiction's law against a First Amendment challenge;[65] as have other courts when evaluating restrictions on activities, including sexting, and finding their laws have not been overbroad under the First Amendment.[66] However, some courts have struck down their jurisdiction's laws on this subject, finding them beyond the sweep of legitimate prohibitions based on the statutes' construction.[67]

In New York, one of the most recent states to enact legislation making revenge pornography a criminal offense—Governor Andrew Cuomo signed the legislation on July 23, 2019—knowledgeable commentators have expressed their belief that the state's law will survive scrutiny in the courts.[68]

Time will tell how successful states and the federal government will be in addressing and countering revenge pornography through legislation and criminal prosecution. But, in the words of Dean Prudenti—revenge pornography, when you understand it at its base, is "wrong by any standard of decency."[69]

7.5 – Wrongful Social Media Posts Can Have Associated "Costs"

Social media, as addressed in other segments of this book, is prevalent across society. It is a ubiquitous and omnipresent part of modern life. So much so it has been reported that as much

---

[65] *VanBuren*, 214 A.3d 791, 800 ("For the reasons set forth below, we conclude that 'revenge porn' does not fall within an established categorical exception to full First Amendment protection, and we decline to predict that the U.S. Supreme Court would recognize a new category. However, we conclude that the Vermont statute survives strict scrutiny as the U.S. Supreme Court has applied that standard").

[66] *U.S. v. Streett*, — F.Supp.3d —, 2020 WL 231688 (D. N.M. Jan. 15, 2020); *People v. Austin*, 2019 WL 5287962, at *22 ("As a final matter, we observe that [720 I.. Comp. Stat. Ann.] section 11-23.5 is 'regarded as the country's strongest anti-revenge-porn legislation yet' (internal quotation marks omitted)... and has been proposed as the model for a federal statute targeting the nonconsensual dissemination of private sexual images.... Indeed, section 11-23.5 is regarded as 'a model for all state revenge porn laws.'... Based on the foregoing, we find that section 11-23.5 does not unconstitutionally restrict the rights to free speech and due process on the grounds asserted by defendant") (citations omitted).

[67] *See State v. Casillas*, 938 N.W.2d at 90-91; *Ex parte Jones*, — S.W.3d —, 2018 WL 2228888 (Tex. Ct. Apps., Tyler, May 16, 2018); *but cf. Ex parte Lopez*, 2019 WL 1905243 (Tex. Ct. Apps., Beaumont, Mar. 27, 2019) (unpublished) (reaching a different conclusion, and noting that the decision in *Ex parte Jones* will be reviewed by the Texas Court of Criminal Appeals).

[68] Gail Prudenti, *NY State's New Revenge Porn Law Will Likely Be Effective*, N.Y. L.J. at 6 (Oct. 2, 2019) (letter to the editor; noting that states have had both successes and failures in designing legislation to keep up with ever-changing technological activities without running afoul of constitutional protections) (Gail Prudenti is the Dean and Executive Director of the Center for Children, Families, and the Law at Hofstra University Maurice A Dean School of Law, and the former Presiding Justice of the New York State Supreme Court, Appellate Division, Second Department, and former Chief Administrative Judge of the Courts of the State of New York).

[69] *Id.*

as seventy percent (70%) of the U.S. adult population use at least one form of social media, and that sixty-nine percent (69%) of the U.S. adult population use Facebook (with seventy-four percent (74%) of those using it at least daily, if not more frequently).[70] Seventy-three percent (73%) of the U.S. adult population utilize YouTube.[71]

For Americans in the age bracket 18-24, preferences tend more toward Instagram and Snapchat, with seventy-five percent (75%) and seventy-three percent (73%) of 18-24 year-olds preferring those platforms, respectively.[72]

Aside from revenge pornography, and additional matters raised in this and other chapters, the pervasiveness of social media in our society, and the ease with which individuals often post without thinking twice, has caused a number of concerns and problems. For example, in 2020, with the rise of the COVID-19 virus, and restrictions put into place to combat infection-spread, individuals in society seemed to form into two "camps," for lack of a better description. In one camp resided those who believed it important to social distance, wear masks and gloves in public, and obey government restrictions imposed for public health, safety, and welfare. In the other camp resided those who resented the restrictions, thought them violative of constitutional rights, and/or disputed the scientific basis behind the government's actions. From this intellectual dispute arose a very real physical dispute, where people in the latter camp would resort to physical violence if confronted by those in stores or elsewhere requesting that they wear masks and/or gloves, and socially distance.[73]

As one would expect, social media abounded with posts in support of both camps at the same time. However, at least one post garnered extra attention—and earned its author a pink slip from their employer. During the controversial time, an administrative employee with a law firm in Texas made offensive posts on social media, including a threat referencing a handgun, because he did not wish to wear a mask at a Dallas Whole Foods.[74] The employer terminated the employee for having violated the values of the firm[75] (understand that the First Amendment does not apply with the same force to private employers versus government/

---

[70] Inset Box, ABA Journal at 47 (Feb./Mar. 2020); A. Perrin & M. Anderson, *Share of U.S. Adults Using Social Media, Including Facebook, is Mostly Unchanged Since 2018*, FactTank (Apr. 10, 2019), *available at* https://www.pewresearch.org/fact-tank/2019/04/10/share-of-u-s-adults-using-social-media-including-facebook-is-mostly-un-changed-since-2018/.

[71] *Id. See also* Christina M. Jordan, *Discovery of Social Media Evidence in Legal Proceedings*, ABA Litigation, vol. 45, no. 1, at 2 (Fall 2019).

[72] *See* A. Perrin & M. Anderson, *Share of U.S. Adults Using Social Media, supra.*

[73] *See* Bill Hutchinson, *"Incomprehensible:" Confrontations Over Masks Erupt Amid COVID-19 Crisis*, ABC News (online May 7, 2020), *available at* https://abcnews.go.com/US/incomprehensible-confrontations-masks-erupt-amid-covid-19-crisis/story?id=70494577.

[74] Debra Cassens Weiss, *Firm Fires Staffer for "No More Masks" Social Media Post That Referred to Glock Pistol*, ABA Journal (online May 11, 2020), *available at* https://www.abajournal.com/news/article/firm-fires-staffer-for-no-more-masks-post-that-referred-to-glock-pistol; Brenda Sapino Jeffreys, *Thompson & Knight Fires Staffer Over "Threatening" Social Media Post*, Law.com (online May 11, 2020), *available at* https://www.law.com/texaslawyer/2020/05/11/thompson-knight-fires-staffer-over-threatening-social-media-post/.

[75] *Id.*

public employers)[76]—the employee made public statements online that did not comport with the principles and ideals of the employer.[77] A similar reason was provided by entertainment and news giant CBS when it terminated an attorney in 2017 for an inappropriate social media post in response to the mass shooting in Las Vegas.[78]

Employment is not the only realm where "costs" may exist in response to wrongful or improper social media posts. Another example again stems from the COVID-19 era, this time related to state laws requiring quarantine. In May of 2020, a tourist from New York State went to the State of Hawaii for vacation, in the midst of the COVID-19 pandemic. Hawaii at the time had a 14-day mandatory quarantine requirement for those people coming from the mainland.[79] The 23-year-old tourist posted pictures of himself enjoying the beach on Instagram—during the time period he was supposed to be in quarantine in his hotel room.[80] To make matters worse, authorities in Hawaii saw the photos after being alerted by citizens, and noted that the tourist had disobeyed documentation that acknowledged violations of quarantine were criminal offenses. Officials in Hawaii subsequently arrested the individual.[81]

However, just as posting on social media could have its associated "dangers" or "costs," as clearly illustrated above, in some circumstances posting may also find itself protected by the First Amendment to the U.S. Constitution. Thus, consider the following two cases.

First, on June 30, 2020, the Third Circuit U.S. Court of Appeals issued a decision that commentators such as the American Civil Liberties Union called a "landmark."[82] In *B.L. by and through Levy v. Mahanoy Area School District*,[83] a Pennsylvania high school student posted a short rant on Snapchat about her school and school cheer team, among other things, using the F-word because she was upset about making the junior varsity cheerleading squad instead of the varsity squad. The school district and coaches took action disciplining the

---

[76] *Alterio v. Almost Family, Inc.*, 2019 WL 7037789, at *9 (D. Conn. Dec. 20, 2019) (citing *Garcetti v. Ceballos*, 547 U.S. 410 (2006)); *Schumann v. Dianon Systems, Inc.*, 304 Conn. 585, 43 A.3d 111 (2012).

[77] D. Weiss, *Firm Fires Staffer, supra*; B. Jeffreys, *Thompson & Knight Fires Staffer, supra*.

[78] *See CBS Fires Lawyer for Social-Media Comment on Las Vegas Shootings* (Associated Press), The Seattle Times (online Oct. 2, 2017) (the employee posted "she wasn't sympathetic because 'country music fans often are Republican gun-toters'"), *available at* https://www.seattletimes.com/business/cbs-fires-lawyer-for-social-media-comment-on-las-vegas/. Candidates for elected office might suffer consequences to their political fortunes as a result of inappropriate social media posts and use, as well. *See* Ally Mutnick, *Top Republican Yanks Endorsement of House Candidate Who Made Offensive Social Media Posts*, Politico (online May 25, 2020), *available at* https://www.politico.com/news/2020/05/25/kevin-mccarthy-pulls-ted-howze-endorsement-281418?cid=apn.

[79] Madeline Holcombe, *New York Tourist is Arrested in Hawaii After Posting Beach Pictures on Instagram*, CNN (online May 17, 2020), *available at* https://www.cnn.com/2020/05/16/us/hawaii-arrest-coronavirus-trnd/index.html.

[80] *Id.*

[81] *Id.*

[82] *See* Debra Cassens Weiss, *'Landmark Decision' Backs Cheerleader Kicked Off Squad for Snapchat F-word Post*, ABA Journal (online July 2, 2020), *available at* https://www.abajournal.com/news/article/federal-appeals-court-rules-for-cheerleader-kicked-off-squad-for-snapchat-message?utm_source=salesforce_233916&utm_medium=email&utm_campaign=weekly_email&utm_medium=email&utm_source=salesforce_233916&sc_sid=01165356&utm_campaign=&promo=&utm_content=&additional4=&additional5=&sfmc_j=233916&sfmc_s=45062262&sfmc_l=1527&sfmc_jb=29&sfmc_mid=100027443&sfmc_u=7490980.

[83] — F.3d —, 2020 WL 3526130 (3d Cir. June 30, 2020).

student and removing her from the JV squad.[84] Student sued under the First Amendment, and the U.S. District Court for the Middle District of Pennsylvania held in her favor, finding the school district violated the First Amendment by overreaching its authority and disciplining the student for off-campus activity (the social media postings) that did not create an actual or foreseeable substantial disruption when it came to the environment in defendant's school or district.[85] The Third Circuit affirmed, examining and distinguishing *Bethel School District No. 403 v. Fraser* and the Vietnam War-era Supreme Court case of *Tinker v. Des Moines Independent Community School District,*[86] finding *Fraser* and *Tinker* do not apply to off-campus speech that does not cause disruption of the school environment, and that the holdings of other courts "distort *Tinker's* narrow exception into a vast font of regulatory authority."[87] Given the importance of this "landmark" decision, large portions of the opinion are provided at length below, for analysis.

B.L. by and through Levy v. Mahanoy Area School District (U.S. Court of Appeals, Third Circuit)

Krause, U.S. Circuit Judge

Public school students' free speech rights have long depended on a vital distinction: We "defer to the school[ ]" when its "arm of authority does not reach beyond the schoolhouse gate," but when it reaches beyond that gate, it "must answer to the same constitutional commands that bind all other institutions of government." *Thomas v. Bd. of Educ.*, 607 F.2d 1043, 1044–45 (2d Cir. 1979). The digital revolution, however, has complicated that distinction. With new forms of communication have come new frontiers of regulation, where educators assert the power to regulate online student speech made off school grounds, after school hours, and without school resources.

This appeal takes us to one such frontier. Appellee B.L. failed to make her high school's varsity cheerleading team and, over a weekend and away from school, posted a picture of herself with the caption "fuck cheer" to Snapchat. J.A. 484. She was suspended from the junior varsity team for a year and sued her school in federal court. The District Court granted summary judgment in B.L.'s favor, ruling that the school had violated her First Amendment rights. We agree and therefore will affirm.

I. BACKGROUND

B.L. is a student at Mahanoy Area High School (MAHS). As a rising freshman, she tried out for cheerleading and made junior varsity. The next year, she was again placed on JV. To add insult to injury, an incoming freshman made the varsity team.

---

[84] *Id.*

[85] *Id.*

[86] *Fraser*, 478 U.S. 675, 106 S.Ct. 3159 (1986); *Tinker*, 393 U.S. 503, 89 S.Ct. 733 (1969).

[87] — F.3d —, 2020 WL 3526130, at *12.

B.L. was frustrated: She had not advanced in cheerleading, was unhappy with her position on a private softball team, and was anxious about upcoming exams. So one Saturday, while hanging out with a friend at a local store, she decided to vent those frustrations. She took a photo of herself and her friend with their middle fingers raised and posted it to her Snapchat story. The snap was visible to about 250 "friends," many of whom were MAHS students and some of whom were cheerleaders, and it was accompanied by a puerile caption: "Fuck school fuck softball fuck cheer fuck everything." J.A. 484. To that post, B.L. added a second: "Love how me and [another student] get told we need a year of jv before we make varsity but that's [sic] doesn't matter to anyone else? [*Upside down smiley face emoji*]."...

One of B.L.'s teammates took a screenshot of her first snap and sent it to one of MAHS's two cheerleading coaches. That coach brought the screenshot to the attention of her co-coach, who, it turned out, was already in the know: "Several students, both cheerleaders and non-cheerleaders," had approached her, "visibly upset," to "express their concerns that [B.L.'s] [s]naps were inappropriate." J.A. 7 (citations omitted).

The coaches decided B.L.'s snap violated team and school rules, which B.L. had acknowledged before joining the team, requiring cheerleaders to "have respect for [their] school, coaches, ... [and] other cheerleaders;" avoid "foul language and inappropriate gestures;" and refrain from sharing "negative information regarding cheerleading, cheerleaders, or coaches ... on the internet." J.A. 439. They also felt B.L.'s snap violated a school rule requiring student athletes to "conduct[ ] themselves in such a way that the image of the Mahanoy School District would not be tarnished in any manner." J.A. 486. So the coaches removed B.L. from the JV team. B.L. and her parents appealed that decision to the athletic director, school principal, district superintendent, and school board. But to no avail: Although school authorities agreed B.L. could try out for the team again the next year, they upheld the coaches' decision for that year. Thus was born this lawsuit.

B.L. sued the Mahanoy Area School District (School District or District) in the United States District Court for the Middle District of Pennsylvania. She advanced three claims under 42 U.S.C. § 1983: that her suspension from the team violated the First Amendment; that the school and team rules she was said to have broken are overbroad and viewpoint discriminatory; and that those rules are unconstitutionally vague.

The District Court granted summary judgment in B.L.'s favor. It first ruled that B.L. had not waived her speech rights by agreeing to the team's rules and that her suspension from the team implicated the First Amendment even though extracurricular participation is merely a privilege. Turning to the merits, the Court ruled that B.L.'s snap was off-campus speech and thus not subject to regulation under *Bethel School District No. 403 v. Fraser*, 478 U.S. 675, 106 S.Ct. 3159, 92 L.Ed.2d 549 (1986). And, finding that B.L.'s snap had not caused any actual or foreseeable substantial disruption of the school environment, the Court ruled her snap was also not subject to discipline under *Tinker v. Des Moines Independent Community School District*, 393 U.S. 503, 89 S.Ct. 733, 21 L.Ed.2d 731 (1969). The Court therefore concluded that the School

District had violated B.L.'s First Amendment rights, rendering unnecessary any consideration of her overbreadth, viewpoint discrimination, or vagueness claims. It entered judgment in B.L.'s favor, awarding nominal damages and requiring the school to expunge her disciplinary record. This appeal followed.

II. DISCUSSION

The First Amendment provides that "Congress shall make no law ... abridging the freedom of speech." U.S. Const. amend. I. Over time, those deceptively simple words have spun off a complex doctrinal web. The briefs here are a testament to that complexity, citing a wealth of cases involving not only student speech but also public employee speech, obscenity, indecency, and many other doctrines.

At its heart, though, this appeal requires that we answer just two questions. The first is whether B.L.'s snap was protected speech. If it was not, our inquiry is at an end. But if it was, we must then decide whether B.L. validly waived that protection. Although navigating those questions requires some stopovers along the way, we ultimately conclude that B.L.'s snap was protected and that she did not waive her right to post it.

A. B.L.'s Speech Was Entitled to First Amendment Protection

We must first determine what, if any, protection the First Amendment affords B.L.'s snap. To do so, we begin by canvassing the Supreme Court's student speech cases. Next, we turn to a threshold question on which B.L.'s rights depend: whether her speech took place "on" or "off" campus. Finally, having found that B.L.'s snap was off-campus speech, we assess the School District's arguments that it was entitled to punish B.L. for that speech under Fraser, Tinker, and several other First Amendment doctrines.

. . . .

3. The punishment of B.L.'s off-campus speech violated the First Amendment

We next ask whether the First Amendment allowed the School District to punish B.L. for her off-campus speech. The District defends its decision under (i) *Fraser*, (ii) *Tinker*, and (iii) a series of First Amendment doctrines beyond the student speech context. We address each in turn.

. . . .

ii. Nor can B.L.'s punishment be justified under *Tinker*

The School District falls back on *Tinker*, arguing that B.L.'s snap was likely to substantially disrupt the cheerleading program. But as we have explained, although B.L.'s snap involved the school and was accessible to MAHS students, it took place beyond the "school context," *J.S.*, 650 F.3d at 932. We therefore confront the question whether *Tinker* applies to off-campus speech.

That is a question we have avoided answering to date. In *Layshock*, the school defended its decision to punish the student only under *Fraser*. *See* 650 F.3d at 216. And in *J.S.*, we were able to "assume, without deciding," that *Tinker* applied to speech like J.S.'s, 650 F.3d at 926, because we held that the school had not "reasonably forecast[ ] a substantial disruption of or material interference with the school," *id.* at 931. But the question is once again squarely before us, and for three reasons we conclude we must answer it today.

First, our choice to sidestep the issue in J.S. adhered to the maxim that, where possible, we should avoid difficult constitutional questions in favor of simpler resolutions. There, it was sensible to avoid the issue because we could resolve the case by applying well-settled precedent addressing the substantial disruption standard in the context of the school environment. *See, e.g., Sypniewski v. Warren Hills Reg'l Bd. of Educ.*, 307 F.3d 243, 254–57 (3d Cir. 2002); *Saxe*, 240 F.3d at 211–12. But that is not the case here. The School District's defense of its decision to punish B.L. focuses not on disruption of the school environment at large, but on disruption in the extracurricular context—specifically, the cheerleading program B.L. decried in her snap. And, as the parties' and amici's dueling citations reveal, the question of how to measure the potentially disruptive effect of student speech on particular extracurricular activities has bedeviled our sister circuits, and it is not one we have addressed to date. So were we to leapfrog *Tinker's* applicability in favor of substantial disruption analysis, we would still face complex and unresolved constitutional questions.

Second, when we decided *J.S.*, the social media revolution was still in its infancy, and few appellate courts had grappled with *Tinker's* application to off-campus online speech. In avoiding the issue, we afforded our sister circuits the chance to coalesce around an approach and the Supreme Court the chance to resolve the issue. Nearly a decade later, however, we see not only that social media has continued its expansion into every corner of modern life, but also that no dominant approach has developed. All the while, we have relegated district courts in this Circuit to confronting this issue without clear guidance, prompting them to turn elsewhere for support, *see, e.g., Dunkley v. Bd. of Educ.*, 216 F. Supp. 3d 485, 492–94 (D.N.J. 2016), and to voice their growing frustration. As one of our district judges put it, "a district court in this Circuit takes up a student off-campus speech case for review with considerable apprehension and anxiety." *R.L. ex rel. Lordan v. Cent. York Sch. Dist.*, 183 F. Supp. 3d 625, 635 (M.D. Pa. 2016).

Finally, while legal uncertainty of any kind is undesirable, uncertainty in this context creates unique problems. Obscure lines between permissible and impermissible speech have an independent chilling effect on speech. *See, e.g., Ashcroft v. Free Speech Coal.*, 535 U.S. 234, 244, 122 S.Ct. 1389, 152 L.Ed.2d 403 (2002) (reasoning that the "uncertain reach" of a law punishing speech would "chill speech within the First Amendment's vast and privileged sphere"). And because local officials are liable for constitutional violations only where "every reasonable official would understand that what he is doing is unlawful," *Russell v. Richardson*, 905 F.3d 239, 251 (3d Cir. 2018) (internal quotation marks and citation omitted), the unresolved issue of *Tinker's* scope has left a significant obstacle in the path of any student seeking to vindicate her free speech rights through a § 1983 suit. *See, e.g., Longoria ex rel. M.L. v. San Benito Indep. Consol. Sch. Dist.*, 942 F.3d 258, 267 (5th Cir. 2019) (holding that because the court had "declin[ed] to adopt a 'specific rule,'" its case law applying *Tinker* to off-campus speech "does not constitute clearly-established binding law that should have placed the defendants on notice about the constitutionality of their actions").

The time has come for us to answer the question. We begin by canvassing the decisions of our sister circuits. We then consider the wisdom of their various approaches, tested against *Tinker's* precepts. Finally, we adopt and explain our own, concluding that Tinker does not apply to off-campus speech and reserving for another day the First Amendment implications of off-campus student speech that threatens violence or harasses others.

. . . .

iii. None of the School District's remaining arguments justifies its punishment of B.L.

. . . .

For these reasons, we hold that B.L.'s snap was not subject to regulation under *Tinker* or *Fraser* and instead enjoyed the full scope of First Amendment protections.

. . . .

B. B.L. Did Not Waive Her Free Speech Rights

We therefore hold that B.L.'s snap was not covered by any of the rules on which the School District relies and reject its contention that B.L. waived her First Amendment rights.

* * *

The heart of the School District's arguments is that it has a duty to "inculcate the habits and manners of civility" in its students. Appellant's Br. 24 (citation omitted). To be sure, B.L.'s snap was crude, rude, and juvenile, just as we might expect of an adolescent. But the primary responsibility for teaching civility rests with parents and other members of the community. As arms of the state, public schools have an interest in teaching civility by example, persuasion, and encouragement, but they may not leverage the coercive power with which they have been entrusted to do so. Otherwise, we give school administrators the power to quash student expression deemed crude or offensive—which far too easily metastasizes into the power to censor valuable speech and legitimate criticism. Instead, by enforcing the Constitution's limits and upholding free speech rights, we teach a deeper and more enduring version of respect for civility and the "hazardous freedom" that is our national treasure and "the basis of our national strength." *Tinker*, 393 U.S. at 508–09, 89 S.Ct. 733.

III. CONCLUSION

For the foregoing reasons, we will affirm the judgment of the District Court.[88]

---

[88] *Id.*

Finally, consider this question: can posting false statements on social media result in *criminal* charges? This is not a question of civil tort claims for libel or slander, but rather criminal charges that carry with them concomitant criminal penalties. Such was an issue addressed in the New York case of *People v. Burwell*.[89] In that case, defendant was charged with, among other things, four counts of making a false statement when she claimed she was the victim of a race crime on a bus on the way to the State University of New York at Albany campus.[90] The false statements of the defendant as to the alleged race crime were circulated by her on social media. Defendant, after conviction, moved to set aside the conviction, arguing, *inter alia*, that the conviction concerning the social media statements violated her First Amendment rights.[91] The trial court denied the motion, but the New York Supreme Court, Appellate Division, Third Department reversed and held that the First Amendment did provide protection to the defendant with regard to the social media activity. The appellate court specifically held, in detail:

> Nevertheless, because we agree with defendant's contention that Penal Law § 240.50(1) as applied here is unconstitutional, the judgment of conviction as to count 7 must be reversed. Specifically, defendant argues that the "public alarm" standard in Penal Law § 240.50(1) is insufficient to criminalize public, noncommercial speech, even if false. To begin, inasmuch as this statute criminalizes a certain type of speech, namely false speech, the restrictions on speech are content-based, rather than time, place or manner limitations. To that end, content-based restrictions are "presumed invalid, and ... the Government bear[s] the burden of showing their constitutionality."… Absent certain historical categories which do not apply here…, even false speech is considered protected… and, in that context, content-based restrictions are subject to "the most exacting scrutiny"…. Under this exacting, or strict, scrutiny standard, governmental regulation of speech "is enforceable only if it is the least restrictive means for serving a compelling government interest"…. "The First Amendment requires that the [g]overnment's chosen restriction on the speech at issue be actually necessary to achieve its interest. There must be a direct causal link between the restriction imposed and the injury to be prevented"….

> We have no trouble finding that Penal Law § 240.50(1) is designed to address at least two compelling governmental interests—preventing public alarm and the waste of public resources that may result from police investigations predicated on false reports. However, when examining whether the statute uses the least restrictive means for serving those purposes, as applied to defendant, we reach the conclusion that the statute is impermissibly broad…. More particularly, neither general concern nor the Twitter storm that ensued following defendant posting the false tweets are the type of "public alarm or inconvenience" that permits defendant's tweets to escape protection under the First Amendment (Penal Law § 240.50[1]), and, therefore, the speech at issue here may not be criminalized.

---

[89] 122 N.Y.S.3d 419 (App. Div. 3d Dep't 2020).
[90] *Id.*
[91] *Id.*

To that end, although it was "not unlikely" that defendant's false tweets about a racial assault at a state university would cause public alarm (Penal Law § 240.50[1]), what level of public alarm rises to the level of criminal liability? Indeed, *United States v. Alvarez*,… informs us that criminalizing false speech requires either proof of specific harm to identifiable victims or a great likelihood of harm. Certainly, general concern by those reading defendant's tweets does not rise to that level, nor does the proof adduced at trial, which established that defendant's tweets were "retweeted" a significant number of times. In fact, because these "retweets" led to nothing more than a charged online discussion about whether a racially motivated assault did in fact occur, which falls far short of meeting the standard set forth in *United States v. Alvarez*…, we reach the inescapable conclusion that Penal Law § 240.50(1), as applied to defendant's conduct, is unconstitutional.

Indeed, Penal Law § 240.50(1) is a "[b]lunt [t]ool for [c]ombating [f]alse [s]peech"… and its "alarming breadth"… is especially on display with respect to social media. Notably, "[t]he remedy for speech that is false is speech that is true"… and "social media platforms are information-disseminating fora. By the very nature of social media, falsehoods can quickly and effectively be countered by truth, making the criminalizing of false speech on social media not 'actually necessary' to prevent alarm and inconvenience"…. This could not be more apparent here, where defendant's false tweets were largely debunked through counter speech; thus, criminalizing her speech by way of Penal Law § 240.50(1) was not actually necessary to prevent public alarm and inconvenience….

Based upon this conclusion, defendant's remaining contentions as to count 7 are rendered academic. Defendant's remaining argument, that her conviction as to count 4 is repugnant to her acquittal of count 5, is not preserved because defendant failed to raise this claim prior to Supreme Court's discharge of the jury….

…

ORDERED that the judgment is modified, on the law, by reversing defendant's conviction of falsely reporting an incident in the third degree under count 7 of the indictment; said count dismissed and the sentence imposed thereon vacated; and, as so modified, affirmed.[92]

Thus, we see that courts, when addressing online activity, often must parse First Amendment challenges—with those challenges sometimes cutting in favor of defendants, and sometimes not. All-in-all, care should be taken on social media, to avoid some of the potential "costs" that may not be immediately or readily foreseeable if one presses "send" in the heat of a moment or as a reflex without consideration.

---

[92] *Id.* (citations omitted) (noting in Footnote 5 that "Defendant's tweets constitute speech rather than conduct") (citing *People v. Marquan M.*, 24 N.Y.3d 1, 994 N.Y.S.2d 554 (2014); *compare People v. Shack*, 86 N.Y.2d 529, 535, 634 N.Y.S.2d 660 (1995)).

7.6 – Conclusion

What is our take-away from all of the above? It certainly is no easier to be a teenager today—spreading one's wings and learning how to be an adult—than it was for generations past. Indeed, it appears it is much harder. It falls on adults—parents, teachers, and even police officers and attorneys performing public service by speaking in schools and to student groups—to educate teenagers in the law and to advise them about the activities that could result in their prosecution as criminal defendants, or, worse, their deaths. Adult users, as is clear from the above text, must be no less cautious.

Most importantly, teens—and, as we see, really anyone utilizing social media and postings on the Internet—must understand that not only is there no privacy on the Internet, but in America there is also no right to be forgotten when one posts on the Internet. While the European Union's Court of Justice has held that Google had an obligation to consider requests of individuals to remove personal information from postings (particularly information that is incorrect or outdated),[93] courts in the United States have regularly rejected such a "right to be forgotten."[94] There is, further, generally no cause of action against the search engines/providers on which the material is posted.[95]

Texting and driving, or the embarrassment and social ostracism that could result from cyberbullying or "sexting" or revenge pornography, have led to the deaths or psychological damage of countless people, including teens. And, when it comes to teens in particular, they need to understand the reality of the world in which they will soon, it is hoped, be adults who contribute in a positive way.

As the *Nash* decision in Alabama concluded on one of the very real legal problems facing our society through the growth of technology:

> Although young people may not listen to this court, they need to know that sexting is a very real problem that can have very real, unexpected consequences. . . . The message of the damages of sexting and the egregious consequences it can bring is one that needs to be shared with teenagers and young people, legislative bodies, and members of the justice system.[96]

---

[93] *Google Spain SL v. Agencia Española de Protección de Datos (AEPD)*, Case C–131/12, ECLI:EU:C:2014:616 (May 13, 2014); although the right to "de-referencing" is by no means absolute, *Google v. Comm'n Nationale de L'Informatique et des Libertés*, C507/17 (CJEV Sept. 24, 2019). *See also* European Union General Data Protection Regulation (GDPR), Article 17, *Right to erasure ("right to be forgotten")*; GDPR Recitals 65 ("Right of rectification and erasure") and 66 ("Right to be forgotten"). *See* https://gdpr-info.eu/art-17-gdpr/. *See also Mosha v. Yandex, Inc.*, 2019 WL 5595037, at *2 (S.D.N.Y. Oct. 30, 2019) (discussing Article 152 of the Civil Procedure Code of the Russian Federation – the "Right to be Forgotten").

[94] *Garcia v. Google, Inc.*, 786 F.3d 733, 745–746 (9th Cir. 2015).

[95] *See Manchanda v. Google*, 2016 WL 6806250, at *3 & n.2 (S.D.N.Y. Nov. 16, 2016) ("The Court is sensitive to the deep personal harms that can result from hurtful information posted on the internet. . . . But the CDA [Communications Decency Act] prevents individuals from 'su[ing] the messenger.' . . . As such, [plaintiff's] defamation and related claims must be dismissed").

[96] *Nash*, 1 F.Supp.3d at 1249–1250.

Chapter End Questions

Note: Answers may be found on page 263.

1. What percentage of teenagers recognize that driving while texting is dangerous?

    a. 27%
    b. 52%
    c. 79%
    d. 94%

2. What percentage of teenagers admit to having driven while texting?

    a. 10%
    b. 20%
    c. 32%
    d. 35%

3. How many states in the U.S. have made it illegal to text using a mobile device while driving?

    a. 15
    b. 27
    c. 39
    d. 48

4. Which of the following is not a major legal issue for teens and their social media use today?

    a. Cyberbullying
    b. "Sexting"
    c. Driving while using mobile devices
    d. All of these selections are major issues

5. According to statistics, how many people were killed on U.S. roads in the year 2015 due to drivers who were distracted - including those on their mobile devices?

    a. 1200
    b. 1950
    c. 2650
    d. 3500

6. According to statistics, how many teenagers die each day because of distracted drivers - including those on their mobile devices?

    a. 1
    b. 5
    c. 9
    d. 11

7. True or False: If you are only 18, the courts will not put you in jail if you are convicted of texting and driving, causing an accident and killing someone in another car.

   a. TRUE
   b. FALSE

8. True or False: If one is in a legal, consensual romantic relationship, where one party is over 18 and one party is under 18, it is legal to text naked photographs of the under 18 person between those two people in the relationship.

   a. TRUE
   b. FALSE

9. True or False: The First Amendment of the U.S. Constitution prevents schools from disciplining students for their off-campus social media communications that bully other students.

   a. TRUE
   b. FALSE

10. True or False: What is discussed as "cyberbullying" is generally harmless joking.

   a. TRUE
   b. FALSE

CHAPTER 8

A Specific Societal Concern and Social Media: Texting to Drivers

8.1 – Introduction

No, this chapter does not concern the liability of drivers who text while driving and cause accidents. The laws of the states, and case decisions on liability, are very clear on that issue. And we all know the very real dangers of texting and driving (as discussed in Chapter 7), which is now the leading cause of death among teenagers and young adults, resulting in more deaths than caused by alcohol-related DUI accidents.[1] This chapter, though, addresses whether a non-passenger, non-driving sender of a text to a driver can be held liable if the driver who receives the message subsequently causes an accident. Yes, that's right; this next issue we will address in this chapter is what liability, if any, does the non-passenger *sender* of a text message face if the recipient driver subsequently looks at the text message and causes an accident. Before you denounce the ridiculousness of such a proposition, and bemoan academic exercises, think about other contexts. A person gets drunk at a pub or a private home, the bartender or homeowner knows the patron or friend is drunk and continues to serve the ossified individual, and then the drunk person drives home. Dram Shop liability and other provisions in the law of torts speak to that scenario.[2] What about a passenger riding in a car who decides to reach over and put his or her hands or a piece of paper in front of the driver's eyes, without the driver's consent, and the driver careens off the road and into a house? Negligence and tort liability speak to that scenario.[3]

---

[1] M. Taggart & B. Robinson, *The Very Real Dangers of Texting While Driving*, HUFFINGTON POST (Apr. 17, 2017), available at: https://www.huffingtonpost.com/entry/the-very-real-dangers-of-texting-while-driving_us_58eeddd4e4b0156697224c60 (citing the Centers for Disease Control (CDC)).

[2] *See, generally, Oursler v. Brennan*, 67 A.D.3d 36, 884 N.Y.S. 2d 534 (4th Dep't 2009); N.Y. GEN. OBLIG. LAW § 11–101; *Place v. Cooper*, 35 A.D.3d 1260, 827 N.Y.S.2d 396 (4th Dep't 2006).

[3] *See, generally, Sartori v. Gregoire*, 259 A.D.2d 1004, 688 N.Y.S.2d 295 (4th Dep't 1999); *Enclarde v. Roach*, 397 So.2d 855 (La. Ct. App. 4th Cir. 1981).

Therefore, it is not unheard of for non-drivers to be held at least partially liable for tort injuries suffered by others in motor vehicle accidents, depending upon the duty owed by those non-drivers. Now, however, we consider the *non-passenger, non-driver sender of a text message* facing potential liability because the driver recipient of a message causes an accident while viewing the text sent to him or her. New York and New Jersey courts, addressing the issue, have utilized differing analyses, which should give pause to some depending on jurisdiction.

8.2 – In the State of New Jersey, Senders Must Be Wary

In 2013, a New Jersey appellate court articulated, for what appears to be the first time, a standard that could result in the sender of a text message (one who is not even in the automobile with the driver) being held liable along with the driver for damages from an accident.[4] In the case, two individuals were seriously injured when the driver of a truck crossed the centerline of a road and caused an accident while he was texting and driving. The injured parties sought to receive damages from the driver, which is typical, but also from the driver's girlfriend who had been texting the driver—even though she *was not* in the vehicle at the time of the accident. The lower court granted summary judgment, dismissing the case against the girlfriend. The appellate court sustained the lower court's decision. *However*, what makes the case noteworthy is that the appellate court held: "we do not adopt the trial court's reasoning that a remote texter does not have a legal duty to avoid sending text messages to one who is driving."[5]

The New Jersey court examined the duties imposed by common law and the standard of conduct one person must have toward another. The Court recognized the injured parties wanted a duty of care imposed upon the girlfriend under a theory that she had aided and abetted the driver's violation of law when the driver used the cell phone to text while driving. Furthermore, there is caselaw holding passengers in a motor vehicle have an affirmative duty to fellow passengers and the driver in terms of a standard of care and behavior in the vehicle—*if* there is either a special relationship or the passenger "actively encourages" the driver to commit a negligent act.[6] Of course, "mere failure to prevent wrongful conduct by another is ordinarily not sufficient to impose liability."[7]

In the *Kubert* case, the New Jersey appellate court held that the driver's girlfriend did not have a special relationship with the driver by which she could control his conduct in the vehicle, nor was there evidence in the record that the driver's girlfriend had actively encouraged him to text while he was driving. "We view [precedent] as appropriately leading to the conclusion that one should not be held liable for sending a wireless transmission simply because

---

[4] *Kubert v. Best*, 75 A.3d 1214, 432 N.J. Super. 495 (App. Div. 2013).

[5] *Id.* at 503.

[6] *See Champion ex rel. Ezzo v. Dunfee*, 398 N.J. Super. 112, 939 A.2d 825 (App. Div.), *certif. denied,* 195 N.J. 420 (2008); *Lombardo v. Hoag*, 269 N.J. Super. 36, 634 A.2d 550 (App. Div. 1993), *certif. denied,* 135 N.J. 469 (1994). Cited by *Kubert*, 432 N.J. Super. at 511–512.

[7] *Kubert*, 432 N.J. Super. at 512.

some recipient might use his cell phone unlawfully and become distracted while driving. Whether by text, email, Twitter, or other means, the mere sending of a wireless transmission that unidentified drivers may receive and view is not enough to impose liability."[8] The court held: "the evidence in this case is not sufficient for a jury to conclude that [driver's girlfriend] took affirmative steps and gave substantial assistance to [driver] in violating the law. Plaintiffs produced no evidence tending to show that [driver's girlfriend] urged [driver] to read and respond to her text while he was driving."[9] The claims against the defendant girlfriend were dismissed.

Nevertheless, of interest for this chapter is the fact that the Court did not stop there. It continued, expanding potential liability in a new area. Examining the foreseeability of the risk of harm, the Court reasoned it is not always foreseeable that a driver would immediately review a text message or other message sent to his or her mobile upon receipt. However,

> if the sender knows that the recipient is both driving and will read the text immediately, then the sender has taken a foreseeable risk in sending a text at that time. The sender has knowingly engaged in distracting conduct, and it is not unfair also to hold the sender responsible for the distraction. [...]"When the risk of harm is that posed by third persons, a plaintiff may be required to prove that defendant was in a position to 'know or have reason to know, from past experience, that there [was] a likelihood of conduct on the part of [a]third person[]' that was 'likely to endanger the safety' of another."[10]

The New Jersey court cited, with approval, several cases from other jurisdictions in which causes of action against cell phone manufacturers were dismissed, with courts holding they were not negligently responsible and had not defectively designed mechanisms that could be viewed while a driver was driving.[11] Despite that conclusion, the Court still found there is a limited duty that can be imposed on a sender—"[w]hen the sender knows that the text will reach the driver while operating a vehicle, the sender has a relationship to the public who use the roadways similar to that of a passenger physically present in the vehicle. As we have stated, a passenger must avoid distracting the driver. The remote sender of a text who knows the recipient is then driving must do the same."[12]

Again, the New Jersey court did not find liability against the driver's girlfriend in this particular case for having sent the text from a remote location. The Court did state, though:

> we do not hold that someone who texts to a person driving is liable for that person's negligent actions; the driver bears responsibility for obeying the law and maintaining safe control of the

---

[8] *Id.* at 514.

[9] *Id.* at 513.

[10] *Id.* at 517 (emphasis added) (citing cases).

[11] *Kubert*, 432 N.J. Super. at 514 (citing *Durkee v. C.H. Robinson Worldwide, Inc.*, 765 F.Supp.2d 742 (W.D.N.C. 2011), *aff'd sub nom. Durkee v. Geologic Solutions, Inc.*, 502 Fed. Appx. 326 (4th Cir. 2013); *Estate of Doyle v. Sprint/Nextel Corp.*, 248 P.3d 947 (Okla. Civ. App. 2010); *Williams v. Cingular Wireless*, 809 N.E.2d 473 (Ind. Ct. App., *app. denied*, 822 N.E.2d 976 (Ind. 2004)).

[12] *Kubert*, 432 N.J. Super. at 517–518.

vehicle. We hold that, when a texter knows or has special reason to know that the intended recipient is driving and is likely to read the text message while driving, *the texter has a duty to users of the public roads to refrain from sending the driver a text at that time.*[13]

The majority decision of the Court did provide an indication as to its concerns and a potential reason for the expansion of this window on liability. The Court was concerned that text messaging may pose as great a risk as drunk driving. "Just as the public has learned the dangers of drinking and driving through a sustained campaign and enhanced criminal penalties and civil liability, the hazards of texting when on the road, *or to someone who is on the road*, may become part of the public consciousness when the liability of those involved matches the seriousness of the harm."[14]

Therefore, a note of caution to those who may text others they know to be driving at the time of the text. If it is also known that the driver will not be able to resist looking at his or her phone or device while driving, *do not* hit "send" if they are in New Jersey, because that may just result in shared liability in the event of an accident. Know to whom you are texting and what it is they are doing when you are texting them, and take the appropriate caution and care.

8.3 – In the State of New York, It Is Drivers Who Must Beware

In 2018, a New York intermediate appellate court, addressing the same issue, articulated a clearly different standard. The case, *Vega v. Crane*[15], arose from a two-vehicle accident wherein one driver died and one was seriously injured when the deceased driver crossed the centerline of a roadway. The plaintiff brought suit, including claims that the deceased driver's girlfriend, who had been texting him, had a duty not to text a person known to be driving, and therefore contributed to the plaintiff's serious injuries. The Supreme Court granted summary judgment to defendant girlfriend and dismissed the claim.

The Appellate Division, Fourth Department, unanimously affirmed, holding: "The [trial] court properly concluded that defendant had no duty to refrain from sending text messages to decedent."[16] Citing, *inter alia*, the seminal case of *Palsgraf* (which needs no introduction to anyone who attended law school), the Court stated: "It is well established that a defendant may not be held liable for negligence unless he or she owes a duty to the plaintiff."[17]

---

[13] *Id.* at 519 (emphasis added).

[14] *Id.* (emphasis added).

[15] 162 A.D.3d 167, 75 N.Y.S.3d 760 (4th Dep't 2018) (lower court decision at 49 N.Y.S.2d 264 (Sup. Ct. Genesee County 2017)).

[16] *Id.* at 762.

[17] *Id.* (citing *Palsgraf v. Long Is. R.R. Co.*, 248 N.Y. 339, 342, 162 N.E. 99 (1928), *rearg. denied* 249 N.Y. 511, 164 N.E. 564 (1928)).

The Fourth Department did reiterate settled law that a passenger in a vehicle may be liable if he or she acts to distract a driver before, or causing, an accident.[18] However, the New York appellate court also acknowledged an absolute distinction between a passenger distracting a driver and a sender of a text message from a remote location. In its analysis, which disagreed with *Kubert v. Best*, and cited to the concurring opinion of Justice Espinosa (who joined only in the result in *Kubert*, not the analysis), the Fourth Department found a clear line to be drawn. An unruly passenger who yells or is otherwise a force of distraction may not be controlled by the driver, but the driver has the absolute ability to control if and when he or she looks at a phone or device while in a vehicle. Indeed, the Court held: "Although the remote sender has the ability to refrain from sending the driver a text message, he or she is powerless to compel the driver to read such a text message at an imprudent moment, *and has no duty to prevent the driver from doing so*. Rather, it is the duty of the driver to see what should be seen and to exercise reasonable care in the operation of his or her vehicle to avoid a collision with another vehicle."[19]

The Fourth Department, furthermore, seemed rightly concerned with creating a slippery slope, recognizing that had this decision resulted in a different holding, the flood gates might be opened and numerous innocent actors, exercising no control over drivers, might be exposed to claims of liability.

Vega v. Crane (N.Y. Supreme Court, Appellate Division, Fourth Department)

Troutman, Associate Justice

If a person were to be held liable for communicating a text message to another person whom he or she knows or reasonably should know is operating a vehicle, such a holding could logically be expanded to encompass all manner of heretofore innocuous activities. A billboard, a sign outside a church, or a child's lemonade stand could all become a potential source of liability in a negligence action. Each of the foregoing examples is a communication directed specifically at passing motorists and intended to divert their attention from the highway.

To be sure, cellular telephones and other electronic devices present unique distractions to motorists. For that reason, the legislature passed laws specifically to regulate the use of cellular telephones and other electronic devices by those *operating* motor vehicles. . . . The legislature did not create a duty to refrain from communicating with persons known to be operating a vehicle. To the contrary, those laws place the responsibility of managing or avoiding the distractions caused by electronic devices squarely with the driver. . . . The driver has various means available for managing or avoiding such distractions, such as a hands-free device to handle incoming

---

[18] *Vega*, 75 N.Y.S.3d at 763 (citing *Sartori v. Gregoire*, 259 A.D.2d 1004, 688 N.Y.S.2d 295 (4th Dep't 1999); *Dziedzic v. Thayer*, 292 A.D.2d 845, 739 N.Y.S.2d 802 (4th Dep't 2002); Restatement [Second] of Torts § 303, Comment d, Illustration 3).

[19] *Vega*, 75 N.Y.S.3d at 763 (emphasis added) (citing *Deering v. Deering*, 134 A.D.3d 1497, 21 N.Y.S.3d 801 (4th Dep't 2015); *Zweeres v. Materi*, 94 A.D.3d 1111, 942 N.Y.S.2d 625 (2d Dep't 2012)).

> calls . . . or a setting for temporarily disabling sounds or alerts. Or, the driver can simply pull over to the side of the highway to engage in any communications deemed too urgent to wait. The remote sender of a text message is not in a good position to know how the driver will or should handle incoming text messages.[20]

8.4 – A Concluding Word on the Matter

New York's Fourth Department has not traveled the same road as the New Jersey appellate court. In New York, the sender of a text message to a driver generally bears no liability because there is no duty to refrain from texting the driver. The duty falls squarely on drivers to remain in control of their vehicle and to mitigate or ignore any distractions, as required by the law of the jurisdiction.[21] In New Jersey, to the contrary, while there is generally no liability for merely sending a communication that may be viewed in the future, *if* the sender knows or has reason to know that the recipient is driving a vehicle and is likely to look at and be distracted by the text message, the sender faces the potential for shared liability in New Jersey for any resulting accident. It should now be clear to the reader why the specific internal section headings in this chapter were chosen—in New Jersey, it is senders of text messages who have to be wary. In New York, drivers are the ones who must beware and who must take care.[22]

Chapter End Questions

Note: Answers may be found on page 264.

1. What is the name of the law providing that owners of bars/taverns might be held responsible for continuing to serve an intoxicated patron, and then allowing the patron to drive drunk, causing an accident?

   a. Drug Shop
   b. Dram Shop
   c. Drunk Shop
   d. Drop Shop

---

[20] *Vega*, 75 N.Y.S.3d at 763–764 (emphasis added) (citing N.Y. VEHICLE AND TRAFFIC LAW §§ 1225–c, 1225–d).

[21] It is generally accepted (with one potential disagreement out of the First Department) that the separate Appellate Divisions in the State of New York exist for administrative convenience, and the holding of one Division is binding across the State absent a contrary decision by another Division or the Court of Appeals. *See People v. Turner*, 5 N.Y.3d 476, 482 (2005) (Smith, J.); *Mountain View Coach Lines, Inc. v. Storms*, 102 A.D.2d 663, 664, 476 N.Y.S.2d 918 (2d Dep't 1984) (citing, *inter alia*, *Waldo v. Schmidt*, 200 N.Y. 199, 202 (1910), and collecting other cases); *see also* M. Gordon, *Which Appellate Division Rulings Bind Which Trial Courts?*, N.Y.L.J. (Online, Sept. 8, 2009) (citing cases, and discussing status of the statewide binding nature of Appellate Division decisions).

[22] *See also* 8A N.Y. JUR. 2D Automobiles §697.

2. What is the name of the NJ case in which the Court held a non-driver, non-passenger might be liable for an accident in certain circumstances?

   a. *Kubert v. Best*
   b. *Vega v. Crane*
   c. *Best v. Mediocre*
   d. None of these selections

3. Why did the NJ Court find that there might be occasions when a non-driver, non-passenger might be held responsible for a car accident?

   a. The texter could have told the driver, in the text, to "read this later."
   b. The texter could have ridden in the car with the driver to just speak with them in person.
   c. If sender/texter knows recipient is driving, and knows that recipient will not resist reading the text immediately and responding, then sender/texter has reason to wait, and therefore should wait and not distract driver. That is in texter's control.
   d. None of these selections contain the reasoning of the Kubert case.

4. What is the name of the case in NY that distinguished/disagreed with the case from NJ?

   a. *Marbury v. Madison*
   b. *Vega v. Crane*
   c. *Kubert v. Best*
   d. *Kramer v. Vega*

5. What was the reasoning, generally, for why the NY Court distinguished the NJ decision?

   a. Drivers are responsible for controlling their vehicles, and avoiding distractions, particularly those distractions that drivers alone can control - like looking at a phone.
   b. Non-passengers can only be liable for accidents if and when they scream at a driver.
   c. Non-passengers can only be liable for accidents when the driver is driving at night or in bad weather.
   d. None of these selections.

6. What NY law/statute places responsibilities on drivers not to use their devices while driving in a way that causes distractions?

   a. NY V&T § 1225-c
   b. NY V&T § 1225-d
   c. NY CPLR 3126
   d. Both a. & b.

7. True or False: Non-drivers can never be held liable for automobile accidents.

   a. TRUE
   b. FALSE

8. True or False: If someone texts a driver, the driver is then not responsible if they answer the text while driving, because someone else distracted them.

   a. TRUE
   b. FALSE

9. True or False: In NY, a plaintiff injured in an accident can still sue a non-passenger sender of a text message to the defendant driver, simply by citing to the NJ caselaw.

   a. TRUE
   b. FALSE

10. True of False: Regardless of whether a non-passenger can be held liable for a car accident in a particular state, texting and driving is not an important issue in our society any longer.

    a. TRUE
    b. FALSE

CHAPTER 9

Authentication of Electronic Evidence for Trial*

9.1 – Introduction

Electronically stored information and documentation, once it is disclosed through the discovery process and evaluated by attorneys and their experts, becomes the subject of that next hurdle in litigation—the proffer of evidence at trial. That is the subject of this chapter—and, particularly, the issue of authentication of that electronically stored information when proffered as evidence in a courtroom.

> **Evidence** – The body of material, including oral testimony, documents, and other physical items that under the evidentiary rules, are permitted to be seen/heard by the jury as they make a decision (deliberate), and which are part of the official case record.

There is no one, uniform standard across our great Nation. In fact, from state to state, the courts have disagreed regarding what factors and methods to apply when parties seek to authenticate electronically stored or retrieved evidence. States apply their own procedural and evidentiary rules, many of which mirror federal rules. The federal system took a large step toward resolving the issue across its courts with the specific rule change that added provisions to F.R.E. 902 in 2017.

9.2 – Authentication versus Admissibility, Judicial Notice

First, a distinction should be made: we are speaking here about the *authentication* of proffered evidence, and not necessarily its *admissibility*. As experienced

---

*The author of this text published a prior version of this discussion. *See* M. Fox, "*I Show You Exhibit E for Identification*"—*Differing Approaches for Authentication of Electronic Evidence at Trial*, 22 NY LITIGATOR 14 (Spring 2017).

Authenticated – An item is/is shown to be what it is claimed to be; it is authentic.

attorneys are well aware, just because a piece of proffered evidence may be authenticated (it is what it claims to be) does not mean that the evidence has overcome all hurdles to admissibility into evidence (meaning it may then be considered by a jury, or by a judge in a bench trial, in rendering a decision). Often the material must overcome a second evidentiary burden, such as relevance or hearsay.[1] The court is ultimately the gatekeeper.[2] In fact, some courts have particularly stated: "[i]nitially, authentication social media evidence [*sic*] is to be evaluated on a case-by-case basis to determine whether or not there has been an adequate foundational showing of its relevance and authenticity."[3]

Furthermore, courts certainly cannot take judicial notice of most websites and the content of most private Web pages.[4] After all, courts and commentators alike have recognized the possibilities and dangers in the fact that anyone can create a fake profile on social media, or a fake website, thus endangering the trustworthiness of trial evidence if care is not taken with authentication.[5]

---

[1] *See U.S. v. Bertram*, 259 F.Supp.3d 638, 642–643 (E.D. Ky. 2017); *U.S. v. Way*, 2018 WL 2470944, at *2 (E.D. Cal. June 1, 2018) (motion *in limine* and pre-trial certification; "Even with pretrial certification, '[t]he Government still bears the burden to prove the relevance of the business record and the absence of another form of hearsay in order for the record to be formally entered into evidence'") (citing *U.S. v. Kahre*, 610 F.Supp.2d 1261, 1266 (D. Nev. 2009)). Note also that some courts and commentators have been addressing whether and when the Business Records exception applies to e-mails. *See* D. Lender, L. Barrington &. J. Rausch, *Are E-mails Actually Business Records? It Depends.*, N.Y. L.J. at 9 (Feb. 3, 2020).

[2] *See, generally, U.S. v. Vayner*, 769 F.3d 125 (2d Cir. 2014); *Hernandez v. Apple Auto Wholesalers of Waterbury LLC*, 2020 WL 2543785 (D. Conn. May 18, 2020); *U.S. v. Ulbricht*, 79 F.Supp.3d 466 (S.D.N.Y. 2015); *State v. Sample* —A.3d—, 2020 WL2316709, (Md. Ct. Apps. May 11, 2020); *Sublet v. State*, 442 Md. 632, 113 A.3d 695 (Md. Ct. Apps. 2015); *Walker v. State*, No. 1030, 2015 WL 8579806, at *5 (Md. Ct. Spec. Apps. Dec. 11, 2015); *State v. Goodwin*, 2020 WL1551149 at *4 (Iowa Ct. Apps. Apr. 1, 2020); *People v. Hernandez*, 31 Misc.3d 208, 915 N.Y.S.2d 824 (Rochester City Court 2011); Fed. R. Evid. 401, 402, 801–803, 901–902; N.Y. CPLR 4518, 4532-b.

[3] *Commonwealth v. Mangel*, 181 A.3d 1154, 1162 (Pa. Super. Ct. 2018).

[4] *See Veronica Foods Co. v. Ecklin*, 2017 WL 2806706 at *4 (N.D. Cal. June 29, 2017) (court took notice of one Web page because plaintiff itself referred the public to the third-party site); *McGown v. Silverman & Borenstein, PLLC*, Case No. 13-cv-748-RGA/MPT, 2014 U.S. Dist. LEXIS 12823 (D. Del. Feb. 3, 2014); *Victaulic Co. v. Tieman*, 499 F.3d 227, 236 (3d Cir. 2010). *See* M. Hutter, *Judicial Notice of Website Information*, N.Y.L.J. at 3, 9 (June 2, 2016) (if a court takes judicial notice of facts on the Internet, it is "necessarily seeking to determine whether that fact is not subject to reasonable dispute because of the unquestioned accuracy of its source, the website"); *PROF-2013-S3 Legal Title Trust IV by U.S. Bank Nat'l Ass'n v. SFR Invest. Pool 1, LLC*, 2018 WL 2138526, at *2 (D. Nev. May 8, 2018) (citing F.R.E. 201(b) for the same principle). *See also Rivera v. Marriott Int'l, Inc.*, —F. Supp. 3d— 2020 WL 1933968 (D.P.R. Apr. 22, 2020); 29 Am. JUR. 2D EVIDENCE §95.

[5] *See Rabkin v. Lion Biotechnologies, Inc.*, 2018 WL 905862 (N.D. Cal. Feb. 15, 2018); *Commonwealth v. Mangel*, 181 A.3d at 1162 ("Social media evidence presents additional challenges because of the great ease with which a social media account may be falsified, or a legitimate account may be accessed by an imposter"); *Griffin v. State*, 419 Md. 343, 354, 19 A.3d 415, 421–422 (Md. Ct. Apps. 2011) (citing *U.S. v. Drew*, 259 F.R.D. 449 (C.D. Cal. 2009)); *Sublet v. State*, 442 Md. at 662–663, 113 A.3d at 713; J. Cole, *The Brave New World of Internet Evidence*, ABA LITIGATION JOURNAL Vol. 42, No. 4, at 37, 39 (Summer 2016) (citing G. Joseph, *What Every Judge and Lawyer Needs to Know About Electronic Evidence*, 99 JUDICATURE 49, 51 (2015)).

When it comes to authentication of electronic materials—such as e-mails, for example—courts have stated: "[t]o properly authenticate evidence, 'the proponent must produce evidence sufficient to support a finding that the item is what the proponent claims it is.'... 'The burden of proof for authentication is slight.'... In the context of e-mail authentication, 'the key consideration... is not simply whether the witness on the stand was a sender or recipient of the email, but whether the testifying witness can speak to the email's unique characteristics, contents, and appearance.'"[6]

9.3 – State Standards for Authentication of Electronic Evidence

Now let us discuss the standards for authentication applied in several states, through some seminal cases on this topic to date. A three-method proposal was set forth in the Maryland case of *Griffin v. State*.[7] In *Griffin*, the trial court permitted the admission of selected printouts from a MySpace page as evidence in a criminal prosecution. On appeal, the Court of Appeals reversed the conviction, holding that the MySpace printout, utilized to show that a key witness had been threatened, was not properly authenticated. A new trial was ordered. The Court stated:

> A number of authentication opportunities come to mind. . . . The first, and perhaps most obvious method would be to ask the purported creator if she indeed created the profile and also if she added the posting in question, i.e. "[t]estimony of a witness with knowledge that the offered evidence is what it is claimed to be." . . . The second option may be to search the computer of the person who allegedly created the profile and posting and examine the computer's internet history and hard drive to determine whether that computer was used to originate the social networking profile and posting in question. . . . A third method may be to obtain information directly from the social networking website that links the establishment of the profile to the person who allegedly created it and also links the posting sought to be introduced to the person who initiated it.[8]

---

[6] *Cincinnati Holding Co., LLC v. Fireman's Fund Ins. Co.*, 2020 WL 635655, at *3 (S.D. Ohio Feb. 11, 2020) (citing, *inter alia*, Fed. R. Evid. 901(a); *U.S. v. Lundergan*, 2019 WL 4125618, at *1 (E.D. Ky. Aug. 29, 2019) (quoting *U.S. v. Bertram*, 259 F.Supp.3d at 641)). *See also Dalgic v. Misericordia Univ.*, 2019 WL 2867236, at *16 & n. 11 (M.D. Pa. July 3, 2019) ("Federal Rule of Evidence 901(b) provides a non-exhaustive list of ways in which the authentication requirements of 901(a) can be met, and '[a] number of courts have... suggested the key factor in the Rule 901(b) list when it comes to email authentication is Rule 904(b).'... That subsection explains that records may be authenticated by the introduction of testimony regarding their unique characteristics: i.e., the 'appearance, contents, substance, internal patterns, or other distinctive characteristics of the item, taken together with all the circumstances.'") (citing *U.S. v. Bertram*, 259 F. Supp. 3d 638, 640 (E.D. Ky. 2017); Fed. R. Evid. 901(b)(4)).

[7] 419 Md. 343, 19 A.3d 415 (Md. Ct. Apps. 2011). In Maryland, as in New York, the Court of Appeals is the highest Court in the State.

[8] *Griffin*, 419 Md. at 363–364, 19 A.3d at 427–428 (citations omitted).

As to the third method, the Maryland Court cited to a decision of the Third Department of the New York State Supreme Court, Appellate Division, in *People v. Clevenstine*[9].

Four years later, in *Sublet v. State*,[10] the Maryland Court of Appeals consolidated three cases, and both "eludicat[ed] and implement[ed] [its] opinion in *Griffin*."[11] Holding that one trial court did not err in excluding admission of pages from a Facebook conversation, that the second trial court did not err in admitting "direct messages" and "tweets" into evidence, and that the third trial court did not err in admitting Facebook messages authored by the defendant, the *Sublet* Court stated: "We shall hold that, in order to authenticate evidence derived from a social networking website, the trial judge must determine that there is proof from which a reasonable juror could find that the evidence is what the proponent claims it to be."[12] That is, really, the only burden when it comes to authentication. Courts have held the requirements to be quite minimal and the main goal to be one of evaluating whether a jury could determine the evidence is what it is claimed to be.[13]

The courts of other states have disagreed with regard to the methods set forth for authentication in Maryland. In a December 20, 2016, decision, the Appellate Division in New Jersey held that the three methods of authentication proposed by the *Griffin* Court were too strict and that they were not the only methods available for application.[14] "[T]he rules of evidence already in place for determining authenticity are at least generally 'adequate to the task,'" the New Jersey court stated.[15] Furthermore, the New Jersey Appellate Division has held:

"Authenticity can be established by direct proof—such as testimony by the author admitting authenticity[.]" . . . Moreover, the requisite showing "'may be made circumstantially.'" . . . "The requirement of authentication or identification as a condition precedent to admissibility is satisfied by evidence sufficient to support a finding that the matter is what its proponent claims." . . . "Authentication "'does not require absolute certainty or conclusive proof'—only

---

[9] 68 A.D.3d 1448 (App. Div. 3d Dep't 2009).

[10] 442 Md. 632, 113 A.3d 695 (Md. Ct. Apps. 2015), distinguished by *State v. Sample*, 2020 WL 2316709.

[11] *Id.* at 637, 697.

[12] *Id.* at 638, 698. *See also Walker v. State*, No. 1030, 2015 WL 8579806, at *6 (Md. Ct. Spec. Apps. Dec. 11, 2015) (distinguishing *Sublet*; "Appellant relies on the holding in *Sublet* that 'when a witness denies having personal knowledge of the creation of the item to be authenticated, that denial necessarily undercuts the notion of authenticity.' . . . In *Sublet*, however, the purported author of an electronic communication denied authoring a particular writing on social media, though she admitted writing previous posts on the same account. . . . In the present case, however, there is no denial of authorship that must be overcome by evidence, as Appellant did not testify at trial.").

[13] *Jackson v. State*, 460 Md. 107, 116, 188 A.3d 975, 980 (2018) ("a 'Court need not find that the evidence is necessarily what the proponent claims, but only that there is sufficient evidence that the jury ultimately might do so.' . . . The threshold of admissibility is, therefore, slight") (citing *U.S. v. Safavian*, 435 F.Supp.2d 36, 38 (D.D.C. 2006)). *See also Phillips v. Maryland*, 2020 WL 2931744, at *3 (Md. Ct. Spec. Apps. June 3, 2020) (citing *Jackson; Johnson v. State*, 2020 WL 917278, at *8 (Md. Ct. Spec. Apps. Feb. 26, 2020) (Same).

[14] *State v. Hannah*, 448 N.J. Super. 78, 85–87, 151 A.3d 99, 104–105 (N.J. App. Div. 2016). *See also Roy v. State*, 279 So. 3d 238, 243-244 & n.9 (mem.) (Fla. Ct. App. 5th Dist 2019) (distinguishing *Hannah*; collecting cases from across Nation).

[15] *Hannah*, 448 N.J. Super. at 85-87, 151 A.3d at 104–105 (citations omitted).

"a prima facie showing of authenticity" is required.'" . . . [The] evidence could have been "sufficient to meet the low burden imposed by our authentication rules."[16]

> ***Proffer*** – To offer something for admission into evidence in a court proceeding; if the item violates a rule of evidence, the proffer fails, and the item cannot be considered in deciding the case.

The Louisiana Fourth Circuit Court of Appeal, in a 2016 opinion, reversed a trial court decision that denied a defense motion to exclude evidence in a criminal case in the form of Facebook posts offered by the New Orleans police department—posts allegedly made by the defendant.[17] The appellate court held:

> Finding the State failed to present any evidence at all to authenticate the purported social media evidence it wishes to introduce at trial, we find the trial court abused its discretion in ruling the evidence admissible. . . . [W]e remand the matter to the trial court for an evidentiary hearing— outside the presence of the jury—in order for the State to present evidence pursuant to La. C.E. art. 901 to demonstrate the authenticity of the social media posts and for the trial court to rule on their admissibility at trial.[18]

The Louisiana Court continued:

> [a]ccordingly, we find the proper inquiry is whether the proponent has "adduced sufficient evidence to support a finding that the proffered evidence is what it is claimed to be. . . ."[19]

Proffered evidence needs at least circumstantial indicia of authenticity, and in *Smith* because the prosecution did not have the option of presenting the testimony of the alleged creator (the defendant), and therefore direct indicia of authenticity, it needed to present the circumstantial indicia of authenticity. That could include testimony of others, such as a forensic expert, or utilizing other information and identifying characteristics within the posts to authenticate the creator. Software can assist such an endeavor.

In 2017, however, the Louisiana Fourth Circuit Court of Appeal returned to the issue of authentication of Internet/social media/YouTube evidence.[20] In *Gray*, the defendant cited to *Smith* in support of the argument that the State had failed to properly authenticate YouTube videos utilized in the defendant's criminal conviction.[21] The Court reviewed its decision and holding in *Smith* and held that, with regard to YouTube social media evidence in a litigation, "we find [defendant's] argument that there was no authentication evidence as to when the three YouTube videos were recorded and posted or who posted the videos addresses the reliability and the weight of the video evidence, not the authenticity. As noted, the testimony of a

---

[16] *Angeles v. Nieves*, 2018 WL 3149551, at *5–*6 (N.J. App. Div. June 28, 2018) (citing N.J.R.E. 901; *Hannah*, 448 N.J. Super. at 89–91, 151 A.3d at 99).

[17] *State v. Smith*, 192 So.3d 836 (La. Ct. App. 4th Cir. 2016).

[18] *Id.* at 837.

[19] *Id.* at 842 (citing *Sublet*, 113 A.3d at 717 (quoting *U.S. v. Vayner*, 769 F.3d 125, 131 (2d Cir. 2014)).

[20] *State v. Gray*, — So.3d —, 2017 WL 3426021, at *13–*15 (La. Ct. App. 4th Cir. June 28, 2017).

[21] *Id.*

witness with personal knowledge may provide the authentication of evidence necessary for its admission."[22] Thus the burden, while it exists, is often held to be extremely low when it comes to the authentication of electronic evidence.

In New York, we can look at a 2012 decision in the case of *People v. Agudelo*.[23] The First Department of the Appellate Division declined to extend, and distinguished, the Third Department's decision in *People v. Clevenstine*[24] (which, recall, was cited by the Maryland court in *Griffin*). The First Department, in affirming a conviction, held that the victim in the case could testify about text messages exchanged with the defendant that were cut and pasted into a document, because the victim was a party to the conversation, and she was corroborated by a detective who had seen the messages on the victim's phone. The *Agudelo* Court cited to a decision of the U.S. Court of Appeals for the Second Circuit,[25] where it was similarly held that authenticity can be shown through the testimony of a participant to a conversation that the document depicting a cut-and-paste of messages is a fair and accurate representation of the conversation. This method of authentication requires a witness with knowledge that the matter is what it is claimed to be—a permutation of the first method outlined by the *Griffin* Court in Maryland.

The same reasoning and result were seen in the case of *People v. Khan*,[26] a decision of the Criminal Court of the City of New York. In that case, the defendant argued in a CPL § 330.30 application that his conviction should be set aside or modified because of the improper authentication of text messages during trial. The defendant argued both that the messages should not have been authenticated and admitted, and that the court improperly assisted in the authentication of the electronic evidence.[27] With regard to the court assisting in authentication, the judge held that:

> [i]n our current judicial system, it is not uncommon for the courts to take on a more active role in the presentation of evidence, it is when the Court assumes the parties' traditional roles of deciding what evidence or witnesses to call that it will be considered an abuse of discretion. . . .
> The Court of Appeals had held that there is no absolute bar to a trial judge asking a particular number of question [sic] of a seated witness or [even] recalling a witness to the stand.[28]

In *Khan*, the Court stated that it asked approximately eight questions, well within its discretion.

With regard to authentication itself, the court found that only three text messages were admitted, one as a party admission (thus resolving any second-hurdle hearsay issue) and two providing back-up context, while the rest "were precluded by the Court, finding that they were

---

[22] *Id.* at *16 (citing La. C.E. art. 901(B)(1); *Jordan v. State*, 212 So.3d 836, 845 (Miss. Ct. App. 2015)).

[23] 96 A.D.3d 611 (1st Dep't 2012).

[24] 891 N.Y.S.2d 511 (3d Dep't 2009).

[25] *U.S. v. Gagliardi*, 506 F.3d 140 (2d Cir. 2007).

[26] 60 Misc.3d 1216(A), 2018 WL 3552472 (Table) (Crim. Ct. Queens County July 23, 2018).

[27] *Id.*

[28] *Id.* at *3.

self-serving, prone to speculation, and the defendant did not make any admissions of guilt on the texts."[29] Furthermore, just as in *Agudelo*, the complainant/victim testified that the messages proffered were photographs of her phone screen, and a detective testified that he saw the phone screen and took the screen shot. The court also looked to caselaw holding that part of the assessment is whether, in light of the testimonial evidence, the text messages would also not make sense unless they were sent by the defendant.[30] Here, the messages on the victim's phone were held to make no sense unless they could be attributed directly to the defendant. The court further considered that "[w]hen determining admissibility, the person seeking [to] introduce the evidence need not rule out all possibilities inconsistent with authenticity."[31] The defendant's conviction was ultimately upheld, and the CPL 330.30 motion denied.

One of the biggest factors in *Khan* was the fact that the victim and detective testified concerning the text messages and their having seen them on the victim's phone. Think along the lines of a witness who authenticates photographs at trial.[32] Indeed, when the information from the social media pages of a party is proffered as evidence, we can analogize to authentication of photographs, as the New York Court of Appeals (the highest Court in the State of New York) did in *People v. Price*.[33] In *Price*, though, the prosecutor actually argued that testimony from a witness with personal knowledge of the scene of the photograph was not necessary because the photo came from the defendant's Internet profile page.[34] The Court, however, held that:

> the People point out that courts of several other jurisdictions have adopted a two-pronged analysis for authenticating evidence obtained from Internet profiles or social media accounts. This approach allows for admission of the proffered evidence upon proof that the printout of the web

---

[29] *Id.* at *1–*2.

[30] *Id.* at *2 (citing *People v. Pierre*, 41 A.D.3d 289, 291 (App. Div. 1st Dep't 2007)).

[31] *Id.* (citing *U.S. v. Pluta*, 176 F.3d 43, 49 (2d Cir. 1999)). Further in the criminal context, see also *People v. Cotto*, 79 N.Y.S.3d 535 (Mem) (App. Div. 2d Dep't Aug. 22, 2018) ("Photographs of text messages between the defendant and the complainant were properly admitted into evidence. . . . The complainant's testimony that the photographs of the text messages fairly and accurately depicted the text message conversation between her and the defendant was sufficient to authenticate the photographs") (citing, *Pierre*; and *People v. Green*, 107 A.D.3d 915, 916–917 (App. Div. 2d Dep't 2013)).

[32] *Commonwealth v. Martin*, 2018 WL 3121766, at *9 (Pa. Super. Ct. June 26, 2018) ("photographic evidence may be authenticated by testimony that the photograph fairly and accurately depicts the incident or object portrayed") (citing *Commonwealth v. McKellick*, 24 A.3d 982, 986–87 (Pa. Super. Ct. 2011) (citing *Nyce v. Muffley*, 119 A.2d 530, 532 (Pa. 1956))). *See also De Vera v. 243 Suydam, LLC*, 60 Misc.3d 1224(A), 107 N.Y.S. 3d 818 (Table) (Sup. Ct. Kings County Apr. 24, 2018) ("As a preliminary matter, the Court finds that the text messages . . . are admissible into evidence in the case at bar. . . . In [*Agudelo*], the Court admitted into evidence text messages that the victim received on her phone and compiled into a single document. . . . Similarly to *Agudelo* and *Javier*, here, [the party] proffers his email to his attorney containing snapshots of the text messages with the Tenant [ ], showing the date and time of the messages, and also showing the Tenant identifying herself as such. . . . Thus, the Court finds that [the party], by his affidavit, sufficiently authenticated his text message conversation with the Tenant [ ]") (citing, *inter alia, Agudelo*; *People v. Javier*, 154 A.D.3d 445, 445 (1st Dep't 2017)).

[33] 29 N.Y.3d 472, 80 N.E.3d 1005 (2017).

[34] *Id.* at 478, 1010.

page is an accurate depiction thereof, and that the web page is attributable to and controlled by a certain person, often the defendant . . . The courts that have adopted this approach have generally held that circumstantial evidence, such as identifying information and pictures, may be used to authenticate a profile page or social media account as belonging to the defendant. Relying on these out-of-state cases, the People contend that the detective's testimony identifying and describing the profile page she found on BlackPlanet.com, combined with her testimony that the printout was an accurate representation of the photograph displayed thereon, provided sufficient authentication evidence to allow admission of the photograph. We disagree.[35]

In *Price*, the prosecution was held to a higher burden than that which they posited for authentication. However, the *Price* Court did not establish a single, unifying standard to be applied in circumstances like those presented in the case. The law is still developing and, according to the Court, is unsettled, requiring elucidation in light of numerous different factual scenarios that may present themselves. The Court noted:

> We disagree with the assertion of our concurring colleagues that we should not decide this appeal without conclusively adopting a general and comprehensive test for authentication to be applied, not only in this case, but in all cases involving authentication of photographs found on a social network web page. Because we conclude that the proffer was insufficient under any potential standard for authentication—whether it be the traditional method of authenticating a photograph or the standard offered by the People (or some variation thereof)—we need not go any further than deciding the case presently before us. . . . "We reject the premise that we must now declare that one test would be appropriate for all situations, or that the proffered tests are the only options that should be considered". . . . In our view, it is more prudent to proceed with caution in a new and unsettled area of law such as this. We prefer to allow the law to develop with input from the courts below and with a better understanding of the numerous factual variations that will undoubtedly be presented to the trial courts. Because we necessarily decide each case based on the facts presented therein, it would be premature to decide whether the People's proffer would have been sufficient had the prosecution, hypothetically, established that the web page was controlled by defendant. At this time, it is sufficient and appropriate for us to hold that, based on the proffer actually made, the photograph was not admissible.[36]

Thereafter, in *People v. Franzese*,[37] New York's Appellate Division, Second Department, took its own turn at the issue of authentication of YouTube evidence. Citing to and distinguishing *Price*, the *Franzese* Court held:

---

[35] *Id.* (citing, *inter alia*, *Tienda v. State*, 358 S.W.3d 633, 642 (Tex. Crim. App. 2012); *U.S. v. Bansal*, 663 F.3d 634, 667 (3d Cir. 2011)).

[36] *Price*, 29 N.Y.3d at 478 & n.3 (citations omitted) (citing, *inter alia*, *Matter of Brooke S.B. v Elizabeth A.C.C.*, 28 N.Y.3d 1, 27 (2016)).

[37] 154 A.D.3d 706, 61 N.Y.S.3d 661 (App. Div. 2d Dep't 2017). *Cf. People v. Y.L.*, 64 Misc. 3d 664, 104 N.Y.S.3d 839 (County Ct. Monroe County 2019).

"[A]uthenticity is established by proof that the offered evidence is genuine and that there has been no tampering with it," and "[t]he foundation necessary to establish these elements may differ according to the nature of the evidence sought to be admitted". . . . Here, the YouTube video was properly authenticated by a YouTube certification, which indicated when the video was posted online, by a police officer who viewed the video at or about the time that it was posted online, and by the defendant's own admissions about the video made in a phone call while he was housed at Rikers Island Detention Center. . . . The video was further authenticated by its appearance, contents, substance, internal patterns, and other distinctive characteristics. . . . The quantum of authenticating evidence is greater here than what the Court of Appeals found to be inadequate in [*People v. Price*]. . . .

The Supreme Court properly permitted a police officer to testify that, in her opinion, the defendant was the person depicted in a surveillance video. The officer testified, inter alia, that she knew the defendant from her patrols of the neighborhood and from interacting with him on several occasions. Thus, under the circumstances of this case, her testimony served to aid the jury in making an independent evaluation of the videotape evidence.[38]

Compare that result, though, with another of the Appellate Division, Second Department's 2017 decisions: *Lantigua v Goldstein*.[39] In *Goldstein*, the defendant sought to introduce social media (Facebook) posts that allegedly contradicted the plaintiff's claims in the case. At a fourth deposition of plaintiff in the case, the defendant's counsel presented the Facebook pages and questioned the plaintiff about them. The plaintiff acknowledged having a Facebook account in 2010 but denied that the pages and the text on the pages were made by him.[40] The plaintiff sought information about the person who provided the material to the defense so the person could be deposed, and asked the trial court to preclude the proffered Facebook evidence. The trial court denied the motion, but the Appellate Division reversed. The appellate court held:

> the Supreme Court improvidently exercised its discretion in denying that branch of the plaintiff's motion which was to preclude the defendants from offering the printouts as evidence at trial unless the defendants produced the person who obtained the printouts for a deposition, because the plaintiff denied that the printouts were from his Facebook account, and he had no other means to prove or disprove their authenticity.[41]

Finally, in December 2016, New York's Appellate Division, Third Department, in *In re. Colby II (In re. Sheba II)*,[42] reversed a Family Court order that had terminated parental rights

---

[38] 154 A.D.3d at 706–707, 61 N.Y.S.3d at 662–663 (citing, *inter alia*, F.R.E. 901(b)(4); *People v. Clevenstine*, 68 A.D.3d 1448, 891 N.Y.S.2d 511; *People v. Russell*, 79 N.Y.2d 1024, 594 N.E.2d 922 (N.Y. 1992)). *See also People v. Serrano*, 173 A.D.3d 1484, 103 N.Y.S.3d 648 (App. Div. 3d Dep't 2019.)

[39] 149 A.D.3d 1057, 53 N.Y.S.3d 163 (App. Div. 2d Dep't 2017).

[40] *Id.* at 1058, 165.

[41] *Id.* at 1059, 166.

[42] 145 A.D.3d 1271, 43 N.Y.S.3d 587 (App. Div. 3d Dep't 2016).

on the basis of neglect and abandonment. The respondent parent facing loss of rights had proffered Facebook messages in opposition to the petitioner's application, but the Family Court held that a proper foundation was not laid for the proffered messages. The Third Department disagreed and is quoted at length here for analysis and discussion.

In re. Colby II (In re. Sheba II) (N.Y. Supreme Court, Appellate Division, Third Department)

Mulvey, Associate Justice

Respondent's principal contention on appeal is that Family Court erred in its rulings that no proper foundation had been established for the admission of proof that she had communicated with the subject child by Facebook messenger using her adult son's account. The parties stipulated that the child did have contact with respondent through Facebook, and, specifically, that the child was the sender of Facebook messages transmitted under his name. Although the parties so stipulated, Family Court erred in finding that respondent failed to establish a foundation for the proffered document through her testimony and in precluding her testimony regarding the frequency of her communications with the child via Facebook.

A recorded conversation -- such as a printed copy of the content of a set of cell phone instant messages -- may be authenticated through, among other methods, the "testimony of a participant in the conversation that it is a complete and accurate reproduction of the conversation and has not been altered" . . . Notably, "[t]he credibility of the authenticating witness and any motive she [or he] may have had to alter the evidence go to the weight to be accorded this evidence, rather than its admissibility." . . . Respondent testified that she was present when her counsel printed the Facebook messages at his office, and that she reviewed the entire document to ensure that it was a full and complete copy. The aforementioned stipulation and respondent's testimony, when combined with her adult son's testimony confirming that he had provided respondent with his account information, password and permission to use the account for communication with the child, constituted a sufficient foundation for the admission into evidence of the printed messages and her related testimony.

By erroneously precluding this proffered evidence, Family Court deprived respondent of her due process right to a full and fair opportunity to be heard. . . . Accordingly, we reverse the order and remit the matter for a further fact-finding hearing at which the printed Facebook messages are to be admitted into evidence and respondent permitted to testify as to, and be cross-examined on, the nature and extent of her Facebook communications with the child and any other issues related thereto.[43]

---

[43] *Id.* at 1272–1273, 589–590 (citing, *inter alia*, *People v. Agudelo*, 96 A.D.3d at 611; *People v. Green*, 107 A.D.3d at 916–917).

It should be noted that in addition to resolving the issue of authentication, the *Colby II* Court went a step further and held that the proffered messages were to be admitted into evidence by the trial court—determining not only authenticity but also implicitly finding that there was relevancy and an exception to the hearsay rule.[44]

9.4 – Federal Standards for Authentication of Electronic Evidence

In the federal system, one seminal case, especially for Second Circuit litigators and litigants, is *U.S. v. Vayner*.[45] In *Vayner*, the Circuit reversed a conviction because, in its eyes, the

---

[44] N.Y. CPLR 4518(a) also specifically addresses electronic records, as follows in pertinent part: "An electronic record, as defined in section three hundred two of the state technology law, used or stored as such a memorandum or record, shall be admissible in a tangible exhibit that is a true and accurate representation of such electronic record. The court may consider the method or manner by which the electronic record was stored, maintained or retrieved in determining whether the exhibit is a true and accurate representation of such electronic record. All other circumstances of the making of the memorandum or record, including lack of personal knowledge by the maker, may be proved to affect its weight, but they shall not affect its admissibility." Additionally, harkening back to our discussion earlier in this chapter concerning judicial notice of electronic evidence, in 2019 the New York State Assembly and Senate passed, and the Governor signed into law, a provision of the state's Civil Practice Law and Rules repealing an amendment from the prior year (CPLR 4511(c)) and replacing it with a new section CPLR 4532-b (made effective retroactively as of December 28, 2018), which expanded that state's approach to judicial notice for certain categories of material. The new CPLR 4532-b provides as follows:

> An image, map, location, distance, calculation, or other information taken from a web mapping service, a global satellite imaging site, or an internet mapping tool, is admissible in evidence if such image, map, location, distance, calculation, or other information indicates the date such material was created and subject to a challenge that the image, map, location, distance, calculation, or other information taken from a web mapping service, a global satellite imaging site, or an internet mapping tool does not fairly and accurately portray that which it is being offered to prove. A party intending to offer such image or information in evidence at a trial or hearing shall, at least thirty days before the trial or hearing, give notice of such intent, providing a copy or specifying the internet address at which such image or information may be inspected. No later than ten days before the trial or hearing, or later for good cause shown, a party upon whom such notice is served may object to the request to admit into evidence such image or information, stating the grounds for the objection. Unless objection is made pursuant to this subdivision, the court shall take judicial notice and admit into evidence such image, map, location, distance, calculation or other information.

Another thing to consider—some states are beginning to allow litigants to authenticate documents by means of utilizing blockchain technology (likely best known for its connection to Bitcoin and cryptocurrency, but having many other uses). *See* Jason Tashea, *Best Evidence*, ABA Journal at 22-23 (Sept./Oct. 2019) (discussing, in particular, that in the spring of 2019 "the Vermont Supreme Court included blockchain in the state's rules of evidence for the first time"—a digital record registered on a blockchain is self-authenticating and admissible into evidence if a qualified person provides an accompanying declaration/certification; and business records on blockchain are now included for the business records exception to hearsay; the metadata recorded with the entry on the blockchain helps ensure authenticity). According to the *Merriam-Webster Dictionary*, blockchain is defined as: "a digital database containing information (such as records of financial transactions) that can be simultaneously used and shared within a large decentralized, publicly accessible network. [A]lso: the technology used to create such a database. The technology at the heart of bitcoin and other virtual currencies, blockchain is an open, distributed ledger that can record transactions between two parties efficiently and in a verifiable and permanent way." Merriam-Webster Dictionary (online), *available at* https://www.merriam-webster.com/dictionary/blockchain.

[45] 769 F.3d 125 (2d Cir. 2014).

Government's use of a social media account as evidence against the defendant ran afoul of the evidence rules, and the authentication of the proffered material was not sufficient. The Circuit held that if the Government had introduced a paper flyer found on the street, the Government would have had to connect the defendant to that flyer in some way to show involvement or control. However, with regard to the social media account utilized as evidence in the criminal prosecution at issue, the Circuit found the trial judge had permitted use of evidence without a showing that the defendant created, used, or controlled the content of the pages.[46]

In 2015, however, *U.S. v. Ulbricht*[47] distinguished and declined to extend *Vayner*. In *Ulbricht*, the defendant moved, pre-trial, to preclude certain exhibits as insufficiently authenticated under *Vayner* and Federal Rule of Evidence 901. The exhibits at issue included screenshots of websites, forum posts, private Internet messages and chats, and files from the defendant's computer. The District Court held that:

> *Vayner* is not a blanket prohibition on the admissibility of electronic communications. . . . As the Second Circuit observed, "[e]vidence may be authenticated in many ways" and "the 'type and quantum' of evidence necessary to authenticate a web page will always depend on context". . . . *The Second Circuit also expressed skepticism that authentication of evidence derived from the Internet required "greater scrutiny" than authentication of any other record.* . . . Whether the Government can meet Rule 901's authentication standard with respect to the challenged exhibits is a question best answered at trial. There simply is no basis to prejudge the Government's ability to meet that standard.[48]

---

[46] *Id.* Consider, separately, another case: *U.S. v. Brinson*, 772 F.3d 1314, 1320–1321 (10th Cir. 2014). *Brinson* involved a federal criminal prosecution for, *inter alia*, conspiracy to engage in sex trafficking. With regard to the authentication, and non-hearsay issues, the Court held:

> A statement is not considered "hearsay" if it is offered against a party and is the party's own statement. Fed.R.Evid. 801(d)(2)(A). Proponents of the evidence need only show by a preponderance of the evidence that the opposing party had made the statement. . . .
>
> To link Mr. Brinson to the messages, the government had to show by a preponderance of evidence that Mr. Brinson was "Twinchee Vanto." The prosecution sufficiently established that link through five facts:
>
> 1. The "Twinchee Vanto" account was registered to an email address:
> 2. "Twinchee Vanto" identified himself in one message as "Tarran."
> 3. A witness testified that "Twinchee Vanto" had identified himself as "Tarran."
> 4. A phone number on the bill of sale for Mr. Brinson's SUV matched the number that "Twinchee Vanto" had given as a contact number.
> 5. Two witnesses testified that "Twinchee Vanto" was Mr. Brinson's "Facebook name" and that Mr. Brinson was known as "Twin."
>
> . . .
>
> Mr. Brinson presented evidence that other individuals had access to the Facebook account and had posted messages. But the district court could reasonably find by a preponderance of the evidence that Mr. Brinson had authored the messages. . . . Therefore, the district court properly admitted the Facebook messages as statements of a party opponent.

[47] *U.S. v. Ulbricht*, 79 F.Supp.3d 466 (S.D.N.Y. 2015).

[48] *Id.* at 488 (citations omitted) (emphasis added).

The emphasized text in the above quoted language also serves to highlight that federal courts appear to share some of the skepticism of the New Jersey Court in *Hannah*—that is, that the Maryland methods are too strict and that present rules and methods already utilized for non-electronic evidence may suffice for authentication of electronic evidence.

It is important to recognize that federal courts, like state courts, have acknowledged that the rules are not so strict as to require, for instance, authentication of emails only by those having direct knowledge of or involvement with the drafting of the emails. Circumstantial evidence is acceptable for authentication of emails and is very dependent on the context of the case and facts.[49]

Litigators, especially those who practice in the federal courts, should be aware that Federal Rule of Evidence 902 was amended, effective December 1, 2017, with the addition of subsections (13) and (14). The rule amendments created a process whereby electronically stored information that is copied from an electronic device, storage medium or file, may be authenticated by digital identification, accompanied by a certification of a qualified person complying with F.R.E. 902(11) and 902(12).[50] Of course, again, the opposing party may still object at trial on other evidentiary grounds, including hearsay or relevance.[51] However, the rule, at the federal level, addressed and largely resolved uncertainty as to the standard applied when courts and attorneys are first looking to authenticate proffered electronically stored and retrieved information during trial. It also potentially

> *Litigator* – An attorney who specializes in representing clients/trying cases in courts, arbitrations & mediations, and often before administrative bodies.

---

[49] *U.S. v. Gasperini*, 2017 WL 3140366, at *5 (E.D.N.Y. July 21, 2017) ("The court is unaware of any authority that authentication of emails can only come through a witness with direct knowledge of the drafting, and Defendant provides none. Rather, courts considering the admissibility of electronic documents and communications have held that "'evidence may be authenticated in many ways" and "the type and quantum of evidence necessary to authenticate [electronic sources] will always depend on context."' . . . The court concludes that the Government may authenticate the emails through circumstantial evidence and DENIES the motion to exclude those emails"; the Court then reserved on the issue of exclusion of the emails as inadmissible hearsay) (citing *Ulbricht* and *Vayner*). The District Court's decision was affirmed by the Second Circuit in 2018—*U.S. v. Gasperini*, 894 F.3d 482 (2d Cir. 2018).

[50] *See Report of Advisory Committee on Rules of Evidence*, Judicial Conference of the United States, October 21, 2016 (Tab 3), available at: http://www.uscourts.gov/sites/default/files/2016-10-evidence-agenda-book.pdf. *See also* David A. Prange & Benjamin C. Linden, *Explaining the Almost Unexplainable: Source Code Evidence at Trial*, N.Y.L.J. at 4 (May 5, 2020).

[51] For comparison, see N.Y. CPLR 4540-a, added to the New York Rules after Governor Cuomo signed into law on August 24, 2018 (effective January 1, 2019). New York's provision states:

> Rule 4540-a. Presumption of authenticity based on a party's production of material authored or otherwise created by the party. Material produced by a party in response to a demand pursuant to article thirty-one of this chapter for material authored or otherwise created by such party shall be presumed authentic when offered into evidence by an adverse party. Such presumption may be rebutted by a preponderance of evidence proving such material is not authentic, and shall not preclude any other objection to admissibility.

See N.Y.S Assembly website (for actions on Assembly Bill A06048), available at: http://assembly.state.ny.us/leg/?default_fld=&leg_video=&bn=A06048&term=2017&Summary=Y&Actions=Y&Memo=Y&Text=Y. Of course, the provision does not apply just to electronic material, but it could reasonably be read to include same.

alleviates the time and expense required when calling authentication witnesses for electronic evidence at trial,[52] in most cases—although in some cases, where significant dispute still exists, that is still unavoidable.

As the Advisory Committee states at length in the *Committee Note* to the Rule:

> Today, data copied from electronic devices, storage media, and electronic files are ordinarily authenticated by "hash value." A hash value is a number that is often represented as a sequence of characters and is produced by an algorithm based upon the digital contents of a drive, medium, or file. If the hash values for the original and copy are different, then the copy is not identical to the original. If the hash values for the original and copy are the same, it is highly improbable that the original and copy are not identical. Thus, identical hash values for the original and copy reliably attest to the fact that they are exact duplicates. This amendment allows self-authentication by a certification of a qualified person that she checked the hash value of the proffered item and that it was identical to the original. The rule is flexible enough to allow certifications through processes other than comparison of hash value, including by other reliable means of identification provided by future technology.

> A proponent establishing authenticity under this Rule must present a certification containing information that would be sufficient to establish authenticity were that information provided by a witness at trial. If the certification provides information that would be insufficient to authenticate the record if the certifying person testified, then authenticity is not established under this Rule.

> ...

> The opponent remains free to object to admissibility of the proffered item on other grounds—including hearsay, relevance, or in criminal cases the right to confrontation. For example, in a criminal case in which data copied from a hard drive is proffered, the defendant can still challenge hearsay found in the hard drive, and can still challenge whether the information on the hard drive was placed there by the defendant. . . .[53]

One should be aware, though, that electronic records are not automatically self-authenticating.[54] That is not what the cases discussed in this chapter hold. Parties proffering electronic evidence must demonstrate that the materials are authentic, trustworthy, and what they appear to be.

---

[52] For further discussion, see C. Aveni, *New Federal Evidence Rules Reflect Modern World*, 43 Litigation News 10 (ABA Section of Litigation Spring 2018).

[53] F.R.E. 902, Advisory Committee Notes to 2017 Amendments [Paragraph 14].

[54] Consider case holdings even before the creation of Federal Rule 902(13)&(14): *Specht v. Google Inc.*, 747 F.3d 929 (7th Cir. 2014), where the Court held in an infringement action:

> Specht's second procedural challenge concerns evidentiary rulings that the district court made before ruling on Google's summary judgment motion. He contests the exclusion of the screenshots of certain client websites as they supposedly appeared in 2005, bearing the Android Data mark. He argues that the screenshots are admissible because the creators of the sites asserted from memory that the screenshots reflected how those sites appeared in 2005. But the district court reasonably required more than memory, which is fallible; it required authentication by someone with personal knowledge of reliability of the archive service from which the screenshots were retrieved.

However, if a party submits a declaration or affidavit, or other testimony, attesting to qualities of an electronic record but does not address "computer policy and system control procedures, including control of access to the pertinent databases, control of access to the pertinent programs, recording and logging of changes to the data, backup practices, and audit procedures utilized to assure the continuing integrity of the records," the court is within its discretion to reject the proffer.[55]

If parties are not clear about the data's origin, and any changes or modifications thereof, a significant question exists, working against authentication. Particularly so for business records, "a proponent must show that there were no breaks in the electronic chain of custody that could have altered the record from its original state."[56] As summarized by a North Carolina federal district court: "'It is necessary . . . that the authenticating witness provide factual specificity about the process by which the electronically stored information is created, acquired, maintained, and preserved without alteration or change, or the process by which it is produced if the result of a system or process that does so.' . . . '[B]oilerplate, conclusory statements' are insufficient. . . . The *Lorraine* court also noted the importance of cross-examination to test the reliability of such evidence and the fact-specific nature of the required foundation."[57]

Make note, however, that the burden in federal court is similarly low as compared to the state courts—although often that low bar is not surpassed.[58]

> "[T]he proponent must produce evidence sufficient to support a finding that the item is what the proponent claims it is." . . . "Importantly, 'the burden to authenticate under Rule 901 is not high—only a *prima facie* [on its face] showing is required,' and a 'district court's role is to serve as gatekeeper in assessing whether the proponent has offered a satisfactory foundation from which the jury could reasonably find that the evidence is authentic.'"[59]

One federal bankruptcy court in South Carolina set forth an 11-part test for authentication of electronic business records.[60] Courts have stated that a wide range of evidence speaking to

---

See also U.S. v. Bansal, 663 F.3d 634, 667–68 (3d Cir. 2011) (screenshots from Internet archive authenticated via testimony of witness with personal knowledge of how Internet archive works). *Cf. Conyers v.City of Chicago,* 2020 WL 2528534 (N.D.Ill. May 18, 2020) (distinguishing *Specht*) (on appeal to 7th Circuit, No. 20-1934, June 3, 2020).

[55] *See In re Vee Vinhnee*, 336 B.R. 437 (9th Cir. B.A.P. 2005).

[56] *Hose v. Washington Inventory Services, Inc.*, 2016 WL 6427810, at *6 (S.D. Cal. Aug. 30, 2016) (citing *In re Vee Vinhnee*, 336 B.R. at 444).

[57] *Dillon v. BMO Harris Bank, N.A.*, 173 F.Supp.3d 258, 267–268 (M.D. N.C. 2016) (citing *In re. Vee Vinhnee*, 336 B.R. at 447; *Lorraine v. Markel Am. Ins. Co.*, 241 F.R.D. 534, 543–545 (D. Md. 2007)).

[58] *Dillon*, 173 F. Supp.3d at 267 (citing *Lorraine*, 241 F.R.D. at 542).

[59] *U.S. v. Landaverde-Giron*, 2018 WL 902168, at *1 (D. Md. Feb. 14, 2018) (citing F.R.E. 901; *U.S. v. Hassan*, 742 F.3d 104, 133 (4th Cir. 2014)).

[60] *In re McFadden*, 471 B.R. 136, 157 (Bankr. D.S.C. 2016) (citing to an 11-part test advocated by Professor Imwinkelreid).

authentication should be considered for electronic materials, just as for documentary evidence (the traditional analog).[61]

9.5 – Conclusion

Clearly there is no one-size-fits-all method for authentication of electronically created, stored, and retrieved material for use as evidence during trial[62]—although there are a number of common threads that run through the analyses of both the state and federal courts. Several reasonable, workable, and reliable methods have been set forth by the courts of multiple states, by the federal courts, and by the additions to the Federal Rules of Evidence in 2017 and the procedural rules in some of the states. The choice of method to apply will, as with most trial rules, be jurisdiction-specific. If a jurisdiction does not yet have a settled method for authentication of electronic evidence, though, there are a number to choose from—either by stipulation or as the result of motion practice.

Chapter End Questions

Note: Answers may be found on page 264.

1. Which of the following is the newly amended Federal Rule concerning authentication of evidence?
   a. FRE 902(14)
   b. FRE 101
   c. FRCP 56
   d. FRCP 12

2. Which of the following could prevent something that is authenticated from being admitted into evidence?
   a. Hearsay
   b. Lack of relevance
   c. Neither of these
   d. Both of these

---

[61] *U.S. v. Landaverde-Giron*, 2018 WL 902168, at *2 (citing *U.S. v. Browne*, 834 F.3d 403 (3d Cir. 2016)). *Browne* was distinguished by *Commonwealth v. Martin*, 2018 WL 3121766, at *9 (Pa. Super. Ct. June 26, 2018), which also distinguished *Commonwealth v. Mangel*, discussed earlier. In *Martin*, the Court stated: "[the question in *Browne* and *Mangel* was] how to authenticate properly text or chat message evidence, mindful of the unique challenges posed by social media. The question before [the *Martin* Court, however, was] not whether Appellant posted the Instagram images, but whether the Instagram posts accurately portrayed [the assault on the victim]. . . . Here, the Commonwealth presented sufficient direct and circumstantial evidence tending to support the authentication of the photographs of [victim's] assault."

[62] *See People v. Khan*, 60 Misc.3d 1216(A), 2018 WL 3552472 (Table), at *2.

3. Which of the following is not an option for the authentication of electronic or social media evidence in Maryland courts?

   a. Ask the purported creator if she/he indeed created the profile and also if she/he added the posting in question.

   b. Search the computer of the person who allegedly created the profile and posting and examine the computer's internet history and hard drive.

   c. Obtain information directly from the social networking website that links the establishment of the profile to the person who allegedly created it.

   d. All of these selections are options per Maryland caselaw.

4. What is the name of a seminal federal case in the field of authentication of electronic evidence?

   a. *U.S. v. Venture*

   b. *U.S. v. Vayner*

   c. *U.S. v. Vulture*

   d. None of these case names

5. What is the new procedural rule in New York, effective on January 1, 2019, that looks very similar to the new federal rules on authentication of evidence produced by the party that created it?

   a. CPLR 3500

   b. CPLR 1515

   c. CPLR 4540-a

   d. CPLR 3212-f

6. Which of the following cases in NY hold that defendants had to produce for deposition the person who allegedly obtained printouts proffered to be from plaintiff's social media account, because the plaintiff denied the printouts were from his account, and plaintiff would have no other way to prove or disprove authenticity?

   a. *Sherwood v. Walker*

   b. *Lantigua v. Goldstein*

   c. *Gold v. Antigua*

   d. *U.S. v. Vayner*

7. Which of the following is the name of the Louisiana state court case, in the area of authentication, that held: "[a]ccordingly, we find the proper inquiry is whether the proponent has 'adduced sufficient evidence to support a finding that the proffered evidence is what it is claimed to be....'"?

   a. *Smith v. Jones*

   b. *State v. Smith*

   c. *State v. Smythe*

   d. *Smith v. White*

8. True or False: There is no distinction between admissibility and authentication. If something is authenticated, then it will automatically be admitted into evidence in a court proceeding.

   a. TRUE
   b. FALSE

9. True or False: In the field of authentication of electronic evidence, most jurisdictions use similar standards and rules, and apply the same options or rationales for parties to use in the process.

   a. TRUE
   b. FALSE

10. True or False: The standards and processes for authentication of electronic evidence in state courts can differ from the processes and standards applied in the federal courts.

   a. TRUE
   b. FALSE

CHAPTER 10

Electronics in Times of Crisis: COVID-19 and Beyond

10.1 – Introduction

In January/February of the year 2020 (or earlier, depending on the timeline reported), a novel coronavirus emerged—COVID-19. The viral infection appeared to spread rapidly and virulently, causing complications with numerous organ systems—most specifically the respiratory and circulatory systems. Scientific experts attempted to predict death rates while hospitals quickly ran out of beds across states,[1] and morgues ran out of room for the bodies of the deceased.[2] With no known effective treatment, and no single discernable vector of transmission, the winter and spring of 2020 saw many governors across the nation take the unprecedented step of shutting down their states.[3] Stay-at-home orders were issued for all but those deemed essential workers (those including, but not limited to, health care workers; pharmacies; utilities and telecommunications; food processing and grocery stores; trash collection; mail service; and services providing for the health, safety, and welfare of the community).[4] Elementary and secondary

---

[1] *See* Dan Mangan & John W. Schoen, *Coronavirus cases: These States Face Biggest Potential Shortfalls in Hospital ICU Beds*, CNBC (online Apr. 6, 2020), *available at* https://www.cnbc.com/2020/04/06/coronavirus-cases-states-with-biggest-hospital-bed-shortfalls.html.

[2] *See* Dan Mosher, *New York City Hospitals Are Running Out of Room in Their Morgues, But the Flow of Coronavirus Bodies Is Just Starting To Ramp Up*, Business Insider (online Mar. 25, 2020), *available at* https://www.businessinsider.com/coronavirus-covid-19-deaths-bodies-morgue-storage-capacity-maxed-out-2020-3; Alan Feuer & William K. Rashbaum, *'We Ran Out of Space': Bodies Pile Up as N.Y. Struggles to Bury Its Dead*, The N.Y. Times (online Apr. 30, 2020), *available at* https://www.nytimes.com/2020/04/30/nyregion/coronavirus-nyc-funeral-home-morgue-bodies.html.

[3] *See* Alicia Lee, *These States Have Implemented Stay-at-Home Orders. Here's What That Means for You*, CNN (online Apr. 7, 2020), *available at* https://www.cnn.com/2020/03/23/us/coronavirus-which-states-stay-at-home-order-trnd/index.html.

[4] *See, e.g.*, N.Y. Executive Orders 202.6 (Mar. 18, 2020) & 202.8 (Mar. 20, 2020).

schools, colleges, and universities closed their doors, and sent students home to transition to a completely online world of instruction—many turning to platforms such as Zoom—with concomitant cybersecurity concerns.[5] Law firms and courts closed their physical doors but turned to online platforms including Zoom, Microsoft Teams, Google Meet, Skype for Business, and others to open their virtual doors and resume many necessary activities.[6]

However, scores of businesses not deemed essential, and unable to transition to an online world due to the products or services offered, were closed, with workers finding themselves unemployed—resulting in a national unemployment rate of 14.7% in April 2020, and 13.3% in May 2020,[7] and a "real" national unemployment rate of 23.9% as more than 40 million Americans found themselves out of work (according to initial jobless claims/unemployment benefits claims)[8]—economic conditions not seen in the United States since the Great Depression in the 1930's.

Thus, life, not only in the United States but around the world, was forced online into eWorld. Broadway shows, theatres, art shows, performances, and concerts were either cancelled or performers and audiences opted for online and televised assemblies such as the global *One World: Together At Home* concert.[9] Newscasters, television talk-show hosts, and others began broadcasting from their homes.[10] Likewise, numerous professional and social organizations suspended in-person conferences and meetings, and resorted to online gatherings. For instance, the 70,000-plus member New York State Bar Association held an online meeting of its governing House of Delegates, attended by 207 of the members (including the author of this book), in April 2020—the first time in the Association's 143-year history; followed by the

---

[5] *See* Alex Zimmerman & Christina Veiga, *NYC Allows Zoom (Once Again) for Remote Learning*, Chalkbeat New York (online May 6, 2020), *available at* https://ny.chalkbeat.org/2020/5/6/21249689/nyc-schools-education-zoom-ban-reversed. *See also Cybersecurity Alert: Tips for Students during the #Stayathome Semester*, Tech. & the Legal Profession Comm. of the N.Y. St. Bar Ass'n (Mar. 31, 2020), *available at* https://nysba.org/app/uploads/2020/04/NYSBA-Student-Cyber-Alert-FINAL-GG-33120.pdf.

[6] *See* Brandon Vogel, *How New York's Courts Use Skype for Business and What You Need to Know*, NYSBA News (online May 8, 2020), *available at* https://nysba.org/how-new-yorks-courts-use-skype-for-business-and-what-you-need-to-know/.

[7] *See Bureau of Labor Statistics*, U.S. Dep't of Labor, USDL-20-1140 (News Release June 5, 2020), *available at* https://www.bls.gov/news.release/pdf/empsit.pdf.

[8] *See* Lance Lambert, *Over 40 Million Americans Have Filed for Unemployment During the Pandemic—Real Jobless Rate Over 23.9%*, Fortune (May 28, 2020), *available at* https://fortune.com/2020/05/28/us-unemployment-rate-numbers-claims-this-week-total-job-losses-may-28-2020-benefits-claims-job-losses/.

[9] *One World: Together At Home*, Global Citizen, *available at* https://www.globalcitizen.org/en/connect/togetherathome/.

[10] *See, e.g.,* Cynthia Littleton, *'CBS This Morning' Team Anchors From Home Because of Coronavirus Precautions*, Variety (online Mar. 30, 2020), *available at* https://variety.com/2020/tv/news/cbs-this-morning-coronavirus-gayle-king-anthony-mason-tony-dokoupil-1203548602/; *Conan O'Brien To Return To Air, With an iPhone From Home*, AP News (online Mar. 19, 2020), *available at* https://apnews.com/92a470d14c5e91cf3dfd3582fe369978; Cortney Moore, *Coronavirus Causes Stephen Colbert To Broadcast From Home*, Fox Business (online Mar. 25, 2020), *available at* https://www.foxbusiness.com/media/stephen-colbert-late-night-hosts-return-home-coronavirus. *See also Gold/Fox: Non-Billable*, N.Y. St. Bar Ass'n Podcast, Season 2, Episodes 1 & 2 (May 13, 2020 & May 20, 2020).

online Summer 2020 meeting that June.[11] The approximately 400,000-member American Bar Association decided to hold its Annual Meeting in August 2020 entirely virtually.[12] Theme parks, bowling alleys, gyms, indoor malls, movie theaters, carnivals, arcades, amusements parks, water parks, aquariums, and zoos, among other entities, were closed by the Governor of New York State in March 2020,[13] as well as in other states by actions of their respective governors.[14]

As members of the public feared for their health, professional licensing examinations were postponed or cancelled—with some proposed to be held online.[15] "Telehealth" medical visits with physicians, surgeons, and physician assistants took on a whole new meaning.[16] Graduation ceremonies and commencement speeches at many colleges and universities went online—such as former Vice President Joseph R. Biden's address to the Columbia University School of Law Class of 2020.[17] Some new attorneys in New York State took their oaths of admission to the Bar virtually via Skype.[18] It was even reported that a majority of Americans favored things like online voting in elections.[19]

---

[11] *See* Christian Nolan, *NYSBA's Governing Body Holds Its First-Ever All-Virtual Meeting*, NYSBA News (Apr. 8, 2020), *available at* https://nysba.org/nysbas-governing-body-holds-its-first-ever-all-virtual-meeting/; *House of Delegates Meeting Agendas and Materials*, N.Y. St. Bar Ass'n, *at* https://nysba.org/house-of-delegates-meeting-agendas-and-materials/. Other state bar associations, such as those in Ohio and Florida, likewise held virtual meetings. *See Virtual 2020 OSBA Annual Meeting/Rescheduled for July 24*, Ohio St. Bar Ass'n, *at* https://www.ohiobar.org/meetings--events/OSBA-Annual-Meeting/; *Annual Florida Bar Convention, June 15-19, 2020, Virtual*, The Fla. Bar, *at* https://www.floridabar.org/news/meetings/.

[12] *See* Amanda Robert, *ABA Annual Meeting Will Be Held Online, Board of Governors Decides*, ABA Journal (online Apr. 30, 2020), *available at* https://www.abajournal.com/news/article/ABA-Annual-Meeting-will-be-held-online-Board-of-Governors-decides-american-bar-association-chicago.

[13] *See, e.g.,* N.Y. Executive Orders 202.3 (Mar. 16, 2020); 202.5 (Mar. 18, 2020); 202.14 (Apr. 7, 2020); 202.28 (May 7, 2020).

[14] *See, e.g.,* N.J. Executive Order 104 (Mar. 16, 2020); S.C. Executive Order 2020-17 (Mar. 31, 2020).

[15] *See* Stephanie Francis Ward, *California Bar Exam Will Be Postponed and Administered Online*, ABA Journal (online Apr. 27, 2020), *available at* https://www.abajournal.com/news/article/california-looks-to-online-testing-for-state-bar-exam; Michael Marciano, *Bar Exam Dates Finalized*, Law.com & N.Y. L.J. (online Apr. 6, 2020), *available at* https://www.law.com/newyorklawjournal/2020/04/06/bar-exam-dates-finalized/ (New York Bar Exam rescheduled from July 2020 to September 2020).

[16] *See Using Telehealth to Expand Access to Essential Health Services During the COVID-19 Pandemic*, Ctrs. for Disease Contr. & Prev., U.S. Dep't of Health & Human Servs. (Online June 10, 2020), *available at* https://www.cdc.gov/coronavirus/2019-ncov/hcp/telehealth.html.

[17] *See* Karen Sloan, *Biden Set to Address Columbia Law Graduates—Virtually*, N.Y. L.J. at 1 (Apr. 23, 2020); Karen Sloan, *Joe Biden Urges Columbia Law Grads to Protect Democracy*, N.Y. L.J. (online May 20, 2020), *available at* https://www.law.com/newyorklawjournal/2020/05/20/joe-biden-urges-columbia-law-grads-to-protect-democracy/.

[18] *See* Jack Newsham, *Unmute Yourself and Raise Your Right Hand: NY Lawyers Sworn In on Skype*, N.Y. L.J. (online Apr. 28, 2020), *available at* https://www.law.com/newyorklawjournal/2020/04/28/unmute-yourself-and-raise-your-right-hand-ny-lawyers-sworn-in-on-skype/.

[19] *See* Amanda Robert, *More Than Half of Americans Support Online Voting During COVID-19 Pandemic, Second ABA Civics Survey Shows*, ABA Journal (online May 1, 2020), *available at* https://www.abajournal.com/web/article/more-than-half-of-americans-support-online-voting-during-covid-19-pandemic-second-aba-civics-survey-shows.

What do we see from all of the above? Our world is rapidly evolving into a true eWorld. While the material addressed in this chapter could rightfully be incorporated into other chapters within this book, the material is carved-out here to specifically identify how the world has changed, particularly following the health pandemic, and some ways in which attorneys, businesspeople and others must recognize and adapt.[20]

10.2 – Select Laws and Orders in the Face of Pandemic

As the COVID-19 pandemic spread, and states of emergency were declared across the United States of America, state legislatures, executives, and judiciaries took action. For example, in New York, the Legislature amended Section 29-a of the Executive Law to read as follows, in pertinent part:

> The governor, by executive order, may issue any directive during a state disaster emergency declared in the following instances: fire, flood, earthquake, hurricane, tornado, high water, landslide, mudslide, wind, storm, wave action, volcanic activity, epidemic, disease outbreak, air contamination, terrorism, cyber event, blight, drought, infestation, explosion, radiological accident, nuclear, chemical, biological, or bacteriological release, water contamination, bridge failure or bridge collapse. Any such directive must be necessary to cope with the disaster and may provide for procedures reasonably necessary to enforce such directive.[21]

---

[20] Another potential effect from COVID-19 that is worth mentioning for its impact on society has to do with divorces and families. It was reported that among divorced couples with children, who have co-parenting/custody and visitation agreements, the pandemic forced them to work out alternatives and informal compromises, often without needing formal involvement of courts and attorneys. For instance, if visitation was an issue because of social distancing, some parents agreed to longer continuous visits (perhaps a week rather than two weekends) to avoid multiple back-and-forth stays, or parents exchanged real-world time with children for online visits utilizing Zoom, FaceTime and other platforms. *See* Stephanie Zimmerman, *Divorce Lawyers Say Technology Changes May Outlive the COVID-19 Pandemic*, ABA Journal (online June 11, 2020), *available at* https://www.abajournal.com/web/article/divorce-in-the-time-of-coronavirus-attorneys-say-tech-changes-may-outlive-the-pandemic.

[21] N.Y. Exec. Law § 29-a(1) (as amended by L.2020, c. 23, § 2, eff. March 3, 2020) (but provisions in Sections 1 and 2 of the statute expire and will be deemed repealed as of April 30, 2021, unless extended by act of the Legislature). It is also noteworthy that New York's Legislature held remote sessions, where Members attended *and voted* virtually. *See* Denis Slattery, *New York Legislature to Reconvene for Remote Session Next Week*, N.Y. Daily News (online May 22, 2020), *available at* https://www.nydailynews.com/news/politics/ny-legislature-reconvening-remote-session-albany-20200522-eln6w4thmfdq7ls72zj34fbiwm-story.html; Jon Campbell, *New York Lawmakers Passed New COVID-19 Bills. Here's What the Measures Would Do*, Democrat & Chronicle (online May 28, 2020), *available at* https://www.democratandchronicle.com/story/news/politics/albany/2020/05/28/ny-lawmakers-passed-new-covid-19-bills-heres-what-they-would-do/5270110002/. The U.S. House of Representatives in Congress took a similar step, permitting proxy voting in the House for the first time in history by rule change. *See* H. Res. 965 (2020); Nicholas Fandos, *With Move to Remote Voting, House Alters What It Means for Congress to Meet*, The N.Y. Times (online May 15, 2020), *available at* https://www.nytimes.com/2020/05/15/us/politics/remote-voting-house-coronavirus.html.

Thereafter, Governor Andrew Cuomo issued no less than forty-five (45) Executive Orders (as of June 26, 2020) "Continuing Temporary Suspension and Modification of Laws Relating to the Disaster Emergency."[22] In New York, non-essential businesses were forced to have employees work from home or be furloughed, and notarizations and weddings (occurrences previously required to be in-person) were permitted to take place by remote/virtual means, among myriad other changes. However, the New York Governor was not alone. In Connecticut, Governor Ned Lamont limited the size of groups in childcare, and also permitted remote/virtual notarizations of documents (an unprecedented action of the governors, previously unheard of in the United States), among a number of other provisions. In New Jersey, Governor Philip Murphy permitted remote/virtual weddings and remote/virtual issuance of working papers to minors, among numerous additional orders.[23]

> *Executive Order* – A written and signed directive of the President of the United States, or the Governor of a State, having force of law but limited effect—only within the Executive Branch under usual circumstances (the COVID-19 pandemic and N.Y. Exec. Law § 29-a, for example, notwithstanding). They do not require legislative approval, and are not laws/statutes in the traditional sense of "law." Furthermore, they must not violate the Constitution.

In Massachusetts, the State Legislature acted to permit remote/virtual notarizations, balancing health needs with the requirements of law and business.[24]

Additionally, the judiciary in New York State took further action to address the needs of attorneys, litigants, the public, and the courts in the face of the pandemic. Virtual/remote operations began, and electronic filing (already developed and implemented before the pandemic)

---

[22] The Executive Orders were part of Governor Cuomo's "New York State on PAUSE" response to COVID-19 (specifically Executive Order 202.8 (March 20, 2020), which directed the 100% reduction of in-person workforces for all non-essential businesses in the state). *See* https://www.governor.ny.gov/executiveorders.

[23] *See, e.g.,* N.Y. Executive Orders 202.5, 202.6, 202.7 & 202.20, *at* https://www.governor.ny.gov/executiveorders; *see also* **Appendix 6** (CT Executive Order 7Q; N.J. Executive Order 135). *See also* Joseph D. Nohavicka, *Interim Comment on Use of Remote Notarizations, Pursuant to Executive Order 202.7*, N.Y. L.J. at 6 (Apr. 29, 2020) (providing an example of the steps for e-notarization and then the accompanying e-notary certification); Josephine Harvey, *New Yorkers Can Now Legally Get Married Via Zoom*, Huffington Post (online May 5, 2020), *available at* https://www.huffpost.com/entry/wedding-new-york-zoom-legal_n_5e9d08b4c5b6ea335d5df1d9. It should be recognized, though, that constitutional challenges were raised to some of the executive orders issued by governors, such as in Illinois—see David Thomas, *Small-Town Lawyer Squared Off With Illinois' Governor Over COVID-19 Lockdown*, N.Y. L.J. at 5 (May 8, 2020).

[24] Massachusetts Legislature, Chapter 71 of the Acts of 2020 (amending Chapter 222 of the General Laws to address virtual notarizations). *See also* Amanda Robert, *Lawyers Address Problems with Estate-Planning Document Signing during Coronavirus Crisis*, ABA Journal (May 4, 2020), *available at* https://www.abajournal.com/web/article/lawyers-and-problems-with-estate-planning-signing-during-coronavirus-crisis.

was expanded with new procedures.[25] The New York Court of Appeals (the state's highest court) also engaged in remote/virtual work, as did Appellate Divisions in the state.[26]

The federal courts, likewise, expanded their remote/virtual operations—for instance, appellate oral arguments took place via use of technology and remote appearances.[27] Even the Supreme Court of the United States, usually proudly steeped in tradition and decorum, permitted telephonic arguments in a number of cases, and the public was able to listen-in[28] (meaning, for the first time, the world could hear live-action U.S. Supreme Court arguments by counsel in front of the justices—what had previously been possible only for those who managed to get public seating in the Court's relatively small ceremonial courtroom,[29] with other seats reserved for the press, retired justices, law clerks, officers of the Court, and attorneys admitted to practice before the Supreme Court).[30]

The courts remained functioning, to address the needs of our constitutional Republic—although backlogs did still develop, since the courts were not designed for purely remote/virtual workings, particularly when it comes to juries and jury trials.[31] Note, though, that in the Federal Rules, there was a provision pre-dating the pandemic that permitted the transmission of testimony from a remote location when good cause was shown.[32] Rule 43 provides, in pertinent part:

---

[25] *See* **Appendix 7** (May 1, 2020 and May 28, 2020 Administrative Orders of the Chief Administrative Judge of the New York State Courts; May 4, 2020 Message from Hon. Janet DiFiore, Chief Judge of the State of New York; May 11, 2020 Notice to the Bar by Chief Clerk and Legal Counsel to the N.Y. Court of Appeals; and April 21, 2020 Operations Announcement of the N.Y.S. Supreme Court, Appellate Division, First Department).

[26] *See Order of the New York State Court of Appeals* (May 7, 2020) (DiFiore, C.J.) (Amending Rules of Practice), N.Y.S. Register, *Court Notices,* N.Y.S. Dep't of State (May 27, 2020), *available at* https://www.dos.ny.gov/info/register/2020/052720.pdf; *COVID-19 Notice to Parties Appearing for Remote Argument in June/July*, N.Y.S. Sup. Ct., App. Div., Fourth Dep't (June 11, 2020), *available at* https://ad4.nycourts.gov/press/notices/5ee24d8bafb6be4d0c5a999e.

[27] *See* **Appendix 8** (March 16, 2020 Order of the Chief Judge for the United States Court of Appeals for the Second Circuit). *See also* Marcia Coyle, *In the Coronavirus Pandemic, State and Federal Judges Lean on Technology*, N.Y. L.J. at 2 (Apr. 23, 2020); Lisa Margaret Smith, *Settlement Conferences in the Age of COVID-19*, Fed. Bar Counc. Quarterly, vol. 27, no. 3, at 24 (Mar./Apr./May 2020).

[28] *See* April 13, 2020 Press Release, Supreme Court of the United States, *available at* https://www.supremecourt.gov/publicinfo/press/pressreleases/pr_04-13-20; Mike Scarcella & Tony Mauro, *Flush With Embarrassment: Brief History Of Awkward Moments at SCOTUS*, N.Y. L.J. at 5 (May 8, 2020); Adam Liptak, *Supreme Court Hears First Arguments via Phone*, The N.Y. Times (online May 12, 2020), *available at* https://www.nytimes.com/2020/05/04/us/politics/supreme-court-coronavirus-call.html.

[29] *See, generally, Courtroom Seating*, Sup. Ct. of the U.S., *at* https://www.supremecourt.gov/oral_arguments/courtroomseating.aspx.

[30] *See Visitor's Guide to Oral Argument*, Sup. Ct. of the U.S., *at* https://www.supremecourt.gov/visiting/visitorsguidetooralargument.aspx.

[31] *See* Alan Feuer, *et al.*, *N.Y.'s Legal Limbo: Pandemic Creates Backlog of 39,200 Criminal Cases*, The N.Y. Times (online June 22, 2020), *available at* https://www.nytimes.com/2020/06/22/nyregion/coronavirus-new-york-courts.html.

[32] *See* Fed. R. Civ. P. 43(a) ("Taking Testimony").

(a) In Open Court. At trial, the witnesses' testimony must be taken in open court unless a federal statute, the Federal Rules of Evidence, these rules, or other rules adopted by the Supreme Court provide otherwise. For good cause in compelling circumstances and with appropriate safeguards, the court may permit testimony in open court by contemporaneous transmission from a different location.[33]

Thus, not every development concerning law and technology has its roots in a disease pandemic, although such provisions do allow for continuing operations during a pandemic.

10.3 – COVID-19 and Law Practice: How Different Things May Look

As discussed in this chapter, virtual proceedings have taken place across the world, from legislative chambers to offices in executive mansions to courtrooms to business showrooms and boardrooms to schools to the offices of attorneys and physicians. In light of everything, some commentators argue that Law may not be able to turn back from technology and remote work,[34] while others argue that Law is not ready for long-term remote working conditions (particularly because of the risks to cybersecurity).[35] As with most debates, the answer likely lies somewhere in-between. Certainly, though, an answer will not be forthcoming before the second edition of this present book is published!

In the meantime, consider that eDiscovery has most definitely been impacted by the online activities since COVID-19—particularly with so much additional online and electronic content having been created on social media and elsewhere, all subject to eDiscovery demands as we know from earlier chapters.[36] Additionally, the New York State Bar Association's Technology and the Legal Profession Committee issued a Cybersecurity Alert in June 2020, warning that, among other things, the recording function available on so many virtual meeting platforms might make a recording of a business meeting a "business record" under evidence rules—thus making it subject to discovery demands.[37] Always food for thought.

Additionally, while there are no Rules of Professional Conduct that specifically address the activities of attorneys in a virtual world, several rules can be extrapolated to apply—including Rules 1.1, 1.3, 1.6, 1.15, 5.1, 5.2 and 5.3. One must acknowledge, recognize, and prepare for threats that arise from the use of online storage, communication, and conferencing

---

[33] *Id.*

[34] *See* Frank Ready, *COVID-19 Pushed Legal Toward Tech, Remote Work. There May Be No Going Back*, N.Y. L.J. at 6 (Apr. 10, 2020).

[35] *See* Victoria Hudgins, *Don't Get Cocky: Firms May Not Be Prepared for Long-Term Remote Work*, N.Y. L.J. at 3 (May 12, 2020).

[36] *See* Michael Poskanka & Allison Oliver, *Social Media Use Is on the Rise. What Does This Mean for Litigation?*, N.Y. L.J. at 6 (June 19, 2020).

[37] *Cybersecurity Alert: Discovery of Recordings from Virtual Meeting Platforms,* Tech. & the Legal Profession Comm. of the N.Y. St. Bar Ass'n (June 25, 2020).

services.[38] Attorneys must be competent; must be diligent; must safeguard client confidences and property; must communicate securely with clients to preserve the attorney-client privilege and work-product protection (if any); and properly supervise the activities of all subordinate attorneys and non-attorneys to ensure that no violations of ethical proscriptions occur.[39] Of course the foregoing, and more, is true for any day in the practice of law—both in the virtual and real worlds, as they become more intertwined—as well as in response to any disaster, emergency, or pandemic, and not just in response to COVID-19.[40]

Next, consider trials and hearings. Federal Rule 43(a) notwithstanding, pre-COVID-19 it likely did not seem to many that trials could be conducted in any other fashion than in-person, in a courtroom (particularly criminal trials, with their concomitant Sixth Amendment requirements and safeguards). Yet, thousands of hearings were held in state and federal courts via virtual means since the COVID-19 pandemic began, while "Zoom trials" and even virtual grand jury proceedings, with judges and/or jurors viewing proceedings via video from separate rooms (or from home), became new tools for litigators and courts in an age when social distancing and disease created what many refer to as "the new normal."[41] While many

---

[38] *See* Peter Brown, *Internet Fraud: COVID-19 Causes Increased Vulnerability*, N.Y. L.J. at 5 (Apr. 14, 2020) (discussing some common threats and scams about which to be aware, including those stemming from COVID-19).

[39] *See, e.g.*, Ellen Rosen, *The Zoom Boom: How Videoconferencing Tools Are Changing the Legal Profession*, ABA Journal (online June 3, 2020), *available at* https://www.abajournal.com/web/article/ethics-videoconferencing-tools-are-changing-the-legal-profession; Mark A. Berman, *Remote Computing, Cyber Protection From Home*, N.Y. L.J. at 6 (Mar. 18, 2020); *Cybersecurity Alert: Tips for Working Securely While Working Remotely*, Tech. & the Legal Profession Comm. of the N.Y. St. Bar Ass'n (Mar. 12, 2020), *available at* https://nysba.org/app/uploads/2020/03/NYSBA-Cyber-Alert-031220.pdf. *See also* Benjamin Dynkin & Barry Dynkin, *Professional Responsibility In the Age of Zoom*, N.Y. L.J. at 4 (June 16, 2020); Anthony E. Davis & Janis M. Meyer, *Rising to the Ethical Challenges Of Remote Working*, N.Y. L.J. at 3 (May 4, 2020); Devika Kewalramani, *Pandemics and Ethics: How To Sustain Lawyer Quality and Client Service*, N.Y. L.J. at 4 (Apr. 6, 2020); Mark A. Berman, *Cybersecurity 'Hygiene' For Lawyers*, N.Y. L.J. at 5 (Mar. 3, 2020) (discussing need for attorneys to practice good cybersecurity under both the ethical rules and New York's new S.H.I.E.L.D. Act); Am. Bar Ass'n Formal Op. 477; Am. Bar Ass'n Formal Op. 369; N.Y.S. Bar Ass'n Op. 1019; and N.Y.S. Bar Ass'n Op. 842.

[40] *See* Dave Poston & Ioana Good, *Cyber Attacks Are the New Normal. Be Prepared To Respond*, N.Y. L.J. at 5 (Mar. 6, 2020); Jessica L. Copeland, *COVID-19: Now Infecting Cybersecurity*, N.Y. L.J. at 4 (Apr. 13, 2020).

[41] *See* Charles Scudder, *In a Test Case, Collin County Jury Renders Verdict on Zoom for the First Time; Too Risky for a Full Trial?*, The Dallas Morning News (online May 22, 2020), *available at* https://www.dallasnews.com/news/courts/2020/05/22/in-a-test-case-collin-county-jury-meets-on-zoom-for-the-first-time-but-some-lawyers-say-its-too-risky-for-real-trial/ (discussing summary jury trial in Texas—an alternative dispute resolution procedure—where the jury was selected and heard the matter all via Zoom); Catherine Wilson, *"Most Troublesome" Issue: Experiment Tests Remote Jury Trial With COVID-19 Around*, N.Y. L.J. at 6 (May 18, 2020) (discussing, among other things, Broward, Florida, Chief Circuit Judge, attorneys and judicial assistants experimenting with a mock jury trial); Matt Reynolds, *Could Zoom Jury Trials Become the Norm During the Coronavirus Pandemic?*, ABA Journal (online May 11, 2020), *available at* https://www.abajournal.com/web/article/could-zoom-jury-trials-become-a-reality-during-the-pandemic; Madison Alder, *U.S. Courts Try Out Social Distancing, Video for Grand Juries*, Bloomberg News (online June 8, 2020), *available at* https://news.bloomberglaw.com/us-law-week/u-s-courts-try-out-social-distancing-video-for-grand-juries. *But see* Scott Graham, *Judges Get Creative as Patent Jury Trials Resume Next Month*, Texas Law. (online May 21, 2020), *available at* https://www.law.com/texaslawyer/2020/05/21/judges-get-creative-as-patent-jury-trials-resume-next-month/ (discussing plans in several federal

of the skills and techniques employed by attorneys will not change,[42] litigators will have to adapt and hone skills for application in an online world—where witnesses, judges, and jurors are no longer only a few feet away; where in addition to being concerned about having legal precedents, rules, exhibits, and pens at the ready, attorneys and their paralegals must also be prepared for technological/remote evidentiary pre-

> **Zoombombing** – Unauthorized access to an online video conference, particularly when the unauthorized person creates a disturbance, interrupts, or commits a lewd act to disrupt the meeting or call.

sentation procedures, together with glitches and hardware concerns;[43] where new concerns arise over potential "coaching" of witnesses by off-screen individuals during depositions or trial examinations;[44] where medical diagnoses in personal injury, medical malpractice, and other cases could potentially be attacked because medical treatment or the independent (a.k.a. defense) medical examination took place via telemedicine;[45] and where interruptions, "Zoombombing" and other infractions outside the control of the trial or appellate judge may halt proceedings mid-sentence and disrupt the flow of what would otherwise be structured presentations.[46]

In addition to all of the foregoing, another possibility looms in the future as artificial intelligence (AI) continues its development by leaps and bounds—Holographic Judges/Internet Courts. These are courts with AI judges, not human beings, who handle cases at any time—and they already exist in China.[47] Such may present an answer down the line to case backlogs, or might allow courts to shed relatively minor, or small dollar, matters from their dockets to AI courts. However, as was mentioned in Chapter 1 when discussing TAR and predictive coding, AI learns as it goes, based largely on programming and inputs. Since readers are likely

---

district courts in Texas, California and New York to either use social distancing in large courtrooms, with plexiglass dividers and other features, or hybrid in-person/online procedures, to re-start jury trials in June 2020).

[42] *See* William Dorsey, *Zoom Trials: Same Dog, New Tricks*, N.Y. L.J. at 2 (June 1, 2020).

[43] *Id.*; *see also* Graham Smith-Bernal, *The Future Has Arrived: Practical Advice For Conducting Legal Proceedings Digitally*, N.Y. L.J. at 7 (June 4, 2020); David A Lowe, *Trial by Zoom: A Strange But True Story of How One Lawyer Prepared for Court*, ABA Journal (online June 24, 2020), *available at* https://www.abajournal.com/voice/article/the-strange-but-true-story-preparing-for-a-trial-by-zoom (discussing preparations for what was to be a completely virtual bench trial via Zoom in a California federal district court; and mentioning the issue of thousands of jury trials having been delayed by COVID-19, resulting in a substantial backlog of cases, and a potential need for Zoom/virtual trials to resolve matters).

[44] *See* Michael A. Mora, *Is Someone Coaching Your Witness Off-Camera? In a Changing World, Litigators Adjust Strategies*, N.Y. L.J. at 5 (Apr. 24, 2020).

[45] *Id.*

[46] *See* Debra Cassens Weiss, *Federal Judge Shuts Down Remote Hearing After Interruptions by Listening Audience*, ABA Journal (online Apr. 16, 2020), *available at* https://www.abajournal.com/news/article/federal-judge-shuts-down-remote-hearing-after-interruptions-by-public-audience; Rebecca Brazzano, *Zoombombing, Sexting And Revenge Porn, Oh My!*, N.Y. L.J. at 4 (June 11, 2020).

[47] *See* Katherine B. Forrest, *The Holographic Judge*, N.Y. L.J. at 5 (Dec. 31, 2019) (discussing, *inter alia*, the Beijing Internet Court, and the AI court in Hangzhou, China—which as of December 2019, according to the article, had handled three million cases).

> ***Artificial Intelligence*** *– Computer-ized or machine intelligence, as op-posed to or compared with human intelligence.*

familiar with the popular phrase "garbage in, garbage out," the issue of AI courts should be one to ponder with great caution and analysis.

Finally, as we become comfortable with the idea that online proceedings and virtual technology may aid in the continuing work of courts and attorneys, one should also take heed of a note of caution: the rules and electronic practices may not extend so far as to allow remote/virtual proceedings in all *arbitrations*. A 2019 decision of the Eleventh Circuit U.S. Court of Appeals, in *Managed Care Advisory Group, LLC v. CIGNA Healthcare, Inc.*,[48] held that pursuant to the Federal Arbitration Act (FAA), since arbitration is a contractual arrangement arbitrators only have authority over non-parties as limited by the scope of the provisions of the FAA. As the court held, consistent with holdings in the Second, Third, Fourth, and Ninth Circuits, Section 7 of the FAA limits the authority of an arbitrator to summon a non-party to appear or produce documents/information in-person before that arbitrator or panel. However, the summonses issued by the arbitrator in the *Managed Care* case directed, *inter alia*, video conference with non-parties for the taking of evidence. The summonses requiring video, as opposed to in-person, appearances were disallowed by the Court of Appeals, and the district court's determination was reversed and remanded.[49]

The take away point here?—Although the decision at issue above predated the COVID-19 pandemic, and concerns the Federal Arbitration Act and not necessarily arbitrations under state law,[50] be careful if testimony or documents are needed from non-parties in an arbitration, because if such are necessary at a time when in-person appearances are not feasible then the parties may wish to seek a stay or continuance from the arbitrator or arbitral panel until such time as the necessary witness or documents can appear or be delivered live, in-person. This is one situation eWorld appears not to have reached as of yet.

10.4 – Conclusion

Clearly, COVID-19 accelerated what was already a hastening approach of the real world to the virtual world in law and business. While there are aspects that are not all that much different between the two worlds, others are as different as night and day—although fortunately few are like oil and water, so long as the correct practices, ethical procedures, and safety protocols

---

[48] 939 F.3d 1145 (11th Cir. 2019).

[49] *Id.* at 1158-1161 (citing and agreeing with *Hay Group, Inc. v. E.B.S. Acquisition Corp.*, 360 F.3d 404, 407 (3d Cir. 2004) (Alito, J.); *Life Receivables Tr. v. Syndicate 102 at Lloyd's of London*, 549 F.3d 210, 216 (2d Cir. 2008); *COMSAT Corp. v. Nat'l Sci. Found.*, 190 F.3d 269, 275-76 (4th Cir. 1999); *CVS Health Corp. v. Vividus, LLC*, 878 F.3d 703, 708 (9th Cir. 2017); and declining to follow *In re Sec. Life Ins. Co. of Am.*, 228 F.3d 865, 870-71 (8th Cir. 2000)).

[50] Although beware that holdings of the Supreme Court of the United States have struck at certain state law provisions that conflict with, circumvent or are inconsistent with specific provisions of the FAA, on preemption grounds. *See, e.g., Kindred Nursing Ctrs. Ltd. P'Ship v. Clark*, 137 S.Ct. 1421 (2017).

are followed. There are those who believe the virtual eWorld actually helped save much of society during the COVID-19 pandemic—allowing many to stay in touch with family and friends for their mental health, consult with medical professionals for physical health, confer with attorneys for legal representation and protection, and continue their daily practices, work, schooling, and online shopping from home or other remote locations to keep portions of the economy moving. Some even point out that technology could permit contact-tracing to help identify those with whom infected individuals come into contact during a health pandemic—although, at the same time, as always, raising potential privacy concerns and attendant legal issues.[51] Nevertheless, while there are certainly privacy and security issues to ponder, debate, and address,[52] what is certain is that absent a world-wide crash of the Internet and all technology, the merging of real world and eWorld practices will continue, in large part unabated.

Chapter End Questions

Note: Answers may be found on page 265.

1. Which of the following Ethical Rules for Attorneys are implicated in online activities stemming from COVID-19?

   a. 1.1
   b. 1.15
   c. 5.3
   d. All of these selections

2. In which nation of the world have we already seen examples of holographic judges, where cases are decided utilizing AI rather than human judges?

   a. Australia
   b. France
   c. China
   d. United States

3. Which of the following is the number of the New York Executive Order that permitted virtual/online notarizations of documents?

   a. 202.1
   b. 202.2

---

[51] *See* Matt Reynolds, *Contact-Tracing Apps Could Help Contain COVID-19 But Raise Thorny Legal and Privacy Issues*, ABA Journal (online Apr. 23, 2020), *available at* https://www.abajournal.com/web/article/contact-tracing-apps-and-spread-of-covid-19; Scott Pink & John Dermody, *Where Will the Needle Land? COVID-19 Contact Tracing v. Protecting Personal Privacy*, N.Y. L.J. at 5 (June 12, 2020).

[52] *See* Katherine B. Forrest, *Living With the Illusions of Privacy*, N.Y. L.J. at 3 (June 5, 2020); Jeff John Roberts, *The Boss In Your Bedroom: As Workplace Surveillance Spreads, What Are Your Rights?*, Fortune (online May 20, 2020), *available at* https://fortune.com/2020/05/20/work-surveillance-worker-rights-privacy-coronavirus/.

    c. 202.7

    d. 202.8

4. Which of the following is the number of the New York Executive Order that placed New York State "on PAUSE"?

    a. 202.1

    b. 202.2

    c. 202.7

    d. 202.8

5. Which of the following is the number of the New Jersey Executive Order that permitted remote/virtual weddings and remote/virtual issuance of working papers to minors?

    a. 125

    b. 130

    c. 135

    d. 140

6. Which of the following is the New York statute that provides the Governor authority to issue broad executive orders during emergencies?

    a. Executive Law 20-b

    b. Executive Law 29-a

    c. Civil Rights Law 50-c

    d. Governor's Act 40-b

7. True or False: Arbitrators acting under federal law, whether or not in the face of a pandemic, can order both parties and non-parties to appear and provide evidence or testimony either in-person or by electronic/virtual means.

    a. TRUE

    b. FALSE

8. True or False: While some legal matters may be handled online, things like jury trials will likely never be held in any forum other than in a courtroom.

    a. TRUE

    b. FALSE

9. True or False: "Zoombombing" is when an individual gains unauthorized access to an online video conference, and usually creates a disturbance, interrupts, or commits a lewd act.

    a. TRUE

    b. FALSE

10. True or False: During online witness examinations or depositions, "coaching" of witnesses by off-screen individuals is a concern in eWorld legal proceedings.

    a. TRUE

    b. FALSE

CHAPTER 11

The "Final Chapter"—What Happens to Electronic Assets After Death?[*]

11.1 – Introduction

"Of all our possessions, wisdom alone is immortal."[1] One wonders, would Isocrates revise his statement in present day if he knew about social media and online electronic accounts? While he, no doubt, would have pondered his philosophy in light of the modern age, we nevertheless must certainly redefine our idea of what is immortal. Of course, Andy Warhol, ever the modern philosopher, made his concerns clear: "Dying is the most embarrassing thing that can ever happen to you, because someone's got to take care of all your details."[2] The devil in your personal details after death can be the source of great embarrassment (which is why even after death, public policy favors respecting the privacy of the dead, in certain respects, just as if they were still alive—particularly for information that would not be available under FOIA or FOIL—Freedom of Information Act or Law—requests as public records).[3] However, it is not just embarrassment we

---

[*]The author of this text published a prior version of this discussion. *See* M. Fox, *Who Has Access to Decedents' Electronic Assets and Social Media?*, 26 PRETRIAL PRACTICE & DISCOVERY, Issue 1 (Winter 2018) (American Bar Association).

[1]Isocrates, ancient Greek rhetorician (436-338 B.C.), as quoted in DICTIONARY OF QUOTATIONS (CLASSICAL) (1906), edited by Thomas Benfield Harbottle, p. 495. For more on Isocrates, *see* ENCYCLOPAEDIA BRITANNICA, available at https://www.britannica.com/biography/Isocrates.

[2]Andy Warhol, U.S. Pop Artist (1928-1987), as quoted in Victor Bokris, *Warhol,* "Goodbye 1986-7," quoted in THE COLUMBIA DICTIONARY OF QUOTATIONS (Colum. Univ. Press 1993).

[3]*See, generally*, A. Cohen, *Damage Control: The Adoption of the Uniform Fiduciary Access to Digital Assets Act in Texas*, 8 EST. PLAN. & COMMUNITY PROP. L.J. 317 (2015) (general discussion, including the concerns about the privacy of the deceased that intertwines with the access of fiduciaries and the statutory frameworks governing same); *see, generally, Morley v. C.I.A.*, 508 F.3d 1108 (D.C. Cir. 2007); *U.S. Dept. of Justice v. Reporters Comm. for Freedom of Press*, 489 U.S. 749, 109 S.Ct. 1468 (1989) (distinction between FOIA and privacy in tort law); *Swickard v. Wayne County Med. Examiner*, 438 Mich. 536, 475 N.W.2d 304 (Mich. 1991).

address in this chapter; we must also understand what happens to the ownership and control of virtually immortal electronic assets when the mortal owner passes from this life.[4]

Indeed, as we will discuss,[5] social media, electronic communications, and electronic financial accounts have the capacity to live on, eternally, as part of what has been called in other contexts the "vast cosmos of the Internet."[6] We know that "nothing is ever truly deleted from the Internet." In fact, type that exact phrase into Google or a similar search engine and see how many out there have uttered—or written—and repeated that same sentiment on the Internet. Once an owner of electronic assets passes away, who then has access to the assets? Who owns them? As Warhol would ask, "Who has access to all those potentially embarrassing details?" You might say, "The executor of the estate, the personal representative(s), or the heirs." Not so fast! Do they have complete access to all details in the electronic asset? Do they exert complete control over all facets of the electronic asset? The answer is actually more complicated than one might initially expect and requires specific attention, planning, and parsing of statutes and legal rules.

11.2 – Scenarios to Consider

Consider two scenarios—without needing much imagination at all for either.

Scenario 1—A person utilizes, among other digital or electronic assets, an online Google or Hotmail/Outlook email account with calendar and contacts; a Facebook or Twitter account; and perhaps an Amazon, eBay, or other such account. The individual is concerned about estate planning and works with an attorney to formulate an estate plan that incorporates specific direction regarding the electronic assets and accounts.

Scenario 2—We encounter a person similar to the person in Scenario 1. However, this time the person passes away before creating a specific estate plan concerning the electronic assets and accounts. What happens in each scenario?

---

[4] In most instances, the laws and rules addressed in this Chapter can also be applied to circumstances of incapacity (when the owner of a digital asset becomes incapacitated and a guardian or fiduciary is appointed on his or her behalf), as well as digital trust assets and trustees, and digital assets falling under the principal-agent relationship. *See, e.g.*, N.Y. EPTL §§ 13-A-1(m); 13-A-3.3 through 13-A-3.8. However, for purposes of this Chapter we focus on access after the death of the asset's owner.

[5] This chapter focuses on cases and statutes in New York and Massachusetts, but New York follows a revised uniform code adopted in numerous other states—and therefore one can imagine a similar framework and set of decisions governing across those states, as well. Electronic communications and accounts have no regard for geographic boundaries, thus uniformity across jurisdictions will be key in addressing these unique issues.

[6] *See U.S. v. Mayo*, No. 2:13-CR-48, 2013 WL 5945802, at *8 (D. Vt. Nov. 6, 2013) (citing M. Orso, *Cellular Phones, Warrantless Searches, and the New Frontier of Fourth Amendment Jurisprudence*, 50 SANTA CLARA L. REV. 183, 211 (2010)). *See also* Robert Kirk Walker, *The Right to be Forgotten*, 64 HASTINGS L.J. 257, 259 (2012).

11.3 – Relevant Provisions of Law

To begin, it is settled as a matter of law that digital assets such as email accounts and social media accounts are capable of being owned as a form of property.[7]

In 2016, New York State passed its version of the Uniform Law Commission's Revised Uniform Fiduciary Access to Digital Assets Act, codifying it in a new Article 13-A of the Estates, Powers and Trusts Law (EPTL).[8] Article 13-A addresses access to financial accounts, electronic communications, and digital assets.[9] As of the time of this writing, at least 46 states and the U.S. Virgin Islands had passed laws addressing access to digital assets (email accounts, social media accounts, and other electronic assets) after death or incapacity of the owner—most of which enacted the Uniform Fiduciary Access to Digital Assets Act or Revised Uniform Fiduciary Access to Digital Assets Act.[10]

> *Content Material* – The text and body of messages; the content of communications.
>
> *Non-Content Material* – Information but not content; who created a message; when; contact lists; calendar appointments.

Specifically, there is a hierarchy of access to the online items, governed at the first level by the user's own direction.[11] The hierarchy is exactly what one would expect, with ultimate authority and decision-making as to both content and non-content digital material existing in the hands of the user/decedent while they are among the living. Bear in mind for this analysis, "content material" is actual content of communications between persons (user and another) online. "Non-content material" is other information—as the cases discuss, material such as contact lists, screen name, subscriber information, and calendar appointments—having no content of actual communications between user and others, and thus less, if any, impact on the privacy of the user compared with content material.[12] A key issue being non–content information not finding inclusion/protection under the Federal Electronics Communication Privacy Act (ECPA).[13]

At the first level of authority, the user may direct who, if anyone at all, is to have access to the content or non-content material of online communications through use of an online tool

---

[7] *Carpenter v. U.S.*, 138 S.Ct. 2206, 2270 (2018) (citing, *inter alia*, Tex. Prop. Code Ann. § 111.004(12) ("[p]roperty" includes "property held in any digital or electronic medium"); *Ajemian v. Yahoo!, Inc.*, 478 Mass. 169, 170, 84 N.E.3d 766, 768 (2017) (email account "form of property often referred to as a 'digital asset'"), *cert. denied*, *Oath Holdings, Inc. v. Ajemian*, 138 S.Ct. 1327 (Mem) (U.S. 2018); *Eysoldt v. ProScan Imaging*, 194 Ohio App.3d 630, 638, 957 N.E.2d 780, 786 (1st Dist. 2011)). *See also Matter of Coleman*, 96 N.Y.S.3d 515, 518 (Surr. Ct. Westchester County 2019) (defining terms).

[8] *See* P. Skelos, *et al.*, *The Digital Footprint After Death: Who Wears the Shoes?*, N.Y. L.J. at 10 (Sept. 11, 2017).

[9] Note that the statutory framework does not only include fiduciaries/executors of a decedent but also trustees, powers of attorney, and guardians/fiduciaries of incapacitated persons. *See id.*; *see also* N.Y. EPTL Art. 13-A.

[10] *See* National Conference of State Legislatures, *Access to Digital Assets of Decedents* (May 23, 2019), available at: ncsl.org/research/telecommunications-and-information-technology/access-to-digital-assets-of-decedents.aspx.

[11] *See* N.Y. EPTL § 13-A-2.2.

[12] *See, e.g.*, *Freedman v. America Online, Inc.*, 412 F.Supp.2d 174 (D. Conn. 2005); *but cf. Lukowski v. County of Seneca*, 2009 WL 467075 (W.D.N.Y. Feb. 24, 2009).

[13] *Freedman*, 412 F.Supp.2d at 181–182.

with, or provided by, the service provider/custodian. The statutory framework in New York specifically provides that at the first level of analysis, "[i]f the online tool allows the user to modify or delete a direction at all times, a direction regarding disclosure using an online tool overrides a contrary direction by the user in a will, trust, power of attorney, or other record."[14]

However, if the user does not utilize an online tool directive or the service provider does not make one available, then the next level of authority comes from the usual estate planning documents. "[T]he user may allow or prohibit in a will, trust, power of attorney, or other record, disclosure to a fiduciary of some or all of the user's digital assets, including the content of electronic communications sent or received by the user."[15]

Finally, if the decedent executed no estate planning documents while living (*i.e.*, the user dies intestate), one must look to the last level of authority and engage in an evaluation of the Terms of Service Agreement between the user and the service provider.[16] Such Terms of Service Agreements are generally binding contracts between users and service providers and provide rules governing the relationship, and both maintenance and disclosure of information—as long as the "I agree" button and terms make clear to the user that clicks on the button will form a contract – the standard of "reasonably communicated and accepted."[17]

On that note, courts have been clear that the facts and circumstances calling into question the formation of a binding contract *do not include* the failure of a user to carefully read and evaluate the terms before agreeing. In 2018, the U.S. District Court for the Southern District of New York held: "The process by which [plaintiff] assented to the December 2015 agreement provided him reasonable notice of the terms of the operative agreements. This Court therefore finds that [plaintiff] had constructive knowledge of the arbitration provision. The fact that [plaintiff] may have failed to review the contract carefully is not a valid defense."[18] So, be careful what you "click on" electronically, the same as you should be careful concerning what you sign on paper.

---

[14] *See* N.Y. EPTL § 13-A-2.2(a).

[15] *See* N.Y. EPTL § 13-A-2.2(b).

[16] *See* N.Y. EPTL § 13-A-2.2(c).

[17] *See, e.g., Caraccioli v. Facebook, Inc.*, 700 Fed. Appx. 588 (9th Cir. 2017); *Lewis v. YouTube, LLC*, 244 Cal. App.4th 118, 197 Cal.Rptr.3d 219 (Ca. Ct. of Apps. 6th Dist. 2015); *but cf. Applebaum v. Lyft, Inc.*, 263 F.Supp.3d 454 (S.D.N.Y. 2017) ("where, as here, there is no evidence that the [mobile application] user had actual knowledge of the agreement, the validity of the . . . agreement turns on whether the [application] puts a reasonably prudent user on inquiry notice of the terms of the contract"). *See also Ajemian v. Yahoo!, Inc.*, 83 Mass.App.Ct. 565, 987 N.E.2d 604 (Mass. App. Ct. 2013) (discussing "clickwrap" versus "browsewrap" agreements and their binding or non-binding natures) (citing, *inter alia, Hughes v. McMenamon*, 204 F.Supp.2d 178, 181 (D. Mass. 2002)). *See also Cullinane v. Uber Tech., Inc.*, 893 F.3d 53 (1st Cir. 2018) (citing *Ajemian* and holding that "Plaintiffs were not reasonably notified of the terms of the [online Terms of Service] Agreement, [and so] they did not provide their unambiguous assent to those terms"); *cf. Theodore v. Uber Tech., Inc.*, 18 – cv– 12147 – DPW, 2020 WL 1027917, at *5-*6 (D. Mass. Mar. 3, 2020) (citing *Bekele v. Lyft*, 918 F.3d 181, 187 (1st Cir. 2019)); Dan Clark, *Courts Are Asking for More Complex Evidence In Upholding Click – Through Agreements,* N.Y.L.J. at 5 (Mar. 5, 2020).

[18] *O'Callaghan v. Uber Corp. of California*, 2018 WL 3302179 (S.D.N.Y. July 5, 2018) (distinguishing *Applebaum v. Lyft*). And, note that in states such as New York, "an electronic signature 'has the same validity and effect as the use of a signature affixed by hand.'" *Belizaire v. Ahold U.S.A., Inc.*, 2019 WL 280367 (S.D.N.Y. Jan 22, 2019) (quoting *O'Callaghan*, 2018 WL 3302179, at * 7 n. 9).

Under New York's EPTL provisions in Article 13-A, a fiduciary (executor or administrator) who lacks authority from the user, but nevertheless needs access to the non-communication portions of the decedent's electronic accounts—calendar entries, contact lists, and those items that do not have actual communication components to them—may obtain access by court order and appointment, without running afoul of the federal Stored Communications Act[19] (and, thereby, federal criminal law), because there is no access to actual communication elements (*i.e.,* no communication between parties when a calendar appointment is made or when a contact is added to a contacts list).[20] As stated by the *Serrano* Court, in what appears to be New York's first reported case on the matter:

> *Executor* – Representative of an estate of an individual who died with a will naming a representative.
>
> *Administrator* – Representative appointed by the court for an estate of a person who died without a will or without appointing a representative.

> A deceased user's calendar kept electronically is, thus, a digital asset that does not include "content of electronic communications," and, therefore, must be disclosed to a personal representative by a custodian of such a record pursuant to EPTL 13-A-3.2. . . . In the alternative, the court finds that disclosure here is "reasonably necessary for the administration of the estate". . . As previously discussed, disclosure of the requested non-content information is permitted, if not mandated, by Article 13-A of the EPTL and does not violate the EPTL.[21]

> *Decedent* – A person who has passed away.
>
> *Testator* – A person who makes a will.
>
> *Intestate* – Passing away without a will.
>
> *Will* – A formal document directing the distribution of the possessions of the testator after they pass away.

If, however, the decedent's representatives request access to or disclosure of material containing communications between the decedent and others, a different standard will be applied by the court hearing the matter. In such a circumstance, the *Serrano* Court held: "Authority to request from Google disclosure of the content of the decedent's email communications. . .is denied without prejudice to an application by

---

[19] Stored Wire and Electronic Communications and Transactional Records Access Act (Stored Communications Act), 18 U.S.C. §§ 2701, *et seq.* "Like the tort of trespass, the Stored Communications Act protects individuals' privacy and proprietary interests. The Act reflects Congress's judgment that users have a legitimate interest in the confidentiality of communications in electronic storage at a communications facility. Just as trespass protects those who rent space from a commercial storage facility to hold sensitive documents . . . the Act protects users whose electronic communications are in electronic storage with an ISP or other electronic communications facility." *Theofel v. Farey-Jones*, 359 F.3d 1066, 1072-1073 (9th Cir. 2004) (citations omitted).

[20] *See* N.Y. EPTL § 13–A–3.2; 18 U.S.C.A. §§ 2701, 2510(12); *In re. Estate of Serrano*, 56 Misc. 3d 497, 54 N.Y.S.3d 564 (Surr. Ct. N.Y. County 2017); 35B N.Y. Jur. 2d Criminal Law: Principles and Offenses § 1007—*Computer Trespass*.

[21] *In re. Estate of Serrano*, 56 Misc. 3d at 499, 54 N.Y.S.3d at 566. *See also* 2 Harris N.Y. Estates: Probate Admin. & Litig. § 21:103—*Documents-Digital Assets* (6th ed. 2018).

the voluntary administrator, on notice to Google, establishing that disclosure of that electronic information is reasonably necessary for the administration of the estate."[22]

These issues are also important to consider for business concerns. For instance, contemplate a situation where a business owner passes away and perhaps his or her business is a sole proprietorship. Sole proprietorships are the simplest form of business formation, where the business is the person and the person is the business; they are basically one and the same.[23] Or perhaps the decedent is the sole shareholder of a close corporation, managing all business affairs.[24] When the individual passes away, it is vitally important for the estate representative— the fiduciary, the executor or the administrator—to access all assets, including digital assets, to learn about the decedent's business holdings in marshaling the estate. That was the subject in the New York case of *Matter of White*[25] in 2017.

In *Matter of White*, the decedent's appointed fiduciary sought a court order directing Google to provide access to the decedent's electronic mail account—the only way, it seemed, to gain a full picture of decedent's business assets and liabilities.[26] Google had refused such access. The Court evaluated the provisions of NY EPTL Article 13-A. The decedent did not utilize an online tool provided by Google, nor did the decedent use a will, trust, or other document to provide instructions on access to electronic assets after death.[27] Of course, we know that means that the evaluation of the case fell to the lowest level of authority under the EPTL. Because of that, the Court was concerned about allowing complete and unlimited access to the decedent's digital assets, and risking disclosure of details related to personal affairs or confidential information having nothing to do with the decedent's business interests and activities.[28] Thus, the Court had to balance the competing concerns and ultimately decided to grant limited access to the email account of decedent to the extent of Google disclosing only "the contacts information stored and associated with the email account" at issue.[29] For any further access, if necessity was shown, petitioner fiduciary had to return to the court.[30]

Another recent New York case, from 2019, further explored the issue of a decedent passing without a will, and the subsequent difficulty of obtaining digital asset information from Apple. In *Matter of Coleman*[31] the decedent was a 24-year old man, and recent college graduate, who

---

[22] *In re. Estate of Serrano*, 56 Misc. 3d at 499, 54 N.Y.S.3d at 566 (citing N.Y. EPTL § 13-A-3.1(e)(3)(D)).

[23] *Business Law: Text and Exercises*, Chapter 28 (CENGAGE ADVANTAGE BOOKS 9th Ed. 2018).

[24] *Id.* at Chapters 29-30.

[25] *See Matter of White*, N.Y. L.J. p. 25, col. 1, 2017 NYLJ LEXIS 2780 (Surr. Ct. Suffolk County Oct. 3, 2017); *see also* M. Berman, *Electronic Signatures, Inferences and Access to a Decedent's Emails*, N.Y. L.J. at 7-8 (Jan. 2, 2018) (citing *Matter of White*).

[26] *Id.*

[27] *Id.*

[28] *Id.*

[29] *Id.*

[30] *Id.*

[31] 96 N.Y.S.3d 515 (Surr. Ct. Westchester County 2019).

connected his Georgetown University e-mail account (as an alumnus) to his Apple iPhone.[32] After passing away without a will or estate plan on Christmas Day, 2016, his parents were appointed his administrators by letters of administration in 2017.[33] Thereafter, the administrators sought to obtain information from Apple concerning the digital assets contained in/on his iPhone, for which they did not have the passcode.[34] Apple advised the administrators that without the passcode, the information on the device could not be accessed, but if the information was backed-up in the iCloud, then such could be provided. However, Apple demanded that the administrators first obtain a court order—and Apple spelled-out specifically what the court order had to contain.[35] For some reason, more than a year later—in October 2018—petitioners/administrators filed the action in Westchester County, New York, Surrogates' Court, seeking an order.[36]

Following a clear discussion of the statutory framework found in N.Y. EPTL Article 13-A, the *Coleman* court held that:

> [h]ere, [decedent] neither used an online tool to grant his fiduciary access to the content of his digital assets nor had a last will and testament or other document which controlled the disposition of the content of these assets.

> As set forth above, the petitioners seek disclosure of all of [decedent]'s digital assets. As explained by the Apple representative, because the passcode to the iPhone is not known, access to the digital assets would have to be accomplished through the iCloud, providing the data was backed up to the service. According to Apple's Legal Process Guidelines, granting access to a user's iCloud is equivalent to the disclosure of: incoming and outgoing communications such as time, date, sender email addresses, recipient email addresses, email content, photo stream, iCloud photo library, contacts, calendars, bookmarks, Safari browsing history, iOS Device Backups, which may contain photos and videos in the camera roll, device settings, app data, iMessage, SMS, and MMS messages and voicemail.

---

[32] *Id.*

[33] *Id.*

[34] *Id.*

[35] *Id.* at 516 ("On May 23, 2017, the Apple representative sent an email to Adrienne explaining exactly what the court order should contain to insure compliance by Apple in the disclosure of any information sought. The court order must state: (1) the decedent was the user of all accounts associated with the Apple ID; (2) the petitioners are the personal representatives of the decedent; (3) the personal representatives are the "agents" of the decedent, and their authorization constitutes "lawful consent" as those terms are used in the Electronic Communications Privacy Act; and (4) Apple is ordered by the court to assist in the recovery of decedent's personal data from their accounts, which may contain third party personally identifiable information or data, from their accounts").

[36] *Id.* at 516 ("petitioners filed this proceeding seeking a court order to have access to all of Ryan's digital assets associated with his iPhone to: (1) determine whether Ryan had any medical issues that his two younger siblings may also have; (2) determine whether any legal action on behalf of Ryan's estate may be appropriate; (3) identify and collect Ryan's digital and non-digital assets; and (4) marshal any of Ryan's digital assets as part of the estate administration. They also asked that the court make certain findings including that Ryan maintained an iCloud account associated with the Georgetown email address and that he owned a certain iPhone").

Based on the record before this court, which includes the extent of the information available to a personal representative upon accessing the iCloud, and balancing [decedent]'s interests in his not having consented to the disclosure of the content of any of these digital assets, the court finds that the petitioners have not amply demonstrated, at this juncture, the need to access the content of [decedent]'s digital assets for the administration of his estate (EPTL 13-A-3.1[e][3] [D]) or any other reason. Instead, the petitioners may have access to the non-content information disclosable in accordance with EPTL 13-A-3.2.

As the court noted in [*Matter of Serrano*], if once the non-content information is disclosed pursuant to this decision and order to the petitioners by Apple, the petitioners can demonstrate the need for the content-based digital assets, the court will entertain a new petition based on the additional evidence.

Accordingly, the court makes the following findings:

(1) Decedent was the user of an account with Apple, the ID for which is the e-mail account provided by petitioners, individuals with personal knowledge that decedent was the user of that e-mail account;

(2) Petitioners are the fiduciaries of decedent's estate; and

(3) No lawful consent is required for disclosure of the non-content digital assets associated with the e-mail under the Stored Communications Act (18 USC 2701 et seq) or the New York Administration of Digital Assets law (EPTL Article 13-A).

This is the decision and order of the court[37]

It is further worth observing that another 2019 decision, a case out of the New York County (Manhattan) Surrogates' Court—*Matter of Swezey*[38]—addressed the issue of stored photographs contained in both iTunes and iCloud, but by a decedent who died testate (or with a will/ estate plan). In *Swezey*, as summarized by the *Coleman* court:

the fiduciary commenced an SCPA 2103 turnover proceeding against Apple seeking the decedent's photographs which were stored on iTunes and iCloud. There, the decedent did not use an online tool to provide direction for his digital assets and, although he died testate, he did not specifically provide for the disposition of his digital assets. In ordering Apple to disclose the photographs, the court relied on the facts that, in the decedent's last will and testament, he left his personalty and his residuary estate to his surviving spouse and that the decedent and his spouse "gave to each other implicit consent to access each other's digital assets." In doing so, the court quoted EPTL 1-2.15, noting that a decedent's property is "defined as 'anything that

---

[37] *Id.* at 518-519.
[38] N.Y. L.J., Jan. 18, 2019, at 34, col. 2 (Surr. Ct., N.Y. County).

may be the subject of ownership' real or personal'" and that "include[s] assets kept in digital form in cyberspace."[39]

Interestingly, in *Swezey*, because the estate plan included, in the New York court's eyes, the implicit consent of access for each spouse to the other's digital assets, the service provider (there Apple) was ordered to turn over the content information.

Separately, in the State of Massachusetts, the Supreme Judicial Court delivered a landmark decision in this field in October 2017. The Court held that while a service provider is not required to produce electronically stored account information to a fiduciary, absent a greater showing under the estates laws the federal Stored Communications Act,[40] standing on its own, does not act as an absolute bar or prohibition to such disclosure in the first instance—despite what some commentators and litigants have otherwise argued.[41]

In the *Ajemian* case, a man died in an accident, leaving no will to guide his siblings, who were his personal representatives in/after death. The representatives, in trying to access his digital assets, consented to the release of information by Yahoo! for their purposes, but Yahoo! rejected their requests. The lower court had held that Yahoo! could not provide the information because of Stored Communications Act restrictions. The Supreme Judicial Court, however, reversed.

As part of the majority's decision in the *Ajemian* case, though, the matter was returned/remanded to the probate court, with a direction that the probate court make further findings with regard to Yahoo!'s other argument against release of the digital material sought—that the Terms of Service with the decedent also prevented acquiescence to the request of the personal representatives.[42] As discussed earlier in this chapter, because the user died intestate and Yahoo! raised the issue of the relevant provisions of the Terms of Service, it remained for the lower court to decide if the Terms of Service—existing at the lowest level of authority for release of digital information—provided Yahoo! with a basis to reject the decedent's representatives' otherwise valid claims to the non-content material. In any event, the seminal *Ajemian* decision is believed to be the first of its kind to squarely address these issues.[43]

---

[39] *Matter of Coleman*, 96 N.Y.S.3d at 518.

[40] The separate issues of whether a warrant or only a trial subpoena is needed to compel disclosure of electronic material by a service provider in a legal proceeding, and what constitutes "electronic storage" (both assessed by analyzing the Stored Communications Act, among others) is addressed by, *inter alia*, *U.S. v. Weaver*, 636 F.Supp.2d 769 (C.D. Ill. 2009); *Jennings v. Jennings*, 401 S.C. 1, 736 S.E.2d 242 (S.C. 2012); *Sartori v. Schrodt*, 424 F. Supp. 3d 1121, 1133 (N.D. Fla. 2019); *Hately v. Watts*, 917 F.3d 770 (4th Cir. 2019); *Walker v. Coffey*, 956 F.3d 163, 168 & n.1 (3d Cir. 2020).

[41] *Ajemian v. Yahoo!, Inc.*, 478 Mass. 169, 84 N.E.3d 766 (2017), *cert. denied*, *Oath Holdings, Inc. v. Ajemian*, 138 S.Ct. 1327 (Mem) (U.S. 2018). For more on the Stored Communications Act, *see* M. Borden, *Covering Your Digital Assets: Why the Stored Communications Act Stands in the Way of Digital Inheritance*, 75 Ohio St. L.J. 405 (2014) (discussing the provisions of the SCA in detail).

[42] *Ajemian v. Yahoo!, Inc.*, 478 Mass. 169.

[43] For more on the rise of Uniform Laws in this context and discussion of the *Ajemian* case prior to the decision of the Massachusetts Supreme Judicial Court in 2017, *see* N. Banta, *Inherit the Cloud: The Role of Private Contracts in Distributing or Deleting Digital Assets at Death*, 83 Fordham L. Rev. 799 (2014).

11.4 – Review of the Scenarios Posed Earlier to Apply the Law

Given the information contained in this chapter, let us return to the two scenarios mentioned in the introduction and apply what we know.

In *Scenario 1*, you must determine how the estate plan was accomplished. Did the decedent provide authority to estate representatives through an online tool provided by the service provider (in New York under EPTL § 13-A-2.2(a))? If so, the provisions of that tool will govern. If, instead, the decedent worked with an attorney to include specific language in estate planning documents (will, trust, fiduciary appointment) governing content and non-content items in online accounts (in New York under EPTL § 13-A-2.2(b)), then those documents will govern. In either circumstance, the decedent, while living, utilized his or her rights under law to determine the access to his or her digital assets that would exist after one's death—and those decisions are binding.

Attorneys who provide estate planning guidance and drafting for clients should consider including specific language (boilerplate or otherwise) in wills and trust documents, providing fiduciaries with access to digital assets per the testator or grantor's wishes, as set forth in N.Y. EPTL Article 13-A. Should a will document be otherwise silent on specific digital access, but decedent provides that all residuary and personalty are bequeathed to a specific person (particularly if a spouse), then—at least in New York—the determination in *Matter of Swezey* may provide a foothold for the argument that the executor or heir should be granted access to all content and non-content digital materials in decedent's estate. Although specific planning and language are, of course, always best—and the safest course to ensure decedent's last wishes are carried out in full.[44]

---

[44] Although the new Chapter 10 of this book's second edition addresses the ways in which electronics and social media have been utilized by society and the legal profession to address obstacles presented by social distancing restrictions and responses to COVID-19 in the year 2020, this final chapter and the discussion of estate planning and document signing would be incomplete without mention of how will signing ceremonies were altered, at least in New York State, in response to the virus. Under "normal" circumstances, wills that are signed and witnessed must follow an in-person, formalized process. Under the "new normal" created by COVID-19, however, in-person signing and witnessing of estate planning documents was not possible—unfortunately at a time when, one could argue, it was vitally important for there to be a mechanism where one could complete estate planning. Thus, on April 7, 2020, New York Governor Andrew Cuomo signed Executive Order 202.14 (*available at* https://www.governor.ny.gov/news/no-20214-continuing-temporary-suspension-and-modification-laws-relating-disaster-emergency). The relevant portion of that Executive Order provided as follows:

> For the purposes of Estates Powers and Trusts Law (EPTL) 3-2.1(a)(2), EPTL 3-2.1(a)(4), Public Health Law 2981(2)(a), Public Health Law 4201(3), Article 9 of the Real Property Law, General Obligations Law 5-1514(9)(b), and EPTL 7-1.17, the act of witnessing that is required under the aforementioned New York State laws is authorized to be performed utilizing audio-video technology provided that the following conditions are met:
>
> o The person requesting that their signature be witnessed, if not personally known to the witness(es), must present valid photo ID to the witness(es) during the video conference, not merely transmit it prior to or after;
>
> o The video conference must allow for direct interaction between the person and the witness(es), and the supervising attorney, if applicable (e.g. no pre-recorded videos of the person signing);

In *Scenario 2*, however, we know there are no documents that the decedent executed to govern his or her last wishes with regard to online accounts. Therefore, in New York (and other jurisdictions adopting a similar uniform statutory framework), analysis takes place pursuant to EPTL § 13-A-2.2(c)—and one must ask if there are binding Terms of Service, whose terms the decedent understood and which created a known, binding contract. If so, then the Terms of Service will govern the outcome of access to the digital assets. If not, one is left to apply the estate or probate laws of the jurisdiction to determine how, if at all, personal representatives and fiduciaries of the intestate decedent can access online accounts and information contained therein.

Remember, though, that in most cases, *Matter of Swezey* notwithstanding, without specific direction concerning content-based material from the decedent through clear documentation, personal representatives will likely only have access to non-content material absent a showing that content material is in some way necessary and required for the administration of the estate under the statutory framework.

11.5 – Conclusion

Ultimately, it is clear that, with respect to our Greek rhetorician and philosopher ancestors, times have indeed changed. Wisdom is not our only immortal possession. In the vast cosmos of the Internet, there are digital assets and accounts that will live on long beyond our time on this Earth and hence must be carefully considered in any estate plan, or any action to settle an estate. While there is some protection of personal privacy and details even after death—including whether the details are contained in content or non-content elements of online accounts—those private details that are vital and necessary to the administration of

---

o The witnesses must receive a legible copy of the signature page(s), which may be transmitted via fax or electronic means, on the same date that the pages are signed by the person;

o The witness(es) may sign the transmitted copy of the signature page(s) and transmit the same back to the person; and

o The witness(es) may repeat the witnessing of the original signature page(s) as of the date of execution provided the witness(es) receive such original signature pages together with the electronically witnessed copies within thirty days after the date of execution.

By following the strictures set forth in New York E.O. 202.14, individuals could accomplish their estate planning via a method that would have previously been unthinkable. In the same vein, virtual notarizations utilizing similar electronic audio/visual processes, were authorized in some states, for the same reasoning. *See, e.g.*, N.Y. Executive Order 202.7 (Mar. 19, 2020); Massachusetts Legislature, Chapter 71 of the Acts of 2020 (amending Chapter 222 of the General Laws to address virtual notarizations). Times do change when necessary. *See also* Amanda Robert, *Lawyers Address Problems with Estate-Planning Document Signing During Coronavirus Crisis*, ABA Journal (May 4, 2020), *available at* https://www.abajournal.com/web/article/lawyers-and-problems-with-estate-planning-signing-during-coronavirus-crisis. *See also* Brandon Vogel, *Working Face-to-Face but Not in Person: Tips on How to Virtually Execute an Estate Plan*, 92 N.Y. St. B.J. 38 (May 2020).

the estate may be the subject of access by the appointed fiduciary unless another provision is made by the decedent in the hierarchy of authoritarian documents.[45]

Accordingly, we come to our final takeaway point in this book: Take great care in your estate planning, and if you are an attorney or paralegal take great care in advising your clients concerning their planning, when it comes to digital assets and their content.

Chapter End Questions

Note: Answers may be found on page 265.

1. Which of the following is the name of one of the uniform laws that has been adopted in the majority of states regarding access to electronic assets of decedents?
   a. Uniform Access to My Assets Law
   b. Uniform Fiduciary Access to Digital Assets Act
   c. Uniform Email Assets Act
   d. Uniform Fiduciary Email Assets Act

2. Which of the following is New York's statutory codification of the law for access to decedents' digital assets?
   a. EPTL Article 12
   b. EPTL Article 13
   c. EPTL Article 13-A
   d. EPTL Article 15

3. Which of the following is the first level/highest level of authority under law, in jurisdictions like New York and others that have adopted the Uniform law, for access to digital assets of a decedent?
   a. The user may direct who will have access to content or non-content material following user's death, by utilizing a tool with or provided by the service provider.
   b. The usual estate planning documents of the user/owner of the asset - a will or trust.
   c. The terms of service entered into between the user and service provider.
   d. None of these selections are correct.

4. Which of the following is the second level of authority under law, in jurisdictions like New York and others that have adopted the Uniform law, for access to digital assets of a decedent?
   a. The user may direct who will have access to content or non-content material following user's death, by utilizing a tool with or provided by the service provider.
   b. The usual estate planning documents of the user/owner of the asset - a will or trust.

---

[45] Note, however, that the provisions of law, and relevant case decisions, do not apply to digital assets that are utilized by an employee, but which are owned by an employer and/or used "in the ordinary course of the employer's business." 41 N.Y. Jur. 2d. Decedents' Estates §1788 (citing N.Y. EPTL §13-A-1).

    c. The terms of service entered into between the user and service provider.

    d. None of these selections are correct.

5. Which of the following is the lowest level of authority under law, in jurisdictions like New York and others that have adopted the Uniform law, for access to digital assets of a decedent?

    a. The user may direct who will have access to content or non-content material following user's death, by utilizing a tool with or provided by the service provider.

    b. The usual estate planning documents of the user/owner of the asset - a will or trust.

    c. The terms of service entered into between the user and service provider.

    d. None of these selections are correct.

6. Which of the following cases is the seminal Massachusetts case to address access to electronic assets of a decedent, and limitation of access for the heirs/representatives.

    a. *Matter of Whitaker*

    b. *Estate of Serrano*

    c. *Ajemian v. Yahoo!, Inc.*

    d. *Oracle Am., Inc. v. Google Inc.*

7. Which of the following is the federal statute that "protects individuals' privacy and proprietary interests; the Act reflects Congress's judgment that users have a legitimate interest in the confidentiality of communications in electronic storage at a communications facility"?

    a. Storage of Electronic Assets Law

    b. Stored Communications Act

    c. Digital Assets Law

    d. Electronically Stored Information Law

8. True or False: Once a person passes away, there is no regard for their privacy in any situation.

    a. TRUE

    b. FALSE

9. True or False: Special provision must be made in a person's estate plan for electronic assets, such as social media and email accounts, or the heirs or estate representatives may not have full access thereto.

    a. TRUE

    b. FALSE

10. True or False: A decedent utilizes an estate plan consisting of a will that designates access to decedent's digital assets after death, and provides that decedent's heirs will have access to all content and non-content communications. Decedent does not use a tool provided by the service providers. True or false - the law/courts will grant the heirs full access to all of decedent's digital assets.

    a. TRUE

    b. FALSE

UNITED STATES DISTRICT COURT
SOUTHERN DISTRICT OF NEW YORK

Rule 502(d) Order

-- x
:
:
:
:
:
:
-- x

United States Magistrate Judge:

1. The production of privileged or work-product protected documents, electronically stored information ("ESI") or information, whether inadvertent or otherwise, is not a waiver of the privilege or protection from discovery in this case or in any other federal or state proceeding. This Order shall be interpreted to provide the maximum protection allowed by Federal Rule of Evidence 502(d).

2. Nothing contained herein is intended to or shall serve to limit a party's right to conduct a review of documents, ESI or information (including metadata) for relevance, responsiveness and/or segregation of privileged and/or protected information before production.

SO ORDERED.

Dated: New York, New York

[DATE]

_____

Hon.

United States Magistrate Judge

Copies *by ECF* to: All Counsel

Judge _____

Administrative Order of the Chief Administrative Judge of the Courts

Pursuant to the authority vested in me, and upon consultation with and approval by the Administrative Board of the Courts, I hereby promulgate a new subdivision (f) of Rule 11-e of subdivision (g) of section 202.70 of the Uniform Rules for the Supreme Court and County Court (Rules of Practice for the Commercial Division of the Supreme Court), as follows (new matter <u>underlined</u>), effective October 1, 2018:

(f) The parties are encouraged to use the most efficient means to review documents, including electronically stored information ("ESI"), that is consistent with the parties' disclosure obligations under Article 31 of the CPLR and proportional to the needs of the case. Such means may include technology-assisted review, including predictive coding, in appropriate cases. The parties are encouraged to confer, at the outset of discovery and as needed throughout the discovery period, about technology-assisted review mechanisms they intend to use in document review and production.

Chief Administrative Judge of the Courts

Dated: July 19, 2018

AO/242/18

Model

[Date]

[Address]   **[Internal to Client/Employee]**

     Re.: _____

Dear _____:

Please be advised that due to [anticipated or actual] litigation, electronically stored information (ESI) may be at issue, or may prove to contain important information and evidence in the matter of _____. We have determined that you, as an employee of _____, and as a _____ [position and/or department], are one of _____ [client company's] employees affected by litigation hold and preservation requirements under the law. Therefore, from this moment forward until further notice, please begin what is known as a preservation or litigation "hold" on the alteration, destruction, or deletion of any electronic materials in your possession or at your disposal—this hold must include all documents, e-mails, voicemails, data recordings, tapes, hard-drives, computer servers, Internet webpages, and other electronic storage media—with regard to the following subject matter: _____ _____, or any related matters. The breadth and depth of this hold may be cut back by order of a judge, or on advice of counsel, at a later date, but current caselaw and rules make it important that we abide by our preservation obligations with explicit care.

As a part of this litigation and preservation hold that we are instituting, you should immediately suspend any and all automatic deletion programs, any routine document destruction policies, and any other automatic policies designed to routinely purge and destroy/delete documents. Contact our legal counsel, _____, Esq., immediately at _____ should any clarification be required concerning which documents are subject to this requirement in your daily work. This non-deletion policy even reaches any voicemails, text messages, e-mails, or other forms of communication that you send or receive, which are related to the context of this litigation hold.

Any failure on your part or ours with regard to preservation of materials and documents can result in sanctions from the Court. Indeed, if relevant and discoverable material is altered, lost, deleted, destroyed, or otherwise unavailable, sanctions can be severe, and range anywhere from monetary fine to dismissal of claims or an adverse inference—permission for the judge or jury to take the view that any lost or deleted material was not in favor of our position and was therefore evidence against the company in this action.

Finally, this hold concerning electronically stored information is applicable, as well, to any and all paper documentation you may possess or utilize during the course of your employment until further notice. Paper must be preserved just as ESI must be preserved, or the same sanctions may apply.

Again, if you should have any questions concerning this litigation and preservation hold, please contact legal counsel _____ at _____.

Very truly yours,

[Enc.]
cc:

UNITED STATES DISTRICT COURT
SOUTHERN DISTRICT OF NEW YORK

```
------------------------------------------
                              :
                              :
                              :
_____    :   CIV. NO. _____
              Plaintiff(s),   :   **Joint Electronic Discovery**
    -against-                 :   **Submission No. ___ and**
                              :   **[Proposed] Order**     :
                              :
_____    :
              Defendant(s).   :
                              :
------------------------------------------  :
```

One or more of the parties to this litigation have indicated they believe that relevant information may exist or be stored in electronic format, and that this content is potentially responsive to current or anticipated discovery requests. This Joint Submission and [Proposed] Order (and any subsequent ones) shall be the governing document(s) by which the parties and the Court manage the electronic discovery process in this action. The parties and the Court recognize that this Joint Submission and [Proposed] Order is based on facts and circumstances as they are currently known to each party, that the electronic discovery process is iterative, and that additions and modifications to this submission may become necessary as more information becomes known to the parties.

1. Brief Joint Statement Describing the Action, [e.g., "Putative securities class action pertaining to the restatement of earnings for the period May 1, 2017 to May 30, 2017"]:

_____

_____

_____

_____

_____

_____

a. Estimated amount of Plaintiff(s)' Claims:

☐ Monetary (absolute number or range):$ _____

☐ Equitable Relief (if so, specify) _____

☐ Other (if so, specify) _____

b. Estimated amount of Defendant(s)' Counterclaim/Cross-Claims:

☐ Monetary (absolute number or range):$ _____

☐ Equitable Relief (if so, specify) _____

☐ Other (if so, specify) _____

2. Competence. Counsel certify that they are sufficiently knowledgeable in matters relating to their clients' technological systems to discuss competently issues relating to electronic discovery, or have involved someone competent to address these issues on their behalf.

3. Meet and Confer. Pursuant to Fed. R. Civ. P. 26(f), counsel are required to meet and confer regarding certain matters relating to electronic discovery before the Initial Pretrial Conference. Counsel certify that they have met and conferred to discuss these issues.

Date(s) of parties' meet-and-confer conference(s):

_____

4. Unresolved Issues: The following issues concerning discovery of electronic information remain outstanding and/or require court intervention (check all that apply):

☐ Preservation

☐ Search and Review

☐ Sources of Production

☐ Forms of Production

☐ Identification or Logging of Privileged Material

☐ Inadvertent Production of Privileged Material

☐ Cost Allocation

☐ Other (specify): _____

5. Preservation.

a. The parties have discussed the obligation to preserve potentially relevant electronically stored information and agree to the following scope and methods for preservation, including but not limited to: retention of electronic data and implementation of a data preservation plan; identification of potentially relevant data; disclosure of the programs and manner in which data is maintained; identification of computer system(s); and identification of the individual(s) responsible for data preservation, etc. To the extent the parties have reached agreement as to preservation of electronic information, provide details below:

_____

_____

_____

_____

_____

_____

b. State the extent to which the parties have disclosed or have agreed to disclose the dates, contents, and/or recipients of "litigation hold" communications:

_____

_____

_____

_____

_____

c. The parties anticipate the need for judicial intervention regarding the following issues concerning the duty to preserve, the scope, or the method(s) of preserving electronically stored Information:

_____

_____

_____

_____

_____

6. Search and Review.

   a. The parties have discussed methodologies or protocols for the search and review of electronically stored information, as well as the disclosure of techniques to be used. Some of the approaches that may be considered include: the use and exchange of keyword search lists, "hit reports," and/or responsiveness rates; concept search; machine learning, or other advanced analytical tools; limitations on the fields or file types to be searched; date restrictions; limitations on whether back-up, archival, legacy, or deleted electronically stored information will be searched; testing; sampling; etc. To the extent the parties have reached agreement as to search and review methods, provide details below:

_____

_____

_____

_____

_____

b. The parties anticipate the need for judicial intervention regarding the following issues concerning the search and review of electronically stored information:

_____

_____

_____

_____

_____

_____

7. Production.

a. *Source(s) of Electronically Stored Information.* The parties anticipate that discovery may occur from one or more of the following potential source(s) of electronically stored information [e.g., email, word processing documents, spreadsheets, presentations, databases, instant messages, web sites, blogs, social media, etc.]:

Plaintiff(s):

_____

_____

_____

_____

_____

_____

Defendant(s):

_____

_____

_____

_____

_____

_____

b. *Limitations on Production.* The parties have discussed factors relating to the scope of production, including but not limited to: (i) number of custodians; (ii) identity of custodians; (iii) date ranges for which potentially relevant data will be drawn; (iv) locations of data; (v) timing of productions (including phased discovery or rolling productions); and (vi) electronically stored information in the custody or control of non-parties. To the extent the parties have reached agreements related to any of these factors, describe below:

_____

_____

_____

_____

_____

c. *Form(s) of Production*. The parties have discussed and agreed to the following regarding the form(s) of productions (e.g., TIFF, pdf, native, etc.):

_____

_____

_____

_____

_____

d. The parties anticipate the need for judicial intervention regarding the following issues concerning production:

_____

_____

_____

_____

_____

8. Privileged Material.

    a. Identification. The parties have discussed and agreed to the following method(s) for identification (e.g., individual logging, categorical logging, etc.) and redaction of privileged documents:

_____

_____

_____

_____

_____

    b. Inadvertent Production / Claw-Back Agreements. Pursuant to Fed R. Civ. Proc. 26(b)(5) and F.R.E. 502(e), the parties have agreed to the following concerning the inadvertent production of privileged documents (e.g. "quick-peek" agreements, non-waiver agreements or orders pursuant to F.R.E. 502(d), etc.):

_____

_____

_____

_____

_____

c. The parties have discussed a 502(d) Order. Yes ___; No ___ The provisions of any such proposed Order shall be set forth in a separate document and presented to the Court for its consideration.

d. The parties anticipate the need for judicial intervention regarding the following issues concerning privileged material:

_____

_____

_____

_____

_____

9. Cost of Production.

a. *Costs*: The parties have analyzed their client's data repositories and have estimated the costs associated with production of electronically stored information. The factors and components underlying these costs are estimated as follows:

Plaintiff(s):

_____

_____

_____

_____

_____

Defendant(s):

_____

_____

_____

_____

_____

b. *Cost Allocation.* The parties have considered cost-shifting or cost-sharing and have reached the following agreements, if any:

_____

_____

_____

_____

_____

_____

c. *Cost Savings.* The parties have considered cost-saving measures, such as the use of a common electronic discovery vendor or a shared document repository, and have reached the following agreements, if any:

_____

_____

_____

_____

_____

_____

d. The parties anticipate the need for judicial intervention regarding the following issues concerning the costs of production of electronically stored information:

_____

_____

_____

_____

_____

_____

10. Other Issues, if any.

_____

_____

_____

_____

_____

_____

_____

_____

_____

_____

_____

_____

The preceding constitutes the agreement(s) reached, and disputes existing (if any), between the parties to certain matters concerning electronic discovery as of this date. To the extent additional agreements are reached, modifications are necessary, or disputes are identified, they will be outlined in subsequent submissions or agreements and promptly presented to the Court.

Party: _____   By: _____

Party: _____   By: _____

Party: _____   By: _____

Party: _____   By: _____

Party: _____   By: _____

Party: _____   By: _____

The next scheduled meet-and-confer conference to address electronic discovery issues, including the status of electronic discovery and any issues or disputes that have arisen since the last conference or Order, shall take place on _____.

The next scheduled conference with the Court for purposes of updating the Court on electronic discovery issues has been scheduled for _____.

Additional conferences, or written status reports, shall be set for every ___ weeks, as determined by the parties and the Court, based on the complexity of the issues at hand. A joint agenda should be submitted to the Court three (3) business days before such conference indicating the issues to be raised by the parties. The parties may jointly seek to adjourn the conference with the Court no less than 48 hours in advance of the scheduled conference, if the parties agree that there are no issues requiring Court intervention.

Dated: _____, 20__   SO ORDERED:

_____

United States Magistrate Judge

Model

[Date]

[Address] **[Letter to Opposing Counsel/Party]**

RE.: _____

Dear _____:

Please be advised that we have determined that electronically stored information (ESI) may be at issue, or may prove to contain important information and evidence, in [the present/any potential future] litigation. Therefore, [you/your client] should place what is known as a preservation or litigation "hold" on the alteration, destruction, or deletion of any electronic materials—documents, e-mails, voicemails, data recordings, tapes, hard drives, computer servers, Internet Web pages, or other electronic storage media—with regard to the subject matter of the potential matter, or any related matters. While the breadth and depth of this hold may be cut back by order of a judge, or on advice of counsel, at a later date, current case law and rules require preservation. The relevant authorities are clear in their creation and enforcement of preservation obligations when it comes to ESI. Those authorities include, but are not limited to, the provisions of Federal Rules of Civil Procedure 26 and 37; N.Y. CPLR 3126 and 4518; Rule 8 of the Commercial Division of the New York State Supreme Court; 22 N.Y.C.R.R. § 202.12; *McCarthy v. Phillips Electronics NA,* No. 112522/03 (Sup. Ct. NY County June 17, 2005); *Fitzpatrick v. Toy Indus. Assoc., Inc.,* Index No. 116548/2009, 2009 WL 159123 (Sup. Ct. NY County Jan. 5, 2009); *CAT3, LLC v. Black Lineage, Inc.,* 164 F.Supp.3d 488 (S.D.N.Y. 2016); *Dinkins v. Schinzel,* 2017 WL 4183115 (D. Nev. Sept. 19, 2017); *Pegasus Aviation I, Inc. v. Varig Logistica S.A.,* 26 N.Y.3d 543, 46 N.E.3d 601 (2015); and *Douglas Elliman LLC v. Tal,* 156 A.D.3d 583, 65 N.Y.S.3d 697 (1st Dep't 2017).

If relevant and discoverable material is altered, lost, deleted, destroyed, or otherwise unavailable, sanctions can be severe, and range anywhere from monetary fine to dismissal of claims or an adverse inference—permission for the judge or jury to take the view that any lost or deleted material was not in your favor and was evidence against you.

Thus, at this time [you/your client] **are on notice** that a litigation hold must be put in place concerning all relevant or potentially relevant ESI evidence, or any materials that may reasonably lead to relevant evidence. It may be necessary to communicate this hold to all employees (*i.e.,* all individuals with access to the information).

Further be aware that this notice/demand concerning electronically stored information is applicable, as well, to any and all paper documentation. Paper must be preserved just as ESI must be preserved, or the same sanctions may apply.

Thank you for your time and attention, and anticipated cooperation, in this matter.

Very truly yours,

[Enc.]
cc:

STATE OF CONNECTICUT
BY HIS EXCELLENCY
NED LAMONT
EXCUTIVE ORDER NO. 7Q

PROTECTION OF PUBLIC HEALTH AND SAFETY DURING COVID-19 PANDEMIC AND RESPONSE - CHILDCARE SAFETY, REMOTE NOTARIZATION UPDATE

WHEREAS, on March 10, 2020, I issued a declaration of public health and civil preparedness emergencies, proclaiming a state of emergency throughout the State of Connecticut as a result of the coronavirus disease 2019 (COVID-19) outbreak in the United States and confirmed spread in Connecticut; and

WHEREAS, pursuant to such declaration, I have issued seventeen (17) executive orders to suspend or modify statutes and to take other actions necessary to protect public health and safety and to mitigate the effects of the COVID-19 pandemic; and

WHEREAS, COVID-19 is a respiratory disease that spreads easily from person to person and may result in serious illness or death; and

WHEREAS, the World Health Organization has declared the COVID-19 outbreak a pandemic; and

WHEREAS, the risk of severe illness and death from COVID-19 appears to be higher for individuals who are 60 years of age or older and for those who have chronic health conditions; and

WHEREAS, to reduce the spread of COVID-19, the United States Centers for Disease Control and Prevention and the Connecticut Department of Public Health recommend implementation of community mitigation strategies to increase containment of the virus and to slow transmission of the virus, including cancellation of gatherings of ten people or more and social distancing in smaller gatherings; and

WHEREAS, ongoing childcare operations are necessary to support the essential workforce, and it is vital to protect the health and safety of children and staff in childcare facilities and limit the spread of COVID-19; and

WHEREAS, current regulations allow group sizes of as many as 20 children in childcare facilities, increasing the risk of transmission of COVID-19 among staff, children, and their families, and reducing such group sizes is necessary to reduce such risk; and

WHEREAS, certain documents require the in-person services of a Notary Public or Commissioner of the Superior Court and such interactions should be avoided to the maximum extent possible in order to promote social distancing and the mitigation of the spread of the COIVD-19; and

WHEREAS, certain documents, in addition to notarization or acknowledgement, require the presence of in-person witnesses to their signature, which could increase the risk of transmission of COVID-19; and

NOW, THEREFORE, I, NED **LAMONT**, Governor of the State of Connecticut, by virtue of the authority vested in me by the Constitution and the laws of the State of Connecticut, do hereby ORDER AND DIRECT:

1. Limited Group Sizes in Childcare. To limit the spread of COVID-19 and protect the health and safety of children and staff in all child care facilities that are continuing to operate during this civil preparedness and public health emergency, Section 19a-79 and any related regulations, rules, or policies, are modified to require that all child care facilities shall limit group sizes to no more than ten children in one space, and to authorize the Commissioner of Early Childhood to issue any implementing orders she deems necessary. Any childcare operation seeking to caring for more than thirty children in one facility shall seek approval to do so from the Commissioner and demonstrate sufficient separation of groups within the facility. This order applies to all childcare operations, including but not Sited to childcare centers, group childcare homes, family childcare homes, youth camps, and childcare facilities that are exempt from licensing requirements pursuant to Section 19a-77 of the Connecticut General Statues.

2. Enhanced Health Procedures for All Operating Child Care Programs. All children and childcare workers shall be screened before entrance to any childcare operation, as described herein, for any observable illness, including cough or respiratory distress, and to confirm body temperature below one hundred degrees Fahrenheit. All staff shall practice enhanced handwashing and health practices, including covering coughs and sneezes with a tissue or the comer of the elbow and assisting children with such increased handwashing and health practices. Enhanced cleaning and disinfection practices shall be implemented in all facilities to prevent the spread of COVID-19. This order applies to all childcare operations including but not limited to childcare centers, group childcare homes, family childcare homes, youth camps, and childcare facilities that are exempt from licensing requirements pursuant to Section 19a-77 of the Connecticut General Statues. The Commissioner of Early Childhood may issue any implementing orders she deems necessary consistent with this order.

3. Remote Notarization - Amended Procedures. Effective immediately and through June 23, 2020, unless modified, extended or terminated by me, Section 3 of my prior Executive Order 7K concerning remote notarizations is hereby superseded and replaced in its entirety by this Executive Order. All relevant state laws and regulations are hereby modified to permit any notarial act that is required under Connecticut law to be performed using an electronic device or process that allows a notary public commissioned by the Connecticut Secretary of the State pursuant to section 3-94b of the Connecticut General Statutes ("Notary Public") or a Commissioner of the Superior Court as defined by section 51-85 of the Connecticut General Statutes ("Commissioner") and a remotely located individual to communicate with each other simultaneously by sight and sound ("Communication Technology"), provided that the following conditions are met:

a. The person seeking the notarial act ("Signatory") from a Notary Public or Commissioner, if not personally known to the Notary Public or Commissioner, shall present satisfactory evidence of identity, as defined by subsection 10 of section 3-94a of the General Statutes, while connected to the Communication Technology, not merely transmit it prior to or after the transaction;

b. The Communication Technology must be capable of recording the complete notarial act and such recoding shall be made and retained by the Notary Public for a period of not less than ten (10) years;

c. The Signatory must affirmatively represent via the Communication Technology that he or she is physically situated in the State of Connecticut;

d. The Signatory must transmit by fax or electronic means a legible copy of the signed document directly to the Notary Public or Commissioner on the same date it was executed;

e. The Notary Public or Commissioner may notarize the transmitted copy of the document and transmit the same back to the Signatory by fax or electronic means;

f. The Notary Public or Commissioner may repeat the notarization of the original signed document as of the date of execution, provided the Notary Public or Commissioner receives such original signed document, together with the electronically notarized copy, within thirty days after the date of execution;

g. Notwithstanding the foregoing, only an attorney admitted to practice law in the State of Connecticut and in good standing may remotely administer a self-proving affidavit to a Last Will and Testament pursuant to section 45a-285 of the General Statues or conduct a real estate closing as required by Public Act 19-88. Any witnessing requirement for a Last Will and Testament may be satisfied remotely through the use of Communication Technology if it is completed under the supervision of a Commissioner. The supervising Commissioner shall certify that he or she supervised the remote witnessing of the Last Will and Testament

h. All witness requirements on any document, other than a Last Will and Testament, requiring a notarial act are hereby suspended for the duration of this Executive Order.

i. All Remotely Notarized documents pertaining to real property shall be accepted for recording on the land records by all Connecticut Town or City Clerks. A one-page certification confirming the use of Remote Notarization procedures shall be attached to each remotely notarized document submitted for recording on the land records in Connecticut.

Unless otherwise specified herein, this order shall take effect immediately and shall remain in effect for the duration of the public health and civil preparedness emergency, unless earlier modified or terminated by me.

Dated at Hartford, Connecticut, this 30th day of March, 2020.

Ned Lamont

Governor

By is Excellency's Command

By His Excellency's Command

Denise W. Merrill

Secretary of the State

EXECUTIVE ORDER NO. 135

WHEREAS, in light of the dangers posed by Coronavirus disease 2019 ("COVID-19"), I issued Executive Order No. 103 (2020) on March 9, 2020, the facts and circumstances of which are adopted by reference herein, which declared both a Public Health Emergency and State of Emergency; and

WHEREAS, on April 7, 2020, I issued Executive Order No. 119 (2020), the facts and circumstances of which are adopted by reference herein. which declared that the Public Health Emergency declared in Executive Order No. 103 (2020) continues to exist; and

WHEREAS, in accordance with N.J.S.A. App. A:9-34 and -51, I reserve the right to utilize and employ all available resources of State government to protect against the emergency created by COVID-19; and

WHEREAS, as COVID-19 continued to spread across New Jersey and an increasing number of individuals required medical care or hospitalization, I issued a series of Executive Orders pursuant to my authority under the New Jersey Civilian Defense and Disaster Control Act and the Emergency Health Powers Act, to protect the public health, safety, and welfare against the emergency created by COVID-19, including Executive Order Nos. 104-133 (2020), the facts and circumstances of which are all adopted by reference herein; and

WHEREAS, in recognition that the Centers for Disease Control and Prevention ("CDC") has advised that social mitigation strategies for combatting COVID-19 require every effort to reduce the rate of community spread of the disease and that COVID-19 spreads most frequently through person-to-person contact when individuals are within six feet or less of one another, Executive Order No. 107 (2020) ordered steps to mitigate community spread of COVID-19; and

WHEREAS, Executive Order No. 107 (2020) directed all New Jersey residents to remain at home or at their place of residence, unless they qualified under certain defined categories, including reporting to, or performing, their job; and

WHEREAS, Executive Order No. 107 (2020) directed all businesses or non-profits in the State, whether closed or open to the public, to accommodate their workforce, wherever practicable, for telework or work-from-home arrangements; and

WHEREAS, Executive Order No. 107 (2020) stated that businesses or non-profits who have employees who cannot perform their functions via telework or work-from-home arrangements should make best efforts to reduce staff on site to the minimal number necessary to ensure that essential operations can continue; and

WHEREAS, Executive Order No. 104 (2020) directed that all public, private and parochial schools, including charter and renaissance schools, close to the public beginning on March 18, 2020, and Executive Order No. 107 (2020) directed those schools to remain closed; and

WHEREAS, compliance with the social distancing strategies and travel restrictions required by Executive Orders No. 104 and 107 (2020) impact the ability of the residents of this State to comply with certain statutory requirements to appear in person before certain public officials when seeking to obtain certain licenses, certificates, and other benefits; and

WHEREAS, under N.J.S.A. 37:1-2, persons intending to he married or to enter into a civil union must obtain a marriage or civil union license from a licensing officer and deliver it to the person who is to officiate; and

WHEREAS, under N.J.S.A. 37: 1-7 and -8, individuals seeking to obtain a marriage or civil union license must appear personally or through an attorney-in-fact before the licensing officer and, in the presence of the licensing officer, subscribe and swear to an oath attesting to certain facts respecting the legality of the proposed marriage or civil union, which must also be verified by a witness of legal age; and

WHEREAS, under N.J.S.A. 37:1-4, a marriage or civil union license shall not be issued sooner than 72 hours after an application has been made, and a license so issued shall be good and valid only for 30 days after the date of issuance; and

WHEREAS, under N.J.S.A. 37:1-17, the certificate of marriage or civil union must be signed by the officiant by or before whom the marriage or civil union was solemnized, who must also indicate the date and place of the marriage or civil union; and

WHEREAS, under N.J.S.A. 37:1-17, the certificate of marriage or civil union must also be signed by at least two witnesses who were present at the marriage or civil union ceremony; and

WHEREAS, to comply with my directives to implement social distancing strategies and to limit person-to-person interactions in accordance with CDC and New Jersey Department of Health ('DOH") guidance, certain municipal and State offices have implemented telework or work-from-home arrangements, thus reducing staff on site to the minimal number necessary to continue essential operations; and

WHEREAS, given my direction to strictly observe social distancing practices, including my direction that employees work remotely, individuals seeking to marry or enter into a civil union may be unable to appear in person before a licensing official and/or to solemnize the marriage or civil union as required in the presence of an officiant and two witnesses within 30 days of issuance of the license; and

WHEREAS, the requirements that individuals seeking to marry or enter into a civil union must appear in person to effectuate the arrangement may be accomplished for a limited time period through the use of audio-visual technology, also referred to as video conferencing, while nevertheless providing confidence that marriages and civil unions are entered into legitimately and free of duress; and

WHEREAS, for these reasons, among others, strict enforcement of the various statutory requirements to appear in person relating to marriage or civil union licenses is detrimental to the public welfare; and

WHEREAS, in light of the current crisis, individuals may need to expedite their marriage or civil union, so that the 72-hour waiting period between application and issuance of license should be suspended: and

WHEREAS, N.J.S.A. 34:2-21.7 (a) generally prohibits the employment of minors under the age of 18 years of age unless the employer procures and keeps on file an employment certificate or special permit for the minor that is issued by the issuing officer of the school district in which the child resides or, if the child is a nonresident of the State, of the district in which the child has obtained a promise of employment; and

WHEREAS, N.J.S.A. 34:2-21.8 provides that a school district issuing officer shall issue an employment certificate or special permit only upon the application in person of the minor desiring employment; and

WHEREAS, N.J.S.A. 34:2-21.10 provides that, upon issuance of such certificate, it must be signed by the child in whose name it is issued in the presence of the issuing officer: and

WHEREAS, given the closure of all public schools, minors are unable to appear personally before school district issuing officers in order to apply far or sign the employment certificate as required by applicable statutes; and

WHEREAS, the inability of minors to obtain employment certificates as required by law will render them unable to obtain gainful employment; and

WHEREAS, the requirements that minors appear personally before school district issuing officers in order to apply for or sign the employment certificate may be accomplished for a limited time period through the use of audio-visual technology, while nevertheless providing confidence that such arrangements are entered into legitimately; and

WHEREAS, for these reasons, among others, strict enforcement of the various statutory requirements to appear in person relating to employment certificates for such minors is detrimental to the public welfare; and

WHEREAS, while it is critical to ensure that individuals are able to use audio-visual technology in order to obtain marriage or civil union licenses or necessary employment certificates during the Public Health Emergency declared for COVID-19, it is important to recognize that not all such individuals have equal access to or experience with such technology, and thus it is important to ensure that such individuals can still obtain the applicable licenses or employment certificates without audio-visual technology; and

WHEREAS, pursuant to N.J.S.A. App. A:9-47, the Governor is authorized to, among other things, suspend any regulatory provision of law when its enforcement is detrimental to the public welfare during an emergency; and

WHEREAS, the Constitution and statutes of the State of New Jersey, particularly the provisions of N.J.S.A. 26:13-1 et seq. and N.J.S.A. App. A:9-33 et seq., and all amendments and supplements thereto, confer upon the Governor of the State of New Jersey certain emergency powers, which I have invoked;

NOW, THEREFORE, I, PHILIP D. MURPHY, Governor of the State of New Jersey, by virtue of the authority vested in me by the Constitution and by the Statutes of this State, do hereby ORDER and DIRECT:

1. For the duration of the Public Health Emergency declared in Executive Order No. 103 (2020), the provisions of N.J.S.A. 37-1-7 and N.J.S.A. 37:1-8 requiring that individuals who wish to marry or enter into a civil union appear personally before a licensing official may be satisfied through the use of audio-visual technology under the following conditions:

a. The video conference shall be live and must allow for interaction between the couple, the licensing official, and a witness;

☐ During the video conference, the following steps must occur in a way that is visible and audible to the couple, the licensing official, and a witness:

i. Both members of the couple and the witness present valid photo identification and any other documents necessary to allow the licensing official to fulfill their statutory duty:

ii. The licensing official administers the oath:

iii. Each member of the couple and the witness shall sign the license application. To the extent that the members of the couple and/or witness are physically located in the same place, they shall sign the document in the physical presence of the other(s). If one or more of the individuals is located in a different physical location, once signatures are obtained in one location, the member of the couple or witness shall transmit, a legible copy of the document, to the next signatory, until all signatures are obtained; and

iv. The final signatory shall transmit a legible copy of the signed license application directly to the licensing official, who shall confirm receipt of the document on the day of transmission.

2. For the duration of the Public Health Emergency declared in Executive Order No. 103 (2020). the provisions of N.J.S.A. 37:1-17 requiring that the marriage or civil union he solemnized in the physical presence of an officiant and two witnesses, who will then sign the certificate, may be satisfied through the use of audio-visual technology under the following conditions:

a. The video conference shall be live and must allow for interaction between the couple, the officiant, and the two witnesses:

☐ During the video conference, the following steps must occur in a way that is visible and audible to the couple, the officiant, and the two witnesses:

i. The couple shall present their marriage or civil union license to the officiant;

ii. The afficiant, each witness, and both members of the couple shall affirm that they are physically situated in the State, though they are not required to be situated in the same municipality. The officiant shall identify the municipality and address where he or she is physically situated and denote that municipality and address as the place of marriage or civil union on the certificate;

iii. Each member of the couple and each witness shall sign the marriage or civil union certificate. To the extent that the members of the couple and/or witnesses are physically located in the same place, they shall sign the document in the physical presence of the other(s). If one or more of the individuals is located in a different physical location, once signatures are obtained in one location, the couple or each witness shall transmit a legible copy of the document, to the next signatory, until all signatures are obtained; and

iv. Once the certificate bas been signed by both members of the couple and both witnesses, the last signatory shall transmit a legible copy of the signed certificate to the officiant at the conclusion of the conference, who shall sign the document on the date of transmission.

☐ Following the video conference, the officiant shall make copies of the license and certificate and shall distribute the original and copies in the manner and within the time period required by N.J.S.A. 26:E-41 and N.J.S.A. 37:1-17.1; and

☐ In effectuating the transmission contemplated in this section, local registrars shall use a means that is secure and maintains the confidentiality of the documents.

3. For the duration of the Public Health Emergency, the provisions of N.J.S.A. 37:1-4 requiring a 72-hour waiting period between the license application and issuance shall be waived.

4. Notwithstanding N.J.S.A. 37:1-4, any marriage or civil union license issued during the Public Health Emergency shall be valid for 90 days.

5. For the duration of the Public Health Emergency declared in Executive Order No. 103 (2020), the fees imposed by N.J.S.A. 37:1-12 and N.J.S.A. 37:1-12.1 for the issuance of a marriage or civil union license shall he waived if the couple seeks a second license, mirroring the original license, due to the expiration of the original.

6. The New Jersey Office of Emergency Management, in consultation with the Commissioner of DOH, shall have the discretion to make additions, amendments, clarifications, exceptions, and exclusions to the above-outlined process and requirements.

7. For the duration of the Public Health Emergency declared in Executive Order No. 103 (2020), the provisions of N.J.S.A. 34:2-21.8 and N.J.S.A. 34:2-21.10, requiring the personal appearance of the minor, and, under certain circumstances, the minor's parent or guardian, before school district issuing officers in order to apply for or sign employment certificates may be satisfied through the use of audio-visual technology. Each public school district shall develop and implement procedures to satisfy the statutory requirements without requiring in-person contact between the school district issuing official and the minor, under the following conditions:

a. During the application process, the child and the school district licensing officer may transmit a single copy of all required documentation by way of electronic transmission, fax, or any other means of transfer of documents developed by the school district that avoids in-person contact, is secure, and maintains the confidentiality of the documents;

b. The video conference shall be live and must allow for interaction between the child and the school district issuing officer, and when applicable, the parent or guardian. During the video conference, the child shall verify his or her identity, authenticate the document's submitted, and sign the application, in a way that is visible and audible to the school district issuing officer; and

c. Following the video conference, the child shall transmit the signed certificate, by electronic or other means as determined by the school district, to the issuing officer, who shall make the requisite copies and distribute the original and copies as required by <u>N.J.S.A</u>. 34:2-21.7.

8. Nothing in this Order shall be construed to require that the marriage or civil union licensing process or ceremony, or the minor employment certification process, occur through the use of audio-visual technology. Applicable public officials should make all reasonable efforts, consistent with the enhanced social distancing and mitigation practices detailed in this State's Executive Orders, including but not limited to Executive Order No. 107 (2020). to accommodate applicants who may lack the technological resources necessary to engage in these processes through audio-visual means. Where in-person services are offered, public officials must similarly require participants and staff to wear cloth face coverings and must implement sanitization protocols, consistent with Executive Order No. 122 (2020).

9. TO the degree that that they are inconsistent with this Order, the provisions of <u>N.J.S.A</u>. 37:1-2, <u>N.J.S.A</u>. 37:1-4, <u>N.J.S.A</u>. 37:1-7 and -8, <u>N.J.S.A</u>. 37:1-12, <u>N.J.S.A</u>. 37:1-12.1, <u>N.J.S.A</u>. 37:1-17, <u>N.J.S.A</u>. 34:2-21.8, and <u>N.J.S.A</u>. 34:2-21.10 are suspended for the duration of the Public Health Emergency. Any provisions of these statutes that are not inconsistent with this Order remain in full force and effect.

10. This Order shall take effect on Monday, May 4, 2020 and shall remain in effect until revoked or modified by the Governor

GIVEN, under my hand and seal this 1st day of May, Two Thousand and Twenty, and of the Independence of the United States, the Two Hundred and Forty-Fourth.

[seal]
/s/ Philip D. Murphy
Governor
Attest:
/s/ Matthew J. Platkin
Chief Counsel to the Governor

ADMINISTRATIVE ORDER OF THE

CHIEF ADMINISTRATIVE JUDGE
OF THE COURTS

Pursuant to the authority vested in me, and at the direction of the Chief Judge, I hereby promulgate, effective May 4, 2020 and until further order, the following additional procedures and protocols to mitigate the effects of the COW-19 outbreak upon the judicial officers, staff, and users of the Unified Court System.

A. In pending matters, digital copies of (1) motions, cross-motions, responses, replies and applications (including post-judgment applications), (2) notices of appeal and cross-appeal, (3) stipulations of discontinuance, stipulations of adjournment, and other stipulations; (4) notes of issue, and (5) such other papers as the Chief Administrative Judge may direct, shall be accepted for filing purposes by all courts and clerical officers of the Unified Court System (including County Clerks acting as clerks of court) when presented for filing through (1) the UCS New York State Courts Electronic Filing (NYSCEF) system; (2) the UCS Electronic Document Delivery System (EDDS); or (3) such other document delivery method as the Chief Administrative Judge shall approve.

B. Documents filed through the EDDS system shall be served by electronic means, including electronic mail or facsimile. Filing fees required for documents filed through the EDDS system shall be paid by credit card or, where credit card payment is unavailable, by check delivered to the appropriate clerk's office by U.S. Mail or overnight mail service.

C. The provisions of paragraphs A and B above are authorized on a temporary basis, and will be reviewed and circumscribed promptly **at** the conclusion of *the* COVID-19 public health emergency.

D. Problem-solving courts may conduct virtual court conferences with counsel, court staff, service providers, and, where practicable, clients.

E. Judges may refer matters for virtual alternative dispute resolution, including to neutrals on court-established panels, community dispute resolution centers, and ADR-dedicated court staff.

F. The court shall not request working copies of documents in paper format.

Chief Administrative Judge of the Courts

Dated: May 1, 2020
AO/87/20

NOTICE TO THE PUBLIC

May 4, 2020

EDDS: UCS Program for Electronic Delivery of Documents

In response to the COVID-19 public health emergency and the expansion of "virtual" court operations, the Unified Court System has initiated a new program to transmit digitized documents (in pdf format) to UCS courts, County Clerks, and other court-related offices around the State. The Electronic Document Delivery System ("EDDS") allows users, in a single transaction, to (1) enter basic information about a matter on a UCS webpage portal page; (2) upload one or more pdf documents; and (3) send those documents electronically to a court or clerk selected by the user. Upon receipt of the document(s) by the court, the sender will receive an email notification, together with a unique code that identifies the delivery. More detailed instructions for sending or filing documents through EDDS may be found on the EDDS FAQ page.

Users/Senders should keep several important points in mind when using this system:

1. EDDS May be Used to File Papers with Certain Courts: At the direction of the Chief Administrative Judge, during the COVID-19 public health crisis EDDS can be used to deliver documents for filing with certain courts-including some Family Courts, Criminal Courts, Supreme Court, the Court of Claims, Surrogate's Courts, and District Courts, and City Courts. (EDDS is not available in the New York City Criminal Court.)

To use the system for filing, the sender must simply check a box on the sender information screen, complete the sending of the document(s) to the appropriate court through the EDDS system, and pay any required filing fee by credit card. The clerk's office will review the document(s) for sufficiency and, if the clerk determines that filing prerequisites have been met, accept them for filing purposes. In the event that a clerk's office has accepted and filed a document received through EDDS, the sender will be notified of that fact by email or publication on a public database. If no email or published notification is issued indicating that the document has been accepted for filing, the sender should not assume that the filing has occurred. The sender may contact the clerk's office to inquire about the status of a proposed filing.

2. EDDS is Not a Substitute for E-filing or NVSCEF: Please note that, although EDDS may be used for filing in various courts, it does not replace and may not substitute for

filing under the New York State Courts Electronic Filing System (NYSCEF). Therefore, it should not be used in matters where NYSCEF is available on either a mandatory or consensual basis. (Counties and case types where NYSCEF is available are listed on NYSCEF's Authorized for E-Filing page.)

3. EDDS Delivery is not "Service" on Other Parties: Finally, unlike NYSCEF, delivery of a document through EDDS does not constitute service of the document on any other party. If service is required, the sender must serve by some other means.

In sum, EDDS is a document delivery portal that complements the UCS electronic filing system and which, upon completion and together with NYSCEF, will allow remote and immediate delivery of digitized documents throughout the Unified Court System.

ADMINISTRATIVE ORDER OF THE

CHIEF ADMINISTRATIVE JUDGE OF THE COURTS

Pursuant to the authority vested in me, at the direction of the Chief Judge, and consistent with the Governor's determination approving the easing of restrictions on commerce imposed due to the COVID-19 health emergency, I hereby direct that, notwithstanding the terms of any prior administrative order:

1. In the counties and on the dates set forth in Exh. A, in courts and case types approved for electronic filing through the New York State Courts Electronic Filing System (NYSCEF), represented parties must commence new matters or proceed in pending matters exclusively by electronic filing through NYSCEF, and must file and serve papers in such matters (other than service of commencement documents) by electronic means through NYSCEF or, where permitted under NYSCEF court rules, by mail. Unrepresented parties must file, serve and be served in such matters by non-electronic means unless they expressly opt in to participate in NYSCEF.

2. To the extent that NYSCEF electronic filing is unavailable in courts or case types in the counties and on the dates set forth in Exh. A, represented parties must commence new matters exclusively by mail, except where otherwise authorized by the Chief Administrative Judge. Following commencement of a new matter, and in pending matters, represented parties must file papers through the Unified Court System's Electronic Document Delivery System (EDDS) or by mail, and must serve papers (other than commencement documents) by electronic means or by mail. Unrepresented parties must file, serve and be served in such matters by non-electronic means unless they provide written notification to the court and all parties that they wish to file, serve and be served electronically.

3. In the counties and on the date set forth in Exh. B, in courts and case types approved for electronic filing through NYSCEF, represented parties must commence new matters or proceed in pending matters exclusively by electronic filing through NYSCEF, and must file and serve papers in such matters (other than service of commencement documents) by electronic means through NYSCEF. Unrepresented parties must file, serve and be

served in such matters by non-electronic means unless they expressly opt in to partici-pate in NYSCEF.

4. To the extent that NYSCEF electronic filing is unavailable in courts or case types in the counties and on the date set forth in Exh. B, represented parties in pending matters may submit for filing digital copies of (1) motions, cross-motions, responses, replies and applications, (2) notices of appeal and cross-appeal, (3) stipulations of discontinuance, stipulations of adjournment, and other stipulations; (4) notes of issue, and (5) such other papers as the Chief Administrative Judge may direct, to courts and clerical officers of the Unified Court System (including County Clerks acting as clerks of court) through EDDS or such other document delivery method as the Chief Administrative Judge shall approve. Represented parties must serve documents filed through EDDS by electronic means, including electronic mail or facsimile. Unrepresented parties must file, serve and be served in such matters by non-electronic means unless they provide written notifica-tion to the court and all parties that they wish to file, serve and be served electronically.

This order shall not affect procedures for the filing and service of essential matters, and, on the dates that it becomes effective, supersedes administrative orders AO/87/20 (pars. A-C) and AO/114/20.

_____
Chief Administrative Judge of the Courts

Dated: May 28, 2020

AO/115/20

Exhibit A

Region: Counties	Effective Date
Finger Lakes: Orleans, Monroe, Wayne, Genesee, Wyoming, Livingston, Ontario, Yates, and Seneca. Mohawk Valley: Herkimer, Oneida, Otsego, Fulton, Montgomery, and Schoharie. Southern Tier: Steuben, Schuyler, Chemung, Tompkins, Tioga, Broome, and Chenango, and Delaware.	May 18,2020
North Country: Clinton, Franklin, St. Lawrence, Jefferson, Lewis, Hamilton, and Essex. Central New York: Oswego, Cayuga, Cortland, Onondaga, and Madison.	May 20,2020

Region: Counties	Effective Date
Western New York: Allegany, Cattaraugus, Chautauqua, Erie, and Niagara.	May 21,2020
Capital Region: Albany, Columbia, Greene, Rensselaer, Saratoga, Schenectady, Warren, and Washington.	May 26,2020
Mid-Hudson: Dutchess, Orange, Putnam, Rockland, and Westchester.	May 27,2020
Mid-Hudson (remainder): Sullivan and Ulster.	May 28,2020
Long Island: Nassau and Suffolk.	May 29,2020

Exhibit B

Region: Counties	Effective Date
New York City: New York, Bronx, Queens, Kings, and Richmond	May 25,2020

MESSAGE FROM CHIEF JUDGE DIFIORE

May 4, 2020

Thank you for a few minutes of your time as we bring you up to date on the latest COVID-19 developments affecting our courts, and our justice system.

As we turn the page from April to May, I hope that you and your families are doing well and staying safe and healthy. 1 can only imagine how anxious you all are to emerge from the shadow of this pandemic. I know that I am as well. And I know that many of you have been wondering and asking when it might be safe to re-open our courthouses and get back to work.

The short answer is that it's too soon to know, and that we are not yet in a position to formulate any final or even long-term plan.

Governor Cuomo confirmed last week that New York is making progress in the battle against COVID-19, flattening the curve and reducing the number of new cases, hospitalizations and deaths, although the rate of decline in all categories, particularly the still-staggering number of New Yorkers who are losing their lives on a daily basis, has been slower than any of us would like.

The Governor also indicated that the planning and the process for re-opening our state has begun and that New York will follow the CDC guidelines, which provide, as one requirement, that states may consider re-opening after new COVID hospitalizations have declined for 14 days. The Governor also suggested that there is likely to be a staged or gradual re-opening, depending upon all the attendant economic and public health considerations across the state.

What does this mean for our court system? It means that any plan we put in place will track the Governor's safe and prudent reopening of the state. As to the details of our plan, the honest answer is that we don't know yet what those details will entail. What we do know is that our services are absolutely essential, and that as the economy re-starts and businesses re-open, there will be a corresponding action by the courts. And we remain acutely aware that so much of what we do, whether it's conducting jury trials or screening the public at magnetometers, depends on, and requires, face-to-face interaction with many different stakeholders and constituencies, which, of course, poses a risk of community spread.

That said, the time has come, and we have begun to plan. We are working within the CDC re-opening guidelines, and adopting them in ways that are relevant to the court system; we are considering all the possible scenarios and timetables for a safe, phased-in re-start of our work; and we are determined to move forward in ways that are fully consistent with the latest public health guidance and closely coordinated with our partners in government. While we can't predict the future, I can confidently say that the New York State Unified Court System will be ready to safely re-open when called upon to do so.

And the key to our future success in safely re-opening lies in a responsible, patient, flexible and perhaps even bold approach. And our flexibility at every level will perhaps be the most important component of our success, for as we have seen, this virus is an unpredictable and opportunistic enemy that will lead us around so long as we allow that to happen.

So, while we are indeed working every day to plan for the future, today's thoughtful plan may be in tomorrow's scrap pile. And that is why your patience and understanding, along with your help and constructive guidance, is so important to us.

So, we will continue to work on parallel tracks: planning our "re-opening," while also improving and expanding our virtual court operations, which have been extremely busy and productive. With each passing day, our judges and staff are conferencing more matters, settling more cases and making good progress toward our goal of clearing out our entire backlog of undecided motions.

Effective today, we have also expanded our virtual operations to allow new motions, responsive papers to previously filed motions, and new applications (including post-judgment applications) to be e-filed in pending cases, either through our NYSCEF e-filing system or through a new electronic document delivery system created for those courts and jurisdictions where e-filing is not available. This is an important step that will allow lawyers and self-represented litigants to pursue relief in pending matters and enable our judges to move cases closer to final disposition or trial. Information has been posted on our website on how to file motions in those courts and jurisdictions that are not part of our current e-filing system.

Also effective today, judges in our civil, matrimonial and family courts have been authorized to refer matters to virtual ADR, and judges in our problem solving courts have been authorized to conduct virtual court conferences with counsel, court staff and service providers.

On other important fronts, we are also paying close attention to the difficulties facing Spring 2020 law graduates who may be prevented from taking the bar exam in a timely manner because of public health constraints. Last week, we announced that qualified law graduates will be authorized to temporarily engage in the limited practice of law under the supervision of an experienced attorney in good standing. I want to thank our working group, led by my Court of Appeals colleague, Judge Michael Garcia, which has been studying these issues and coordinating our efforts with the four Departments of the Appellate Division and the law school community.

The problem of seating everyone who wants to take the September bar exam has been an agonizing one. Because of the pandemic, seating capacity for the September administration will be sharply limited. As a result, applications to sit for the exam will be available initially to graduates of New York State law schools taking the exam for the first time. After this initial registration period is over, we will assess available seating in accordance with public health guidance, and if it is feasible and safe to do so, we will open the application process to additional candidates. Information about this process is available for your review on the New York Board of Law Examiners' website.

I think that I can safely assume that we are all experiencing similar reactions to this unimagined public health crisis: concern for our families and loved ones, concern for our future, and at times, even concern for our ability to withstand the pressures that this pandemic has caused. Not only can we handle the pressure, we can triumph over this pandemic. And I want to thank all of our judges, our professional staff and the entire legal community for the support and solidarity they have offered to all New Yorkers throughout this challenging period.

We will emerge from this awful pandemic -- and we will emerge better, stronger and more resilient than ever before.

So, please stay disciplined in keeping yourselves and your families safe and healthy. And be patient. I know it is sometimes difficult to practice patience, particularly at a time like this, but patience, along with careful, deliberate forward movement, will be a small price to pay to ensure the best possible outcome in a very uncertain world.

Thank you for your time and attention, and please stay tuned for additional updates.

State of New York
Court of Appeals

John P. Asiello
Chief Clerk and
Legal Counsel to the Court

Clerk's Office
20 Eagle Street
Albany, New York 12207-1095

May 11, 2020

NOTICE TO THE BAR

COMPANION FILING UPLOAD PORTAL

To facilitate virtual office operations of the Court of Appeals, the **Court has** amended its Rules of Practice to require, for motions and responses to Rule 500.10 jurisdictional inquiries, submissions in digital format as companions to the printed papers filed and served in **accordance** with the Court's Rules of Practice. The submission of these companion filings in digital format will be via a Companion Filing Upload Portal, similar to the Court-PASS system used to submit companion digital copies of briefs and record materials on appeals, certified questions, and judicial conduct matters.

The Court has also amended its Rules of Practice to reduce the number of printed copies that must be filed from 6 to 1 for civil motions for leave to appeal, **reargument** motions, and **papers** in opposition to those motions and from 10 to 1 for primary election session motions and **papers** in opposition to those motions. The **Rules** of Practice have **been** also changed to provide that the Appellate Division documents required by Rule 500.22 (c), Rule 500.26 (b) (3) (iv), and 500.26 (b) (4) shall be filed in digital format only.

Motions submitted with proof of **indigency** may still be made on one set of papers. **Parties** can request to be relieved of the **digital** submission **requirements** based on a **showing** of undue hardship.

Uploading digital submissions to the Companion Filing Upload Portal does not satisfy the service or filing requirements of the CPLR or the Court's Rules of Practice. The filer is responsible for meeting any applicable CPLR time limit by **serving** and filing as provided by the CPLR. **The** filer is responsible for meeting the Court's applicable due **dates** by filing the required number of paper documents with the Clerk's Office. Motions, papers in opposition to motions, and responses to Rule 500.10 jurisdictional **inquiries are deemed** "filed" with the Clerk's Office on the date of receipt of the paper document.

Relevant portions of the proposed amended Rules are attached and are effective May 27, 2020. Any responses to Rule 500.10 jurisdictional inquiries requested on or after May 27, 2020 and any motions returnable on or after June 1, 2020 must comply with the amended Rules. Questions may be directed to the Clerk's Office at 518-455-7700.
John P. Asiello
Chief Clerk and Legal Counsel to the Court

Appellate Division of the Supreme Court
First Judicial Department
AD1 2.0 – First Department Expands Operations
as a Virtual Court

Updated 4/21/20

In light of the continuing public health emergency in New York State and the obligations of the courts to respect the Governor's New York State on PAUSE Executive Order, the Appellate Division, First Department, is transitioning to a virtual court until further notice.

With our transition to a virtual court model nearly complete, the First Department is expanding the operations of the Court and *its* ancillary agencies. The Court will resume (1) calendaring appeals and motions, (2) scheduling pre-argument conference~(3) admitting attorneys to the bar, and (4) processing of attorney grievance complaints.

Avveals

The Court has issued an order setting forth two special terms: The May 2020 Special Term, which will commence on May 4, 2020, and end on May 29, 2020, and the June 2020 Special Term, which will commence on June 1, 2020, and end on June 26, 2020.

In the May 2020 Special Term, matters have been calendared for each Wednesday and Thursday, commencing on May 6, 2020. The calendar has been published on the Court's Calendars webpage.

ALL CALENDARED MATTERS SHALL BE HEARD ON SUBMISSION or REMOTELY ARGUED VIA SKYPE. There will be no adjournments. If a party wants to argue remotely, a request for remote argument must be made, regardless of whether oral argument was previously requested.

Requests for remote argument via Skype on all matters scheduled for the May and June Special Terms shall be made in advance, no later than one week prior to the calendar date, by emailing the Court at AD1InterimApp@nycourts.gov, with an e-mailed copy to opposing counsel or self-represented litigant. A completed notice of appearance with the contact information for the attorney who will appear remotely shall be attached to the request. When making requests, please indicate the **time needed** for **oral argument, not the time desired.** Parties will be advised as to whether the Court has granted oral argument. The remote oral arguments will be livestreamed on the Court's website.

Inasmuch as the filing deadlines for the responding and reply briefs for the original June 2020 term have been extended by the Governor's extension order, the responding briefs are now due no later than May 8, 2020, and the reply briefs no later than May 18, 2020.

By order of this Court, the perfection, filing, and other deadlines for the remaining terms of the Court continue to be suspended indefinitely and until further directive of the Court. **However, litigants may consensually perfect appeals and file motions.**

Hard Copy Filing

The requirement that hard wpy records, appendices, and briefs be filed continues to be suspended until further notice. Hard copy filings will not be permitted for the safety of our employees and the public.

Electronic Filing

Matters Subject to Mandatory E-filing. **All** filings (appeals, motions and applications) relating to matters subject to mandatory e-filing must still be filed via NYSCEF in accord& with the procedural and electronic rules of the Court.

Matters Not Subject to Mandatory E-filing

Interim Applications. In matters not subject to mandatory e-filing, submissions for emergency applications shall be filed via email to AD1InterimApp@nycourts.gov with notice, via email, to opposing counsel or self-represented litigant. The submission shall be one bookmarked PDF. Counsel will be notified by email or telephone as to the time and manner by which the application will be heard.

Motions. In matters not subject to mandatory e-filing, motions shall be filed via email to AD-1-clerks-office@nycourts.gov with notice, via email, to opposing counsel or self-represented litigant. The submission shall be one bookmarked PDF.

CPL245.70 *Applications*. CPL245.70 applications shall be filed via email to AD1CPL245.70App@nycourts.gov. The submission shall be one bookmarked PDF. Counsel will be notified by email or telephone as to the time **and** manner by which the application will be heard.

Appeals Not Subject to Mandatory E-filing. All filings made in connection **with** appeals that **are** not subject to mandatory e-filing shall be made electronically as follows:

CML: email to AD1copy-civil@nycourts.gov and AD-1-clerks-office@nycourts.gov;

CRIMINAL: email to AD1copy-criminal@nycourts.gov and AD-1-clerks-office@nycourts.gov;

FAMILY: email to AD1copy-family@nycourts.gov and AD-1-clerks-office@nycourts.gov.

Admission of Attorneys to the Bar

As delineated in the statement of the Presiding Justices, the Court's Committee on Character and Fitness is actively processing attorney admission applications. Candidates for admission will be interviewed remotely and the admission ceremonies will be held remotely via Skype. The Court will also resume the issuance of certificates of good standing. Additional information will be posted on the Committee on Character & Fitness webpage.

Attorney Grievance Complaints

Grievance complaints will be accepted electronically and processed remotely. Additional information will be posted on the Attorney Grievance Committee webpage**.**

UNITED STATES COURT OF APPEALS FOR THE SECOND CIRCUIT

ORDER

The President of the United States, Governors of the States of Connecticut, New York and Vermont and numerous mayors of cities within the jurisdiction of the Second Circuit, including the City of New York, have declared a state of emergency, marshaling all private and government resources in the response to the COVID-19 pandemic. In order to give full effect to these declarations of national and local emergencies and at the same time fulfill the Court's constitutional and statutory responsibilities as the Third Branch,

IT IS HEREBY ORDERED that the United States Court of Appeals for the Second Circuit will remain open to conduct the business for which the Court was established. However, individuals who do not have business with the Court will not be admitted until further order of the Court.

The oral arguments of appeals and motions will continue as scheduled on the Court's public calendar. Arguments will be conducted using technology that enables judges and individuals who argue to appear remotely, if they choose, and livestreams the arguments so that public access to the Court's proceedings is maintained. As an exercise of discretion, a panel may take on submission an appeal that meets the standard set out in the Federal Rules of Appellate Procedure 34(a)(2).

IT IS FURTHER ORDERED that individuals who seek access to the Thurgood Marshall United States Courthouse to transact business with the Court must, in the judgment of the attending security officers, meet the health standards specified at the entrance to the courthouse.

Parties who seek to file paper documents with the Court must file and stamp them in the Court's Night Box, which is located in the courthouse lobby. Individuals who have procedural questions may call the Clerk's Office at 212.857.8500 or a member of the Clerk's Office staff at the appropriate number listed in the Clerk's Office Directory which is found on the Court's website www.ca2.uscourts.gov. The Clerk's Office public counter will remain closed until further order of the Court.

March 16, 2020

s/s Robert A. Katzmann

Robert A. Katzmann

Chief Judge

Answers

Chapter 01

1. d. Rule 72 concerns other matters in civil procedure
2. d. Rule 3126 is the sanctions and penalties rule
3. a. See text
4. c. The producing party bears its costs, unless there is cost-shifting
5. c. DaSilva was the first case to directly address allowing computer-assisted predictive coding and technology assisted review
6. c. Under the amendments to Rule 37(e), only intentional loss and destruction warrants the most severe sanctions
7. d. So long as the method is reasonably calculated to apprise the party of the action, and complies with the rules and any treaties that may govern if the party is international
8. T There are ethical and legal burdens on attorneys to assure preservation of relevant information
9. T Once a party produces, the other party cannot demand an entire reproduction in a different format, format should be requested at the outset
10. F Certain types of electronic information are subject to FOIL & FOIA

Chapter 02

1. b. This is the definition of a public cloud
2. a. This is the definition of a private cloud
3. d. This is the definition of a community cloud
4. c. This is the definition of a hybrid cloud
5. c. Attorneys must take reasonable steps to prevent unauthorized disclosure of client information (RPC 1.6), and must supervise non-attorneys whose work relates to the work of the attorney and client (RPC 5.3)
6. d. See Comment 18 to Model Rule 1.6(c)
7. d. See text
8. c. See text
9. d. See text. All speak to obligations and considerations.
10. F Attorneys may use the Cloud, so long as they comply with ethical restrictions, and safeguard sensitive and trade secret client information

Chapter 03

1. d. See text
2. c. See text, this is the new standard articulated by the N.Y. Court of Appeals, abrogating prior "factual predicate" requirement

3. c. See text

4. c. Use of anything other than a private e-mail address accessible by only the client could risk loss of the attorney-client privilege in the communications with the attorney.

5. b. See text

6. c. See text, there is no 4th Amendment expectation of privacy or warrant protection with regard to travel crossing international borders.

7. F Courts have held that there is no such privacy in social media activity.

8. F Although there may not be privacy, a party generally must show that the information sought is reasonably calculated to be relevant to the claims and defenses in the litigation.

9. F Per the U.S. Supreme Court's decision in Riley v. California, law enforcement must have a warrant for the search under normal circumstances

10. T See text, such agreements, so long as they are clear to the user that a contract is being formed/terms are being set, are binding

Chapter 04

1. d. See text, all of these rules and others govern, addressing communication, truthfulness, misconduct, attorneys' responsibility for the conduct of non-lawyers, and communications with represented and unrepresented persons

2. c. See text, this is the only option an attorney may ethically utilize, absent permission from the party/witness'

counsel for the attorney to use direct contact.

3. d. See text, these selections all contain conduct that would not be ethical under the governing rules of professional conduct and the relevant caselaw.

4. d. See text, this was the holding in Gonzalez in 2016, a cautionary tale.

5. d. See text

6. d. See text

7. T See text

8. F Attorneys may do so, if the witness is not represented by counsel; and they must utilize "truthful friending"

9. F Attorneys may do so, if the opposing party is not represented by counsel; and they must utilize "truthful friending"

10. F Ethical authorities have held that attorneys may so advise their clients, so long as the information posted is truthful.

Chapter 05

1. a. Courts give instructions on the law to the jurors, who are the judges of the facts.

2. d. It is recommended that the trial judge give the instruction early and often during a trial.

3. d. See text, all of these jurisdictions have instructions of some kind on the topic.

4. c. In 2016, legislation was proposed in California on this issue.

5. c. This represents the action attorneys are required to take in this circumstance.

6. F Jurors are not permitted to perform outside research. They are to judge the facts of the case from the evidence presented in the courtroom

7. F Depending on the social media posts made, and the extent of the misconduct, courts might find that there has not been conduct rising to the level of requiring a mistrial, and might uphold the verdict. See U.S. v. Liu

8. F Attorneys may only view the public pages of prospective jurors, with no other contact

9. F Attorneys may have no direct contact with jurors, in person or electronically, until after the verdict is rendered and the trial is over

10. F For example, in the State of Missouri there is the case of Spence v. BNSF Railway Co., holding there is an obligation for "reasonable investigation" at least with regard to prior litigation history of potential jurors, and search of Case.net. See also MISSOURI SUPREME COURT RULE 69.025.

Chapter 06

1. d. All of these Opinions provide guidance for judges in different jurisdictions.

2. c. Judges face several potential sanctions or discipline options for misuse of social media creating an impropriety or appearance of impropriety.

3. d. See text - all of these selections are ethical violations.

4. a. See text. Certainly before the Herssein decision, Florida was

deemed to have the most restrictive guidance.

5. F Judges may have social media accounts, so long as they are mindful concerning their communications and the appearance of propriety concerning their office

6. F In most jurisdictions, "friends" status alone is not sufficient, it is similar to being friendly at Bar Association or civic functions

7. T While the kind of communications may matter in some jurisdictions, generally this category of communication will require recusal of the judge

8. F While attorneys and judges may have social media connections, ethics rules specifically bar attorneys from utilizing social media in order to communicate with or influence a judge on a particular matter

9. F While judges have First Amendment protections the same as any Citizen, the Gwinnett County, Georgia Chief Magistrate Judge said it best: "Judges are held to a more stringent standard by the Judicial Canons"

10. F ABA Op. 478 (citing Model Rule 2.9(c)) speaks to the fact that generally judges may not do so

Chapter 07

1. d. According to sources, up to 94% of teens know and admit to the danger.

2. d. Despite the dangers, sources say that 35% of teens admit to having texted while driving.

3. d. 48 states, as well as Washington, D.C., Puerto Rico, Guam and the U.S. Virgin Islands

4. d. All of these selections are major issues for teenagers and young adults today, requiring better education, and making it all the more important that they have ready access to information on the law.

5. d. See the text.

6. d. See the text. According to the statistics, of the number of people killed in distracted driving accidents each day, 11 are teenagers.

7. F If you commit such a crime, there is potential for jail time if the law provides for that as a punishment, even if you are only 18

8. F The legality of the relationship age is a matter of state law, but sending naked photographs via text messages (electronic communications) is governed by federal law, and sending those photos of someone under 18 years of age is a violation of federal law regardless of whether the two people are in a legal relationship

9. F The First Amendment, according to reported court decisions, does not prevent school administrators from taking disciplinary action

10. F See the text; the issue of cyberbullying goes far beyond any kind of innocent joking. The posts discussed in this book illustrate the heinous nature of conduct that has occurred online. Furthermore, cyberbullying and shaming have harmed the lives of the victims, causing them in some cases to uproot themselves

and move to a new community, and in the most extreme cases to turn tragically to suicide

Chapter 08

1. b. See the text.

2. a. See the text.

3. c. See the text; this is the reasoning of the majority of the Court for why a non-passenger texter/sender might be held liable for an accident in a certain case.

4. b. See the text.

5. a. See text; this is the reasoning of the Court's decision.

6. d. See the text; Chapter 8.

7. F Under tort laws, there are instances where non-drivers, depending on their actions, could be held liable in part for their role in causing an accident

8. F Regardless of the jurisdiction, drivers are always responsible for keeping their vehicles under control, and for avoiding distractions

9. F No, this is an issue of state law, and the NY appellate court specifically disagreed with the NJ court's reasoning

10. F See the text; this issue of those who text and drive continues to be a problem on our roads in the U.S., resulting in many accidents and deaths each year

Chapter 09

1. a. See the text; this is one of the rules that allow for certification and

streamlined authentication procedures in certain circumstances.

2. d. See the text; failure to address these
issues could lead to exclusion of
material even if the material is otherwise authenticated.

3. d. See the text.

4. b. See the text.

5. c. See text; Chapter 9.

6. b. See text. This is an Appellate Division, Second Department case on
the issue.

7. b. See text.

8. F See the text; these are separate
principles

9. F See the text; in fact states often
disagree on the standards to apply.
See the differences between Maryland and New Jersey discussed in
Chapter 9

10. T See the text; this is in fact true, see
the discussion in Chapter 9 concerning federal standards and considerations versus those in some states

Chapter 10

1. d. These activities are no different than
any other online activities.

2. c. See text.

3. c. See text.

4. d. See text.

5. c. See text.

6. b. See text.

7. F Arbitrators' authority over non-
parties is limited by the terms of the
FAA, which only requires in-person
appearances of non-parties

8. F We have seen the development of
online court proceedings, including the potential for some civil jury
trials

9. T This is part of the definition of
"Zoombombing"

10. T This is one of several new concerns
and adjustments that must be addressed as more and more legal
proceedings go online following
COVID-19

Chapter 11

1. b. See the text; this is one of the two
- the other being Revised Uniform
Fiduciary Access to Digital Assets
Act.

2. c. See the text; EPTL Article 13-A
is where New York placed the
codification.

3. a. See the text; this is the provision for
the highest level of authority under
law.

4. b. See the text; this is the provision for
the second highest level of authority
under law.

5. c. See the text; this is the provision for
the lowest level of authority under
law.

6. c. See the text. This is the seminal case
in Massachusetts, believed to be the
first to address the issues discussed
in Chapter 11.

7. b. See the text.

8. F There are circumstances when the
privacy of deceased individuals is
still respected under law, particularly in areas where FOIL or FOIA
would not result in discovering the
information

9. T Indeed, specific provision must be made, or else law will usually limit the access that successors may have to the electronic information

10. T This is a situation under the law, such as NY EPTL 13-A, whereby decedent's use of an estate planning document, without using a tool provided by a service provider, will govern issues of access

About the Author

Professor Michael L. Fox is Assistant Professor of Business Law and Graduate Program (MBA) Coordinator in the School of Business at Mount Saint Mary College, in Orange County, New York. He teaches courses in the undergraduate business and Masters of Business Administration programs. He also serves as the College's Pre-Law Advisor. In addition, he is an Assistant Adjunct Professor of Law in Professional Responsibility in the J.D. and LL.M. programs at Columbia University School of Law in New York City.

Dr. Fox received his Bachelor of Arts (B.A.) degree, Phi Beta Kappa and *summa cum laude*, from Bucknell University, with a major in Economics/minor in Biology. He was elected to Phi Beta Kappa in his junior year. Thereafter, at graduation he received the award for highest class standing in the major of Economics. Dr. Fox received his Doctor of Law (J.D.) degree from Columbia University School of Law, where he was a Harlan Fiske Stone Scholar and Articles Editor on the *Columbia Business Law Review*.

Professor Fox served as a law clerk to Hon. Lawrence E. Kahn, U.S. District Judge, Northern District of New York, in Albany. He has been rated AV-Preeminent by Martindale-Hubbell since 2015, and was selected to the Upstate New York Super Lawyers list from 2013 through 2016 while engaged in the practice of law. He is admitted to practice in New York State, as well as in the Southern, Eastern and Northern Districts of New York, Second Circuit U.S. Court of Appeals, and United States Supreme Court. He worked as a litigation associate at Stroock & Stroock & Lavan LLP, in Manhattan; and was then, sequentially, an associate, senior counsel, and then partner, and litigation managing attorney at Jacobowitz & Gubits, LLP, in New York's Hudson Valley. At Jacobowitz & Gubits his primary practice areas included Federal Practice, Electronic Discovery, Estates Litigation, Employment and Discrimination Law, and Business Entity Disputes. Just prior to academia, he was Special Counsel at the Hudson Valley law firm Catania, Mahon, Milligram & Rider, PLLC. From February 2014 to November 2016, he was Deputy Corporation Counsel and special labor counsel for the City of Port Jervis.

Dr. Fox is currently a Delegate in the New York State Bar Association House of Delegates. He was Vice President for the Ninth Judicial District of the New York State Bar Association, and member of the NYSBA Executive Committee, from 2017-2019. He previously served as a Delegate in the American Bar Association House of Delegates from 2008 through 2014, served as a Delegate in the NYSBA House of Delegates from 2008 to 2014, and was a Member-at-Large of the NYSBA Executive Committee from 2015 to 2017. He is a member of the Commercial & Federal Litigation Section; and a past Chairperson of the Young Lawyers Section (one of the largest NYSBA Sections). He serves on the NYSBA Committee on Professional Discipline and Committee on Legal Education and Admission to the Bar, among others; co-chairs the ComFed Committee on Electronic Discovery; and Chairs NYSBA's Standing Committee on Communications and Publications. Dr. Fox served as a member of the critique faculty of the NYSBA YLS Trial Academy program, held at Cornell Law School, from 2013-2017; in 2017 and 2018 he served as a team leader and lecturer at Trial Academy; and in 2019 he was a lecturer and critique faculty for the 10th Anniversary Trial Academy. Professor Fox serves on the Board of Directors of the Orange County Bar Association, and on the Advisory Board of the Food Bank of the Hudson Valley.

Professor Fox is co-host of *Gold/Fox: Non-Billable*, a podcast of the New York State Bar Association, available on Spotify, Google Play, Apple iTunes/Podcast App, iHeart Radio App, or wherever podcasts are available. He also authored the book *A Guide to Diversity and Inclusion in the 21st Century Workplace* (N.Y. St. Bar Ass'n Publ. 2020); and has authored/co-authored articles and CLEs, and spoken at programs/symposia, on Federal civil procedure, attorney-client privilege and work product, electronic discovery and social media, evidence, professional ethics, employment law, and pre-law advice.